Advances in Intelligent Systems and Computing

Volume 385

Series editor

About this Series

The series "Advances in Intelligent Systems and Computing" contains publications on theory, applications, and design methods of Intelligent Systems and Intelligent Computing. Virtually all disciplines such as engineering, natural sciences, computer and information science, ICT, economics, business, e-commerce, environment, healthcare, life science are covered. The list of topics spans all the areas of modern intelligent systems and computing.

The publications within "Advances in Intelligent Systems and Computing" are primarily textbooks and proceedings of important conferences, symposia and congresses. They cover significant recent developments in the field, both of a foundational and applicable character. An important characteristic feature of the series is the short publication time and world-wide distribution. This permits a rapid and broad dissemination of research results.

More information about this series at http://www.springer.com/series/11156

Stefano Berretti · Sabu M. Thampi
Soura Dasgupta
Editors

Intelligent Systems Technologies and Applications

Volume 2

Editors
Stefano Berretti
Dipartimento di Ingegneria dell'Informazione (DINFO)
Università degli Studi di Firenze
Firenze
Italy

Soura Dasgupta
Electrical and Computer Engineering
The University of Iowa College of Engineering
Iowa
IA
USA

Sabu M. Thampi
School of CS/IT
Indian Institute of Information Tech. and Management - Kerala (IIITM-K)
Trivandrum
India

ISSN 2194-5357 ISSN 2194-5365 (electronic)
Advances in Intelligent Systems and Computing
ISBN 978-3-319-23257-7 ISBN 978-3-319-23258-4 (eBook)
DOI 10.1007/978-3-319-23258-4

Library of Congress Control Number: 2015946580

Springer Cham Heidelberg New York Dordrecht London

Printed on acid-free paper

Springer International Publishing AG Switzerland is part of Springer Science+Business Media (www.springer.com)

Preface

Intelligent systems refer broadly to computer embedded or controlled systems, machines and devices that possess a certain degree of intelligence with the capacity to gather and analyze data and communicate with other systems. There is a growing interest in developing intelligent technologies that enable users to accomplish complex tasks in different environments with relative ease. The International Symposium on Intelligent Systems Technologies and Applications (ISTA) aims to bring together researchers in related fields to explore and discuss various aspects of intelligent systems technologies and their applications. ISTA'15 was hosted by SCMS School of Engineering and Technology (SSET), SCMS Group of Institutions, Kochi, India during August 10–13, 2015. The Symposium was co-located with the Fourth International Conference on Advances in Computing, Communications and Informatics (ICACCI'15).

In response to the call for papers, 250 papers were submitted to the symposium. All the papers were evaluated on the basis of their significance, novelty, and technical quality. Each paper was rigorously reviewed by the members of the program committee. This book contains a selection of refereed and revised papers from three special tracks: Ad-hoc and Wireless Sensor Networks, Intelligent Distributed Computing and Business Intelligence and Big Data Analytics.

There is a long list of people who volunteered their time and energy to put together the conference and who warrant acknowledgment. We would like to thank the authors of all the submitted papers, especially the accepted ones, and all the participants who made the symposium a successful event. Thanks to all members of the Technical Program Committee, and the external reviewers, for their hard work in evaluating and discussing papers. The EDAS conference system proved very helpful during the submission, review, and editing phases.

We are grateful to the General Chairs and members of the Steering Committee for their support. Our most sincere thanks go to all keynote and tutorial speakers who shared with us their expertise and knowledge. Special thanks to members of the organizing committee for their time and effort in organizing the conference.

We thank the SCMS School of Engineering and Technology (SSET), SCMS Group of Institutions, Kochi for hosting the event.

We wish to express our thanks to Thomas Ditzinger, Senior Editor, Engineering/Applied Sciences Springer-Verlag for his help and cooperation.

August 2015

Stefano Berretti
Sabu M. Thampi
Soura Dasgupta

Organization

Honorary Chairs

Lotfi A. Zadeh	Founder of Fuzzy Logic, University of California Berkeley, USA
William A. Gruver	Simon Fraser University, Canada

ICACCI Steering Committee

Ravi Sandhu	University of Texas at San Antonio, USA
Sankar Kumar Pal	Indian Statistical Institute, Kolkata, India
Albert Y. Zomaya	The University of Sydney, Australia
H.V. Jagadish	University of Michigan, USA
Sartaj Sahni	University of Florida, USA
John F. Buford	Avaya Labs Research, USA
Jianwei Huang	The Chinese University of Hong Kong, Hong Kong
John Strassner	Software Labs, Futurewei, California, USA
Janusz Kacprzyk	Polish Academy of Sciences, Poland
Tan Kay Chen	National University of Singapore, Singapore
Srinivas Padmanabhuni	Infosys Labs, India & President at ACM India
Suzanne McIntosh	New York University and Cloudera Inc., USA
Prabhat K. Mahanti	University of New Brunswick, Canada
R. Vaidyanathan	Louisiana State University, USA
Hideyuki TAKAGI	Kyushu University, Japan
Haibo He	University of Rhode Island, USA
Nikhil R. Pal	Indian Statistical Institute, Kolkata, India
Chandrasekaran K.	NITK, India
Junichi Suzuki	University of Massachusetts Boston, USA
Deepak Garg (Chair)	IEEE Computer Society Chapter, IEEE India Council & Thapar University, India

Pascal Lorenz	University of Haute Alsace, France
Pramod P. Thevannoor (Vice Chairman)	SCMS Group of Institutions, Kochi, India
Axel Sikora	University of Applied Sciences Offenburg, Germany
Maneesha Ramesh	Amrita Vishwa Vidyapeetham, Kollam, India
Sabu M. Thampi	IIITM-K, India
Suash Deb	INNS India Regional Chapter
Arun Somani	Iowa State University, USA
Preeti Bajaj	G.H. Raisoni COE, Nagpur, India
Arnab Bhattacharya	Indian Institute of Technology (IIT), Kanpur, India

General Chairs

Soura Dasgupta	University of Iowa, USA
Jayanta Mukhopadhyay	Indian Institute of Technology (IIT), Kharagpur, India
Axel Sikora	University of Applied Sciences Offenburg, Germany

Publication Chair

Sabu M. Thampi	IIITM-K, India

Technical Program Committee

Program Chairs

Juan Manuel Corchado Rodriguez	University of Salamanca, Spain
Stefano Berretti	University of Florence, Italy

TPC Members/Additional Reviewers

Girijesh Prasad	University of Ulster, UK
Hanen Idoudi	National School of Computer Science - University of Manouba, Tunisia
M.V.N.K. Prasad	IDRBT, India
Wen Zhou	Shantou University, P.R. China
Marcelo Carvalho	University of Brasilia, Brazil
Yoshitaka Kameya	Meijo University, Japan

Lorenzo Mossucca	Istituto Superiore Mario Boella, Italy
Rodolfo Oliveira	Nova University of Lisbon, Portugal
Vamsi Paruchuri	University of Central Arkansas, USA
Maytham Safar	Kuwait University, Kuwait
Wei Tian	Illinois Institute of Technology, USA
Zheng Wei	Microsoft, USA
Uei-Ren Chen	Hsiuping University of Science and Technology, Taiwan
Son Doan	UC San Diego, USA
Roman Jarina	University of Zilina, Slovakia
Sanjay Singh	Manipal Institute of Technology, India
Ioannis Stiakogiannakis	France Research Center, Huawei Technologies Co. Ltd., France
Imtiez Fliss	ENSI, Tunisia
Alberto Nuñez	University Complutense of Madrid, Spain
Sandeep Reddivari	University of North Florida, USA
Haibin Zhu	Nipissing University, Canada
Philip Branch	Swinburne University of Technology, Australia
GianLuca Foresti	University of Udine, Italy
Yassine Khlifi	Umm Al-Qura University, KSA, Saudi Arabia
Abdelhafid Abouaissa	University of Haute Alsace, France
Jose Delgado	Technical University of Lisbon, Portugal
Sabrina Gaito	University of Milan, Italy
Petro Gopych	Universal Power Systems USA-Ukraine LLC, Ukraine
Pavel Kromer	VSB - Technical University of Ostrava, Czech Republic
Antonio LaTorre	Universidad Politécnica de Madrid, Spain
Suleman Mazhar	GIK Institute, Pakistan
Hidemoto Nakada	National Institute of Advanced Industrial Science and Technology, Japan
Yoshihiro Okada	Kyushu University, Japan
Hai Pham	Ritsumeikan University, Japan
Jose Luis Vazquez-Poletti	Universidad Complutense de Madrid, Spain
Rajib Kar	National Institute of Technology, Durgapur, India
Michael Lauer	Michael Lauer Information Technology, Germany
Anthony Lo	Huawei Technologies Sweden AB, Sweden
Ilka Miloucheva	Media Applications Research, Germany
Sunil Kumar Kopparapu	Tata Consultancy Services, India
Waail Al-waely	Al-Mustafa University College, Iraq
Robert Hendley	University of Birmingham, UK
Yoshiki Yamaguchi	University of Tsukuba, Japan
Ruben Casado	TreeLogic, Spain
Sanjay Chaudhary	Dhirubhai Ambani Institute of Information and Communication Technology, India
Amjad Gawanmeh	Khalifa University, UAE

Guan Gui	Akita Prefectural University, Japan
Deepa Gupta	Amrita Vishwa Vidyapeetham, India
Dirman Hanafi	Universiti Tun Hussein Onn Malaysia, Malaysia
Adil Kenzi	ENSAF, Morocco
N. Lakhoua	ENIT, Tunisia
Emad Mabrouk	Assiut University, Faculty of Science, Egypt
Yasser Madany	IEEE, Senior Member, Alexandria University, Egypt
Prabhaker Mateti	Wright State University, USA
John Moore	University of West London, UK
Ilaria Torre	University of Genoa, Italy
Sa'ed Abed	Kuwait University, Kuwait
Fahd Alharbi	KAU, Saudi Arabia
Chitti Babu B.	VSB – Technical University of Ostrava, Czech Republic
C.M.R. Prabhu	Multimedia University, Malaysia
Walisa Romsaiyud	Siam University, Thailand
Mujdat Soyturk	Marmara University, Turkey
Belal Abuhaija	University of Tabuk, Saudi Arabia
Badrul Hisham Ahmad	Universiti Teknikal Malaysia Melaka, Malaysia
Lee Gillam	University of Surrey, UK
Petia Koprinkova-Hristova	Bulgarian Academy of Sciences, Bulgaria
Abd Kadir Mahamad	Universiti Tun Hussein Onn Malaysia, Malaysia
Koushik Majumder	West Bengal University of Technology, India
Abdallah Makhoul	University of Franche-Comté, France
Nimushakavi Murti Sarma	JNT University Hyderabad, Hyderabad, India
Imed Romdhani	Edinburgh Napier University, UK
Anderson Santana de Oliveira	SAP Labs, France
Rahmat Sanudin	Universiti Tun Hussien Onn Malaysia, Malaysia
Rostyslav Sklyar	Independent Researcher, Ukraine
Qiang Wu	Juniper Networks, USA
Salman Yussof	Universiti Tenaga Nasional, Malaysia
Mohammad Zia Ur Rahman	K L University, India
Md. Shohel Sayeed	Multimedia University, Malaysia
Jhilik Bhattacharya	Thapar University, India
Lai Khin Wee	Universiti Malaya, Malaysia
Mu-Qing Lin	Northeastern University, P.R. China
Monica Mehrotra	Jamia Millia Islamia, Delhi, India
Shireen Panchoo	University of Technology, Mauritius
Angkoon Phinyomark	University of Calgary, Canada
Kashif Saleem	King Saud University, Saudi Arabia

Yuh-Ren Tsai	National Tsing Hua University, Taiwan
Minoru Uehara	Toyo University, Japan
Bin Yang	Shanghai Jiao Tong University, P.R. China
Shyan Ming Yuan	National Chiao Tung University, Taiwan
Meng-Shiuan Pan	Tamkang University, Taiwan
Mohan Kankanhalli	National University of Singapore, Singapore
Philip Moore	Lanzhou University, P.R. China
Sung-Bae Cho	Yonsei University, Korea
Eraclito Argolo	Universidade Federal do Maranhão, Brazil
Tushar Ratanpara	Dharmsinh desai University, India
Bulent Tavli	TOBB University of Economics and Technology, Turkey
Andre Carvalho	University of Sao Paulo, Brazil
Valentin Cristea	University Politehnica of Bucharest, Romania
Boris Novikov	Saint Petersburg State University, Russia
Prasheel Suryawanshi	MIT Academy of Engineering, Alandi (D), Pune, India
Kenneth Camilleri	University of Malta, Malta
Atilla Elçi	Aksaray University, Turkey
Stephane Maag	TELECOM SudParis, France
Jun Qin	Southern Illinois University Carbondale, USA
Mikulas Alexik	University of Zilina, Slovakia
Ali Hennache	Al-Imam Muhammad Ibn Saud Islamic University, Saudi Arabia
Mustafa Man	University Malaysia Terengganu, Malaysia
Misron Norhisam	Universiti Putra Malaysia, Malaysia
Hamid Sarbazi-Azad	IPM & Sharif University of Technology, Iran
Dhaval Shah	Institute of Technology, Nirma University, India
Luiz Angelo Steffenel	Université de Reims Champagne-Ardenne, France
Ramayah Thurasamy	Universiti Sains Malaysia, Malaysia
Chi-Ming Wong	Jinwen University of Science and Technology, Taiwan
Shreekanth T.	Sri Jayachamarajendra College of Engineering, India
Kazumi Nakamatsu	University of Hyogo, Japan
Mario Collotta	Kore University of Enna, Italy
Balaji Balasubramaniam	Tata Research Development and Design Centre (TRDDC), India
Chuanming Wei	Boradcom Corporation, USA
S. Agrawal	Delhi Technological University (DTU) Formerly Delhi College of Engineering (DCE), India
Mukesh Saini	University of Ottawa, Canada
Ciprian Dobre	University Politehnica of Bucharest, Romania
Traian Rebedea	University Politehnica of Bucharest, Romania
Zhaoyu Wang	Georgia Institute of Technology, USA
Chakravarthi Jada	RGUKT Nuzividu, India
Eduardo Rodrigues	Federal University of Rio Grande do Norte, Brazil

Dinesh Sathyamoorthy	Science & Technology Research Institute for Defence (STRIDE), Malaysia
Tomonobu Sato	Hitachi, Ltd., Japan
Peng Xia	Microsoft, USA
Bei Yin	Rice University, USA
Ahmed Almurshedi	Universiti Teknologi Malaysia, Malaysia
Amitava Das	CSIO, India
Povar Digambar	BITS Pilani Hyderabad, India
Joydev Ghosh	The New Horizons Institute of Technology, India
Son Le	Aston University, UK
Wan Hussain Wan Ishak	Universiti Utara Malaysia, Malaysia
Edward Chu	National Yunlin University of Science and Technology, Taiwan
Shom Das	National Institute of Science & Technology, India
Akshay Girdhar	Guru Nanak Dev Engineering College, Ludhiana, India
Akash Mecwan	Nirma University, India
Prasant Kumar Pattnaik	KIIT University, India
Ramesh R.	Asiet Kalady, India
Mostafa Al-Emran	Al Buraimi University College, Oman
Weiwei Chen	University of Southern California, USA
Adib Chowdhury	University College of Technology Sarawak, Malaysia
Josep Domingo-Ferrer	Universitat Rovira i Virgili, Spain
Ravi G.	Sona College of Technology, India
Govindarajan Jayaprakash	AmritaVishwavidyapeetham University, India
Vinayak Kulkarni	MIT Academy of Engineering Pune, India
Nandagopal Jayadevan Nair Lathika	Amrita School of Engineering, India
Tonglin Li	Illinois Institute of Technology (IIT), USA
N. Mathan	Sathyabama University, India
Hu Ng	Multimedia University, Malaysia
Sindiso Nleya	Computer Science Department, South Africa
Marcelo Palma Salas	Campinas State University (UNICAMP), Brazil
Muhammad Raheel	University of Wollongong, Australia
Mohammed Saaidia	University of Souk-Ahras. Algeria, Algeria
Jose Stephen	Centre for Development of Advanced Computing, India
Mohammed Mujahid Ulla Faiz	Hafr Al-Batin Community College (HBCC), Saudi Arabia
Karthik Srinivasan	Philips, India
Shrivishal Tripathi	IIT Jodhpur, India
Hengky Susanto	University of Massachusetts at Lowell, USA
Haijun Pan	New Jersey Institute of Technology, USA
Bhupendra Fataniya	Sarkhej Gandhinagar Highway, India
Afshin Shaabany	University of Fasa, Iran
Ashutosh Gupta	Amity University, India

Pablo Cañizares	Universidad Complutense de Madrid, Spain
Georgios Fortetsanakis	University of Crete, Greece
Tilahun Getu	École de Technologie Supérieure (ETS), Canada
Filippos Giannakas	University of the Aegean, Greece
Navneet Iyengar	University of Cincinnati, USA
Jamsheed K.	Amrita Vishwa Vidyapeetham, India
Rupen Mitra	University of Cincinnati, USA
Krishna Teja Nanduri	University of Cincinnati, USA
Hieu Nguyen	INRS-EMT, Canada
Joshin Mathew	Indian Institute of Information Technology an Management - Kerala, India
Sakthivel P.	TCS, India
Lili Zhou	China Telecom Inc., P.R. China
Scott Kristjanson	Simon Fraser University, Canada
Yupeng Liu	BROADCOM, USA
Sreeja Ashok	Amrita Vishwa Vidyapeetham, Kochi, India
Parul Patel	Veer Narmad South Gujarat University, India
Mona Nasseri	University of Toledo, USA
Indhu R.	CDAC, India
Pranali Choudhari	Fr. C. Rodrigues Institute of Technology, India
Anastasia Douma	University of the Aegean, Greece
Aswathy Nair	Amrita School of Arts and Science, Kochi, USA
Zakia Asad	University of Toronto, Canada
Kala S.	Indian Institute of Science Bangalore, India
Pınar Kırcı	Istanbul University, Turkey
Lee Chung Kwek	Multimedia University, Malaysia
Azian Azamimi Abdullah	Nara Institute of Science and Technology, Japan
Archanaa Rajendran	Amrita, India
Marina Zapater	Universidad Politécnica de Madrid, Spain
Piyali Das	University of Cincinnati, USA
Divya G.	Asiet Kalady, India
Medina Hadjem	Université Paris Descartes, France
Pallavi Meharia	University of Cincinnati, USA
Priyanka Shetti	Amrita Vishwa Vidyapeetham, India
Xiaoqian Wang	UTA, USA
Mingyuan Yan	Georgia State University, USA

Contents

Part I
Ad-hoc and Wireless Sensor Networks

Analysis of Communication Delay and Packet Loss During Localization Among Mobile Robots

B. Madhevan and M. Sreekumar

Abstract Wheeled mobile robots moving in an unknown environment are made to face many obstacles while navigating in a planned or unplanned trajectory to reach their destination. But, no information is available regarding the failure of a leader robot of a group in both unknown and uncertain environments and the subsequent course of action by the follower robots. As the leader fails, one of the follower robots within the group can be assigned as a new leader so as to accomplish the planned trajectory. The present experimental work is carried out by a team of robot comprises of a leader robot and three follower robots and if the present leader fails, a new leader is selected from the group using leader follower approach. But, the problem of localization among multi mobile robots is subjected to communication delay and packet loss. The problem of data loss is analyzed and shows that it can be modeled as a feedback system with dual mode observers. An algorithm has been developed to compensate this packet loss during communication between controller and server in a wifi-based robotics environment. Further, the simulation results prove that the algorithm developed is efficient compared to single mode observers, for both unknown and uncertain environments among multi robots.

Keywords Multi robots · Leader follower approach · Packet loss · Localization · Wheeled mobile robots

---

B. Madhevan(✉)
Vellore Institute of Technology Chennai, Off Vandalur Kelambakkam Road, Melakkotaiyur, Chennai 600127, India
e-mail: bmadhevan@iiitdm.ac.in

M. Sreekumar
Indian Institute of Information Technology Design and Manufacturing (IIITD&M) Kancheepuram, Off Vandalur Kelambakkam Road, Melakkotaiyur, Chennai 600048, India
e-mail: msk@iiitdm.ac.in

S. Berretti et al. (eds.), *Intelligent Systems Technologies and Applications*,
Advances in Intelligent Systems and Computing 385,
DOI: 10.1007/978-3-319-23258-4_1

1 Introduction

In the available literature, the movements of mobile robots to reach the destination as a group or one-to-one approach (leader along with followers) have been considered. However, no information has been reported regarding failure of the leader robot in an unknown environment and the subsequent course of action by the follower robots to select a new leader among them so as to complete the task assigned. Issues such as identification of a new leader in the group of follower robots or whether the group of follower robots will reach the destination without a leader has not yet been explored. Considering this as a major factor in the present research, a group comprises of four robots with one robot assigned as a leader initially, is made to face many obstacles while navigating in an unknown environment to reach their destination. However, the leader robot might fail due to internal and external disturbances while navigating in environments where the follower robots do not have any prior information about the environment. If the leader fails, one of the follower robots within the group need to be assigned as a new leader so as to accomplish the planned trajectory which is a new avenue to pursue research in the domain of mobile robots. Wheeled mobile robots (WMRs) are more efficient than legged robots on hard and smooth surfaces and find widespread application in industry. Map based navigation is developed to achieve this task, and the current development in this area is presented. Mapping, usually involves encoding information on how to get from one place to another. The problem of robotic mapping is related to cartography, which helps in defining maps for navigation [1,8]. The goal is to construct a map or floor plan and all the robots should traverse as a group with it [12-13]. In [14], a scalable hybrid MAC protocol, consisting of a contention period and a transmission period, have been designed for heterogeneous machine to machine networks. Further, to balance the tradeoff between these periods in each frame, an optimization problem is formulated to maximize the channel utility. The feasibility of real- time motion planning is dependent on the accuracy of the map, and on the number of obstacles. Basically, it consists of three processes: map-learning, localization and path planning which are briefly presented as under [4,5]. The robot develops a complete map about the environment and with this information the robot can travel freely in an unknown environment using these strategies which can be implemented through various models rely on the information from various sensors. The important aspect is to develop an efficient model which can extract maximum information from the available cues [9-11]. Based on time variant synchronization, a state estimation predictor is developed through which the bilateral time-delay is compensated. The same was studied for more general systems and as a result of this generalization, a state predictor based on synchronization for nonlinear systems with input time delay was developed [6,7]. The experimental work consists of one leader robot and three followers investigates on a novel Role Assignment (RA) scheme [2]. The leader robot is equipped with sensors such as laser range sensor, sonar and gyro compared to followers, which are equipped only with laser and sonar sensors. A driver constructed using LabVIEW and ELVIS kit, is common for the leader and followers. The robots, fixed with multi sensors and the common driver have been integrated with a Data Acquisition System (DAQ). Depending upon mismatch in frequency, the delay is estimated around 250 ms and hence communication protocol with packet loss plays an important role in transferring the data among the robots. This plays a major block in transferring the data between leader-server, server-follower and follower-leader. This three way communication is halted numerous times due to frequency mismatch. To resolve the above mentioned problem, the packet dropping algorithm has been proposed. The optimal time taken to reach the goal with role assignment has been determined experimentally. Hence, this scheme is embodied to perform various tasks in an unknown environment along with role assignment. The experiments demonstrate that the results obtained are in close agreement with analytical results.

2 Leader Follower Approach (LFA) and Role Assignment

The main problem with Multi Robot System Coordination (MRSC) occurs when the leader robot fails due to unknown faults. Due to this, for each execution of the leadership function, the algorithm checks for unknown faults by the followers in the group. The overall architecture checks for two fault occurrence strategies. When the fault occurrence is related to an obstacle, the controller updates speed and position and based on this, a new RA is defined. When the fault occurrence is related to Environmental Constraints, it updates angle and similarly computes another new RA. The controller carries this information through Wi-Fi and executes the algorithm. Further, the flowchart of the navigable part of the main program which would guide the entire robot team safely, despite all these fault occurrences in an unknown environment, filled with moving and stationary obstacles has been described. For inspection and integrating, a group of sensors and a computer system are developed respectively, for all these subsystems which are a part of the architecture. The formation involves different kinds of robots equipped with sensors for fault identification. The top event is represented as F which emphasizes the failure of coordination [3]. FTA can be performed by constructing a Fault Tree (FT) shown in Fig. 1(a). Generalized RA algorithm is presented in Fig. 1(b). The leader - follower approach is adopted because it easy to use a trajectory and define the motion of each individual robot relative to this trajectory (for followers). It assumes that one robot in the formation can sense the locations of other robots and further each robot can sense its own location, as each and every fraction of time, the data regarding the environment including obstacles is updated in the generalized flow table.

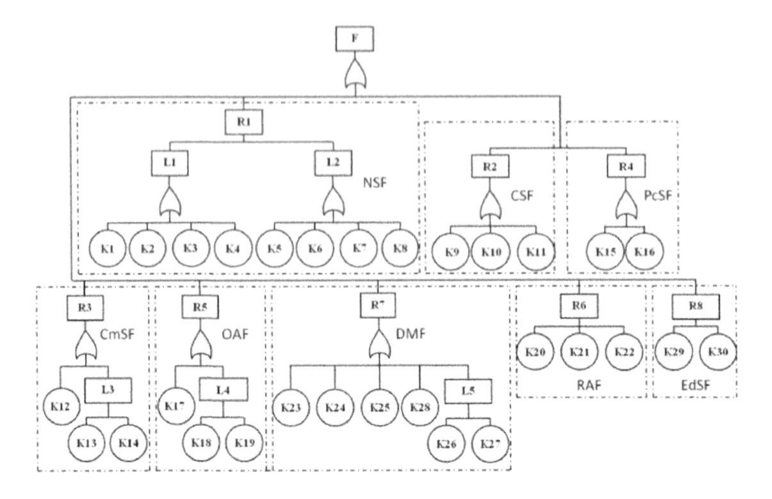

Fig. 1(a) FT structure

Start

Step1 ID: waiting to get information from the server regarding leader

Confirm leader robot is dead.

Step2 RSI:No signal, program not running as next leader yet to be decided.

Step3 SI1:Signal passed, program running. Output ON

d_{goal}: H1, find the destination point

$d_{(F1\text{-}DL)}$, $d_{(F2\text{-}DL)}$, $d_{(F3\text{-}DL)}$: H2, the distance between leader and respective followers

$d_{(Fi)}$: H3, the distance between the robots

$N_{(LF1)O}/N_{(LF1)E}$, $N_{(LF2)O}/N_{(LF2)E}$, $N_{(LF3)O}/N_{(LF3)E}$: H4, update the number of constraints

$t_{(LF1)O}/t_{(LF1)E}$, $t_{(LF2)O}/t_{(LF2)E}$, $t_{(LF3)O}/t_{(LF3)E}$: H5, update time

$N_{(Fi)E}/N_{(Fi)E}$: H6, update the number of constraints

$t_{(L)O}/t_{(L)E}$: H7 update time

N_{Fi}: H8

Leader Selected

Step4 RS2:Signal received, program running. Output OFF

Step5 SI2: Signal reached, program running. Output ON

Step6 SIR: Signal received. Destination Reached

End

Fig. 1(b) RA algorithm

3 Packet Loss

In the conventional dropping algorithm, the flow ID of each arriving packet is compared with a packet from the queue which is selected arbitrarily. If the flow IDs are same, the arriving packet will be lost. By comparing one packet with the one randomly selected, the dropping condition will be reduced. To resolve this problem, a packet dropping algorithm has been proposed. In this packet dropping scheme, comparison is done by comparing incoming packets with multiple number of packets in the queue. The multiple packets are picked from the selected region in the queue.

3.1 Proposed Algorithm

Regional CHOKe is a queue management algorithm that bridges fairness and simplicity. It approximates the fair queuing between TCP and UDP flows at a minimal implementation overhead. In addition, Regional CHOKe is a stateless algorithm which does not require any special data structure. Thus, it is nearly as simple to implement as RED. The basic idea behind regional CHOKe is that the information about the incoming traffic should be used for chastising misbehaving flows.

Algorithm 1 Packet dropping algorithm

Require: n-number of packets
Ensure: Drop the packet which has same flow-id
  Initialize: $avg_q = 0$,$min_{th} \leq avg_q \leq max_{th}$
  and
  $min_{th} \geq 3$
  avg_q=$(1 - w_q) * avg_q + w_q * q$
  Where w_q=0.02 (queue weight), avg_q: average queue size, max_{th}=$3 * (min_{th})$
  q: current queue size
  Get the incoming packets
  for each packet calculate avg_q **do**
    if $avg_q \leq min_{th}$ **then**
      Enqueue the packet into the buffer
    else
      Pick the packet from the selected region and compare the flow-id with incoming packet
    end if
    if matches found **then**
      Drop all the packets
    else
      Enqueue the incoming packet and calculate avg_q
    end if
    if $avg_q \leq max_{th}$ **then**
      Enqueue the packet with drop probability
    else
      Drop the incoming packets
    end if
  end for

The algorithm sets two thresholds (same as RED) on the buffer, min^{th} and max^{th}.

- Average queue size < min^{th}, this implies, every arriving packet is queued.
- Average queue size >max^{th}, this implies, the flow ID of each arriving packet is compared to the flow ID of a selected packet.

If the flow IDs is the same, the arriving packet will be dropped while the randomly selected packet will be marked, and later it will be dropped when it reaches the head of the queue. Based on the dependency on the level of congestion, each arriving packet is divulged into a buffer with a probability scale. The drop probability is computed such that:

- When the average queue size exceeds the maximum threshold value, packets are dropped with probability 1.

Based on this, comparison is done, so that queue occupancy is below maximum threshold. In this packet dropping scheme, the packets are compared with its flow ID's and if they are similar, the compared packets are merely dropped and if not, the packets are kept in their respective places. Comparison is done by comparing the incoming packet with multiple numbers of packets in the queue. The multiple packets are picked from the selected region in the queue. The selected region is measured by the queue length and weight of the queue. In this manner, a particular region in the queue is selected and all packets in that region are compared with an incoming packet. The selected region can be in any part of the buffer, i.e. from head or tail of the queue. It is just an advantage to compare the packets from the head or tail side. As the number of robot increases, there is an increase in the packet loss which results in communication delay. The amount of packet loss for 2, 3, and 4 robots has been obtained analytically. Further, if number of robots increases to 6 or 8, the projected value of communication delay is obtained and effectiveness of delay in packet loss is listed below in Table 1.

Table 1 Packet loss estimation.

S.No.	Number	Packet loss (PL)	Period (ms)	PL rate
1.	2	0	100	0
2.	3	3	400	50
3.	4	7	1000	60
4.	6	17	2000	84
5.	8	25	3000	96

4 Effect of Packet Loss

This section presents simulation results of the proposed algorithm and performance in penalizing misbehaving flows Fig. 2(a) shows the high performance of the proposed Regional CHOKe algorithm. The simulation result is given for three flows, (4,16,30).

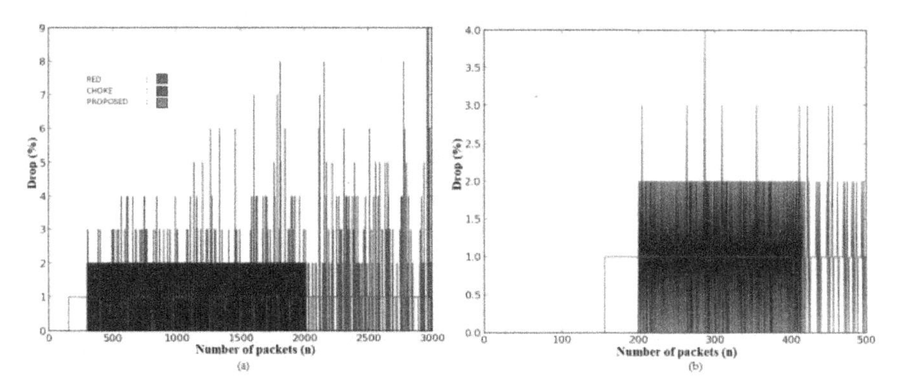

Fig. 2 High performance of proposed regional CHOKe.

The graph states the performance of the RED, CHOKe and proposed algorithm. Compared to the RED and CHOKe, the performance of the packet dropping in the proposed algorithm is high. The packets are dropped by comparing the incoming packet with multiple numbers of packets in the buffer. The multiple packets are chosen in the selected region. The selected region is by dividing the buffer into two parts, and the packets choose from any one of the parts. But for the simulation, the first part is used as selected region. The packets are chosen in that region and compared, then the compared packets are dropped according to the algorithm. Fig. 2(b) shows the high performance of the Regional CHOKe. The total number of packets sent was 3000. Here the regional CHOKe packet dropping algorithm gives high packet dropping. The RED drops maximum one packet at a time, whereas, CHOKe drops two packets at a time because, it compares the flow ID of one packet with another packet. Hence, it can drop two packets at a time and regional CHOKe will drop more than two packets because it compares one packet with more packets. The performance of the algorithms for various numbers of flows is shown in the rest of the figures (Fig. 3 to Fig. 5). Fig. 3(a) & Fig. 3(b) shows the performance of algorithms for 30 and 16 numbers of flows respectively. Fig. 4(a) and Fig. 4 (b) shows the performance of algorithm with 16 numbers of flows. Fig. 5(a) & Fig. 5(b) shows the performance of algorithms for a number of flows 4 and depict the drop of packets. Compared to earlier algorithms RED and CHOKe, the number of packets dropped is high and in certain situation, the proposed algorithm drops 2, 4, 7, 8, 9 packets at a time. The performance analysis of the algorithms for various numbers of flows 4, 16, and 30 are shown. The analysis clearly says the proposed algorithm performance is good for the packet dropping. The algorithm is tested for performance of different flows with variant packets sent. For example: The algorithm is tested for flows such as 4,16 and 30 with variant packets ranging from 100 to 750. Fig. 6 illustrates the performance analysis of the various numbers of flows such as 4, 16, 30. The performance analysis shows that the performance of Regional CHOKe algorithm is better compared to RED and CHOKe. At the number of flow 4, the performance shown is normal to algorithms, but increasing the number of flows offers better performance. The maximum number of flows, the performance was increased as illustrated in Fig. 6. Thus, it shows regional CHOKe algorithm is better compared with RED and CHOKe algorithms.

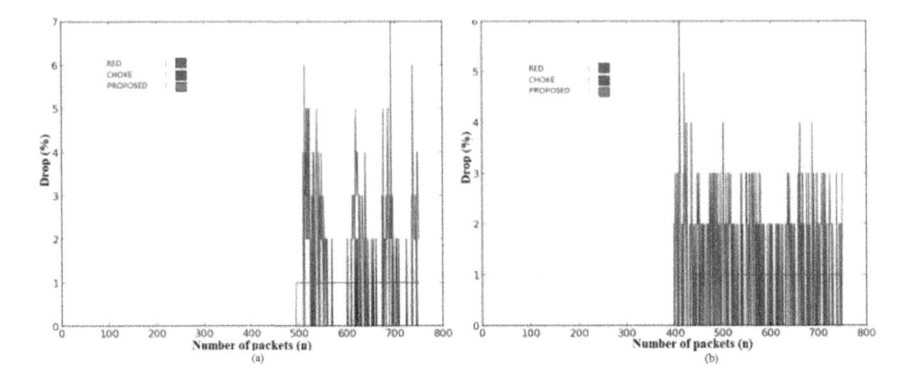

Fig. 3 Performance of 30 flows with 750 packet sent.

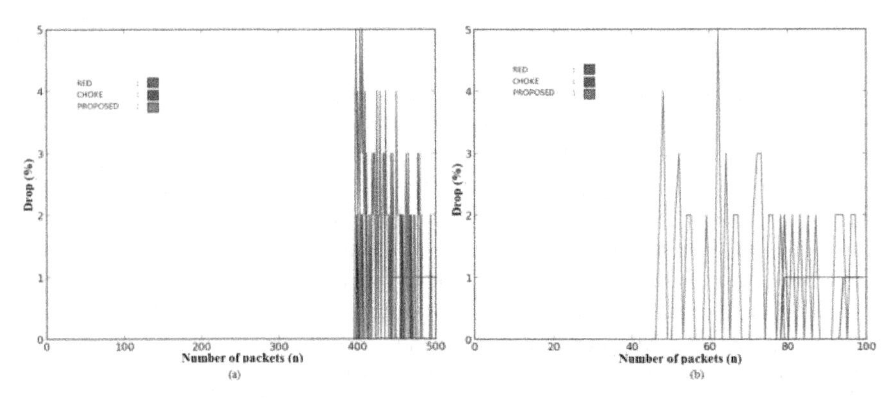

Fig. 4 Performance for 16 flows with 500 packet sent.

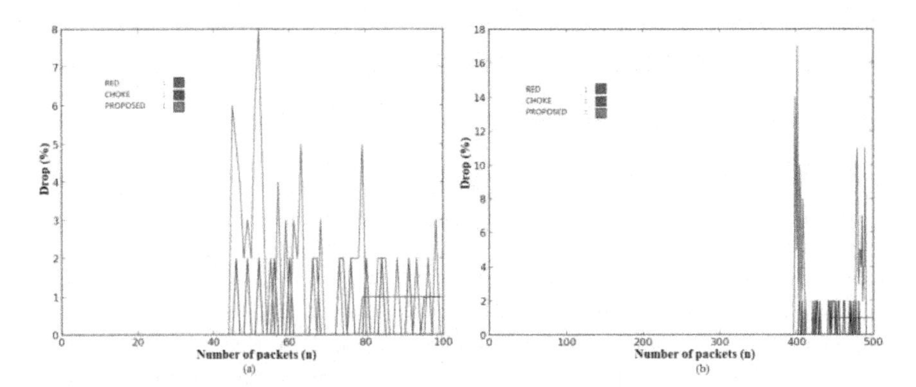

Fig. 5 Performance for 4 flows with 100 packets sent.

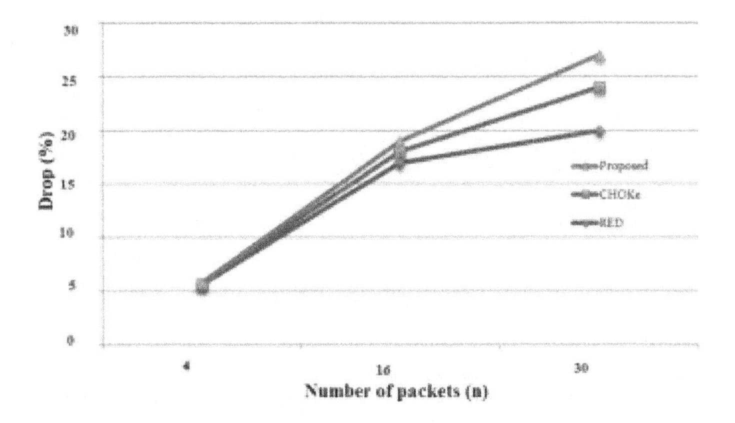

Fig. 6 Performance analysis of 4,16,30 flows.

5 Conclusion

The proposed algorithm can solve endogenous and exogenous faults with multiple models. Further, it can handle model free faults (data driven). In the domain of multirobot navigation, accurate analytical models are not often feasible due to uncertainties in the environment, noisy sensors. The proposed algorithm is developed to tackle faults based on data driven rather than analytical models. Efficiency of the developed algorithm has been checked for three case studies; (i) environment with no constraints in which the leader fails; (ii) environment with constraints in which the leader fails and (iii) environment if more than one follower fails along with the leader. Further, the proposed scheme is a closed loop feedback system, hence, the performance of the algorithm will not be affected by an increase in the number of constraints, and new constraints could be accommodated within milliseconds. As the number of robot increases, there is an increase in the packet loss which results in communication delay. Communication protocol plays a major role in transferring the data among the robots and in the present work; the delay is estimated around 250 ms depending upon mismatch on frequency. This plays a major block in transferring the data between leader-server, server-follower and follower leader. This three way communication was halted many times due to frequency mismatch.

References

1. Austin, D.J., McCarragher, B.J.: Geometric constraint identification and mapping for mobile robots. Robotics and Autonomous Systems **35**, 59–76 (2001). doi:10.1.1.12.2833
2. Madhevan B., Sreekumar M.: A systematic implementation of role assignment in multi robots using leader follower approach: analytical and experimental evaluation. In: 13th International Conference on Control, Automation, Robotics and Vision (ICARCV 2014), pp. 1792–1798 (2014)

3. Madhevan, B., Sreekumar, M.: Tracking algorithm using leader follower approach for multi robots. Procedia Engineering **64**, 1426–1435 (2013). doi:10.1016/j.proeng.2013.09.224
4. Meyer, J.A., Filliat, D.: Map-based navigation in mobile robots: II. A review of maplearning and path-planning strategies. Cognitive Systems Research **4**, 283–317 (2003). doi:10.1016/S1389-0417(03)00007-X
5. Meyer, J.A., Filliat, D.: Map-based navigation in mobile robots: I. A review of maplearning and path-planning strategies. Cognitive Systems Research **4**(4), 243–282 (2003). doi:10.1016/S1389-0417(03)00007-X
6. Niemeyer, G., Preusche, C., Hirzinger, G.: Springer, Handbook of Robotics, chapter 31: Telerobotics, pp. 741–758. Springer-Verlag (2008)
7. Oguchi, T., Nijmeijer, H.: Prediction of chaotic behavior. IEEE Trans. on Circ. and Syst. I. **52**(11), 2464–2472 (2005b). doi:10.1109/TCSI.2005.853396
8. Vasudevan, S.S.: Gachter, Nguyen V., Siegwart, R.: Cognitive maps for mobile robots -an object based approach. Robotics and Autonomous. System **55**, 359–371 (2007). doi:10.1016/j.robot.2006.12.008
9. Wijk, O., Christensen, H.I.: Localization and navigation of a mobile robot using natural point landmarks extracted from sonar data. Robotics and Autonomous Systems **31**, 31–42 (2000). doi:10.1016/S0921-8890(99)00085-8
10. Van den Broek, T.: Formation Control of Unicycle Mobile Robots: Theory and Experiments. (Masters thesis, 2008) Eindhoven University of Technology (2008)
11. Yuan, H., Shim, T.: Model based real-time collision-free motion planning for nonholonomic mobile robots in unknown dynamic environments. Int. J. Precis. Eng. Manuf. **14**(3), 359–365 (2013). doi:10.1007/s12541-013-0050-x
12. http://en.wikipedia.org/wiki/Robotic_mapping
13. http://us.wow.com/wiki/Housenavigationsystem
14. Liu, Y., Yuen, C., Cao, X., Hassan, N.Ul., Chen, J.: Design of a Scalable Hybrid MAC Protocol for Heterogeneous M2M Networks, IEEE Internet of Things Journal, 99-111, February 2014. doi:10.1109/JIOT.2014.2310425

An Encryption Technique to Thwart Android Binder Exploits

Yadu Kaladharan, Prabhaker Mateti and K.P. Jevitha

Abstract Binder handles the interprocess communication in Android. Whether the communication is between the components of the same application or different applications, it happens through Binder. Hence captivating it can expose all the communications. Man-in-the-Binder is one such exploit that can subvert the Binder mechanism. In this paper, we propose an encryption mechanism that can provide confidentiality to app communications to prevent such exploits.

Keywords Android · Binder · Inter process communication · Man-in-the-binder · Encryption · XBRF

1 Introduction

Android has a layered structure. At the bottom lies the Linux kernel. Above the kernel lie the libraries and Android Runtime. Above this lies the Application Framework and at the top are the applications. The most important components of Android Framework include Binder, Zygote, Service Manager, Application Manager etc[12].

Interprocess communication in Android is accomplished by a mechanism known as Binder, that spans over the entire Android Framework. Binder IPC facilitates

Y. Kaladharan(✉)
TIFAC-CORE in Cyber Security, Amrita Vishwa Vidyapeetham, Coimbatore, India
e-mail: yadukaladharan@gmail.com

P. Mateti
Department of Computer Science and Engineering, Wright State University,
Dayton, OH 45435, USA
e-mail: pmateti@wright.edu

K.P. Jevitha
Department of Computer Science, Amrita Vishwa Vidyapeetham, Coimbatore, India
e-mail: kp_jevitha@cb.amrita.edu

S. Berretti et al. (eds.), *Intelligent Systems Technologies and Applications*,
Advances in Intelligent Systems and Computing 385,
DOI: 10.1007/978-3-319-23258-4_2

access to a remote object as if it is a local one. Whether the communication is within the same application or between two different application components, it occurs through the Binder[4]. Hence subverting Binder can expose all application information to an attacker. Man-in-the-Binder is one such exploit and was described by Artenstein *et al*.[2]. It is based on the vulnerability that the application data is sent as plaintext between the application and Binder driver. This transaction is intercepted to get confidential application information.

In this paper, we suggest a technique to encrypt the application data before passing on to the Binder. This is achieved by assigning a shared key between the kernel and each application or service and using it to encrypt the respective communication between them.

The paper first describes the Binder IPC mechanism at various layers of Android. Then, the various attacks on Binder are explained, in which we focus on a man-in-the-middle attack. An encryption scheme to thwart this attack and its effectiveness are described in latter sections.

2 Background

The communication between applications in Android is handled by the Binder mechanism. The Binder framework has a layered structure, which include an API for applications, middleware and kernel driver[10]. The API is implemented in Java, the middleware in C++ and the kernel driver in C. The objective of Binder mechanism is that an application must be able to access a remote object as if it is a local one. The layers, the respective files and methods in each layer are shown in Figure [1].

2.1 Binder API

The Binder API hides the lower level IPC implementation from the applications. For this, it provides two interfaces to each service: proxy and stub. The stub, at the server, provides the actual implementation of methods. The client will call the `bindservice()`, along with the required intent[6]. The `onBind()` method on the server will respond by returning a proxy IBinder interface and will invoke the `onServiceConnected()` with the IBinder object as argument. The client can then call the remote methods directly. On each method call, the API through proxy will invoke the `transact()` of `IBinder.java`[3].

2.2 Binder at the Middleware

The communication between two processes in Android is known as transaction. The data transmitted is in the form of a structure known as Parcel. The call to `transact()` from API will invoke its respective implementation in middleware. The parcels needs to be flattened before being sent. The middleware is responsible for flattening, un-flattening, marshalling, unmarshalling, managing thread pool

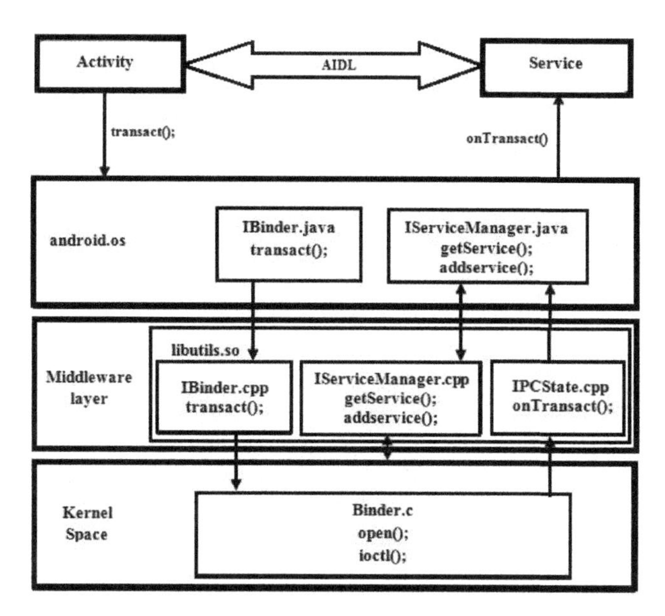

Fig. 1 Binder

etc. The binder middleware functionalities are achieved using a shared library `libbinder.so`[10]. Once the above mentioned operations are done, it will invoke the kernel driver using an ioctl system call. The ioctl will send the Binder command and the data buffer to the driver.

2.3 *Binder at the Kernel*

The ioctl call is of the form: ioctl`(fd, BINDER_WRITE_READ, &bwt)`, where `fd` is a file descriptor to /dev/binder, `BINDER_WRITE_READ` is a command to send a series of transactions and `&bwt` is a reference to the data buffer. The `BINDER_WRITE_READ`, along with some other fields contains a `write_buffer` to hold the request and a `read_buffer` to hold the response and a transaction code. The `write_buffer` holds a protocol tag followed by another data structure `binder_transaction_data`. This data structure contains an enum representing the function to be executed and a reference to the interface, followed by arguments to that method. The tag for Binder trasactions is `BC_TRANSACTION`. The structure of the `BINDER_WRITE_READ` and `binder_transaction_data` are described in [2]. The kernel will then invoke a thread from the thread pool. This thread will copy the parcel to the server's process space using copy from user and copy to user command of the linux kernel. The `onTransact()` of the middleware will do the unmarshalling and the requested operation is then performed. To send the response, the server will then act as the client and the client as the server.

2.4 *Input Method Editor*

For using the standard Input Method Editor (IME), an application will register for its callback.The default IME is usually `com.android.inputmethod.latin`. The application will behave as the client and the IME as server. The response from IME contains the key strokes. The reply buffer from IME when parsed contain data in a format as described in Figure[2]. The first four bytes of the reply represent the protocol code which are defined in enum `binder_driver_ return_protocol`. `BC_REPLY` and `BC_TRANSACTION` are the interested ones here, which indicate the Binder reponse and request respectively.

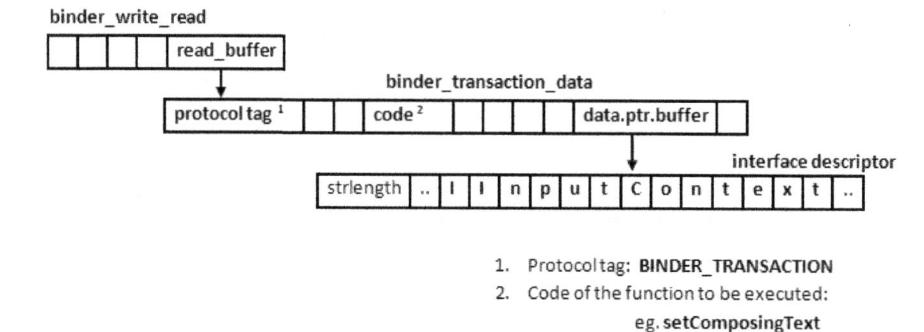

Fig. 2 Binder Transaction between IME and an App

The Parcel which contains the key stroke information is delivered to an internal interface called com.android.internal.view.IInputContext. This interface then sends the data up to the InputContext class for handling the received keyboard inputs for the client process. Each time a key is pressed, another callback to IInputContext is triggered, and a fresh buffer is sent via Binder. Thus, for a key logging attack, the target is IInputContext, and the implementation of which is done using AIDL. There is a direct correspondence between the order of the functions in the AIDL file and the function code as it appears in the transaction.

3 Attacks on the Binder

The Binder mechanism handles all the communication between the applications and services. Hence, an attack on the Binder framework can expose all the critical information to an attacker[1].

3.1 Man-in-The-Binder

The Man-In-The-Binder is one exploit that was described in [2]. This attack is analogous to the man-in-the-middle attack, where the attacker code stands at the middle of a communication.

They performed the attack using a code injection technique on the shared library `libbinder.so`. The attack can only be done on a rooted device. The attack will first hook a method on the target application component. The method hooked is the point where the Binder Framework communicates with the kernel driver. The exploit is based on the vulnerability that the data sent between an application and Binder driver is in plain text. It is in the form of a structure known as Parcel, which is defined publically in parcel.h. The protocol tag `BC_TRANSACTIONS` and `BC_REPLY` can identify a Binder transacton. Further, the action performed on the parcel data can be identified by comparing the enum and the interface implementation. Hence it is possible for an attacker to identify a particular response of a method in a particular transaction. The extent of the attack was explained by an attack scenario in [2]. The attack makes it possible to establish a system wide keylogger.

The applications that use the standard IME are attacked. The target of the attacker is to obtain the Parcel sent by the IME to the client application. More precisely, the data sent between the IME Binder middleware and the kernel driver is the targeted transaction. The only communicaton between the middleware and kernel takes place through `ioctl` call made by the `talkWithDriver()` of `IPCThreadState`[5]. So, inorder to control the `ioctl` transaction, the `talkWith Driver()` is hooked using a library injection technique. The attacker thus replaced the talkWithDriver() with a malicious code. In the hooked method, before the ioctl call is made, the Parcel data is copied to a file on user space. The Parcel, as explained in the earlier section, can be analysed to obtain the relevant data. For a key logger, the response from IME will have the tag `BC_REPLY`, interface as `com.android.internal.view.IInput Context` and the code for `setComposing Text` is 6.

3.2 Cross Binder Reference Forgery(XBRF)

Each application component that uses Binder has a file descriptor to `/dev/ Binder`. It contains a tree structure called `binder_proc`, which maintains the list of Binder references that the respective process can make. This `binder_proc` structure is maintained by the kernel and is updated by sniffing the Parcels. Every time a Parcel is sent through the binder driver, it will analyze it to identify the reference interface mentioned in it. The kernel will then add this reference to a node `node_ref` in `binder_proc` . In addition to this, the `binder_proc` contains the uid and pid, which it checks to ensure the permissions of the application requesting the service. The XBRF[9] is an exploit in which an application having higher privilege passes directly its file descriptor to a lower privileged app. Thus, the less privileged app will get access to the services permitted to the original application.

4 The Encryption Scheme

The Binder mechanism thus handles the entire IPC in Android. Hence, subverting the Binder can expose all critical application data to an attacker[7]. As explained in the previous section, there are a number of ways to do this. Encrypting the application data is a good solution to thwart these attacks. Since, the number of applications and services are very high, getting each communicating entities into a key agreement is not an easy solution. Hence in this paper, we suggest a technique to encrypt the Binder communications through a secret key mechanism. The attackers are usually interested in the Parcel data that is received as `BC_REPLY` from a service. In this encryption scheme, before every application component request a service, it will invoke a particular method `generateSecretKey()`. This method will make the kernel generate a random number, store it respective to the process. The process will then send the request for a service and the server will complete the action. Before transforming the response back to Parcel, the service will invoke a method `encryptAppData()`. This method will take as arguments the server response, and will encrypt it with the client's shared key. Then, the `transact()` method is called and the encrypted data reaches the client. The client after unmarshalling, will decrypt the data using the secret key. The next time, when the same client try to make a Binder call, the `getSecretKey()` is called again and the previous key gets overwritten by the new one. The use of random keys can overcome the problem of secure key storage.

The client after recieving the IBinder object, instead of calling the `transact()` will invoke `getSecretKey(). The getSecretkey()` make use of `/dev/random` for generating a random key. The `/dev/random` is a cryptographically secure mechanism to generate pseudo random numbers. Each application has a file descriptor to `/dev/Binder`. The `/dev/Binder` contains a special structure `binder_proc` respective to each process. In this `binder_proc`, the kernel will store the randomly generated key. The client, then using the Binder IPC will send the request to the server along with the client's file descriptor. The marshalling, unmarshalling, parcelling etc., are all carried the same way. Once the server completes the requested operation, it will then invoke another call `encryptAppData()`. Within this method, another system call is made `getSecretKey()`, which will retrieve the random key corresponding to the file descriptor. The server will thus obtain the client's key and will encrypt the response data using it. Since, the encryption and decryption process should not add much computational overhead, a Lightweight Cryptograhy(LWC)[8] technique must be used. CLEFIA, is one such LWC technique. It is a 128-bit block cipher and support key lengths of 128, 192 and 256 bits. It is efficient both in terms of computation and energy consumption[11].

The server will then invoke the `transact()`. The `transact()` will create the Parcel, but the `write_buffer` data is encrypted. The client on receiving it, through `onTransact()` will unmarshall and unflatten the data. The client, since it knows its own file descriptor to `/dev/Binder`, will use the `getSecretKey()` to retrieve it. It will then decrypt the data. For each communication by the client, the random key will change and get overwritten. The `getSecretKey()` and `generateSecretKey()` must be implemented as an atomic system call. This is to thwart the library injection attack over these methods.

4.1 Communication in Enhanced Binder

The following Figure [3] illustrates the flow of communication in the enhanced Binder. The `generateSceretKey()`, `getSecretKey()` and `encryptAppData()` are the modifications made to the existing Binder mechanism. Calling `bindService()` and obtaining the IBinder object is the first step(1). The client then invokes `generateSecretKey()` (2). The service request is then sent to the driver(3), which inturn send the request and the file descriptor of client to the server(4). After the completion of the request for client's key through `getsecretkey(client_fd)` (5) and obtain the same(6). The encrypted `data_buffer` is parcelled and sent to driver(7) and then to the client(8). The client will then call `getSecretKey(fd)` (9), to get the key(10) to decrypt the data.

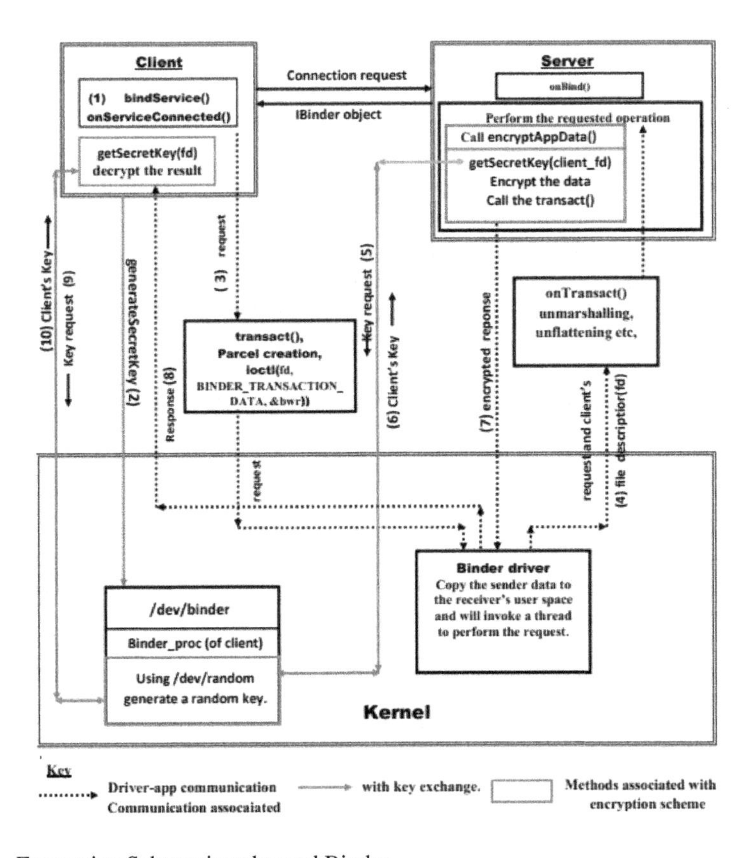

Fig. 3 Encryption Scheme in enhanced Binder

5 Results and Discussion

We created a detailed documentation of the Binder mechanism. The proposed encryption scheme can prevent the man-in-the-middle attacks within the Binder IPC. It makes it impossible to decrypt the parcel data without the secret key, which is known only to the kernel and the respective communicating entities. Hence, in the Man-in-the-Binder attack, even if the attacker perform some library injection, he will only be able to get the encrypted data. Since, encryption is done on the response data alone, no changes are required in the parcel structure. If in addition to this, if the kernel can infer the file decriptor(fd) of a process by itself, then the XBRF can also be prevented. The random key associated with each fd will be different. Then an attacker will not be able to decrypt the response from server as the secret key returned by the kernel will be respective to the actual client application and not with respect to the fd it holds. Hence, the key generated for the encryption and decryption will be different. This will prevent the least privileged process from accessing the higher privileged service. The encryption, decryption will incur additional overhead on the kernel, but the concern here is the security and confidentiality of the inter process communication data.

6 Conclusion

The data transferred between an application and the Binder driver is in plain text. Hence, a man-in-the-middle attack can reveal the communication to an attacker. The proposed encryption scheme, encrypt all the communication from a service to a client, thereby preventing such attacks. For this purpose, a random key is generated by the kernel for each client process and encryption is done using this. Each time the client initiates a communication, a new random key is used. This solves the key storage issue. The additional overhead on kernel due to encryption is a tradeoff between security and performance. Using LWC techniques such as CLEFIA can reduce this overhead.

References

1. Khan, S.J., Cavallaro, L., Fattori, A., Tam, K., Reina, A.: On the reconstruction of android malware behaviors (2014)
2. Artenstein, N., Revivo, I.: Man in the Binder: He who controls IPC, controls the droid (2014). https://www.blackhat.com/docs/eu-14/materials/eu-14-/Artenstein-/Man-In-The-Binder-/He-Who-Controls-/IPC-Controls-The-Droid-wp.pdf
3. Azzola, F.: Android Bound Service: IPC with Messenger. SurvivingWithAndroid.com (2014). http://www.survivingwithandroid.com/2014/01/android-bound-service-ipc-with-messenger.html
4. developer.android.com. Binder (2014)
5. elinux.org. Android binder. *eLinux.org* (2014). http://elinux.org/Android_Binder

6. Gargenta, A.: Deep dive into Android IPC/Binder framework. In: AnDevCon: The Android Developer Conference (2012). https://thenewcircle.com/s/post/1340/deep_dive_into_android_ipc_binder_framework_at_andevcon_iv
7. Jia, P., He, X., Liu, L., Gu, B., Fang, Y.: A framework for privacy information protection on Android. In: 2015 International Conference on Computing, Networking and Communications (ICNC), vol. 2, pp. 1127–1131. IEEE (2015)
8. Katagi, M., Moriai, S.: Lightweight cryptography for the internet of things (2008)
9. Rosa, T.: Android binder security note: On passing binder through another binder
10. Schreiber, T.: Android binder. Master's thesis, Ruhr University, Bochum, Germany, October 2011. http://www.ruhr-uni-bochum.de/. http://www.nds.rub.de/media/attachments/files/2012/03/binder.pdf
11. Shirai, T., Shibutani, K., Akishita, T., Moriai, S., Iwata, T.: The 128-bit blockcipher clefia (2007)
12. Yaghmour, K.: Embedded Android: Porting, Extending, and Customizing. O'Reilly Media Inc., Sebastopol (2013)

Android Smudge Attack Prevention Techniques

M.D. Amruth and K. Praveen

Abstract Graphical patterns are widely used for authentication in touch screen phones. When a user enters a pattern on a touch screen, epidermal oils of his skin leave oily residues on screen called smudge. Attackers can forensically retrieve this smudge which can help them to deduce the unlock pattern. In this paper we analyze some existing techniques and propose new techniques to prevent this attack. We propose Split pattern, Wheel lock, Random PIN lock and Temporal lock to reduce or prevent smudge attack. Usability and shoulder surfing resistance were also considered while designing these techniques. This paper explains how the proposed techniques are effective against smudge attacks.

Keywords Smudge · Smudge attack · Android security · Authentication · Wheel lock · Graphical authentication · Touch screen · Temporal lock · Screen lock · Random PIN lock

1 Introduction

Android is the most widely used OS (Operating System). It appears to be the present and near future of mobile computing. Authentication is a crucial step to Android Security. Authentication can be done based on three factors namely what you know, what you have and what you are. Graphical authentication methods based on what you know are very popular as people can remember pictures better than PINs (Personal Identification Number) [1, 2]. When a user enters a pattern on a touch screen, epidermal oils of his skin leave oily residues on screen called smudge. Attackers can forensically retrieve and analyze this smudge to deduce patterns. Smudge attack (See Figure 1) is one of the most prominent attacks on graphics based

M.D. Amruth(✉) · K. Praveen
Amrita Vishwa Vidyapeetham, Amritanagar, Coimbatore 641112, India
e-mail: {amruthmd,praveen.cys}@gmail.com

S. Berretti et al. (eds.), *Intelligent Systems Technologies and Applications*,
Advances in Intelligent Systems and Computing 385,
DOI: 10.1007/978-3-319-23258-4_3

authentication mechanisms. This vulnerability is brought in by touch screens which are widely used in Android devices. This paper proposes some new screen locking mechanisms which are secure against smudge attacks.

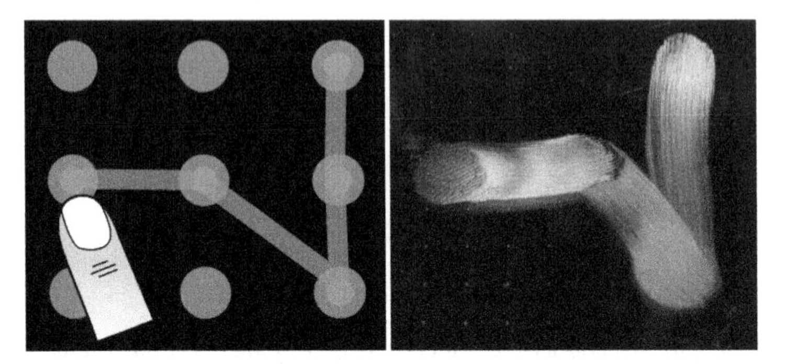

Fig. 1 Sample pattern and its Smudge showing directon of edges

We followed a step by step procedure to come up with the proposal of an implementable model. Smudge attack being a very practical attack has to be solved practically with sound logical backing. Smudge can be wiped off with a good cotton cloth. But such techniques were not considered in this paper as they are not practical. We made a set of ideas in the initial phase. Then they were analyzed and filtered based on usability, implementability, smudge resistance and shoulder surfing resistance to choose four secure techniques to be implemented.

2 Related Work

Android had been subjected to a wide variety of attacks over the years due to its popularity and vulnerablity. It has been found that given physical access an expert attacker can gain privileged access [3] without breaking screen lock. Earlier attackers could install rootkits in phone to spy on user. Android versions from 4.4 supports Verified boot which detects such attacks to a good extent [4]. This has increased importance of screen locking tecniques(to prevent spying) though they are still helpless if attacker takes permanent physical access. Graphical authentication techniques are widely used to lock touch screens. Searchmetric, Locimetric and Drawmetric systems are the three main categories of graphical authentication. These are authentication techniques based on what you know.

In Searchmetric systems user should choose a predefined image from a group of randomly chosen images displayed on screen [5, 6]. Searchmetric systems are usually secure against shoulder surfing [7]. In Locimetric systems user should choose predefined positions from an image. These techniques are prone to smudge attacks on touchscreens. Security of Searchmetric systems depends on its search space which gets limited practically due to small size of touchscreens.

In Drawmetric systems user should draw a predefined pattern on touchscreen. Google has made it the most popular graphical authentication mechanism. But this popular authentication mechanism is prone to smudge attacks. The ordinary pattern technique on 3×3 matrix of vertices used by Google is a broken technique. Though it has pretty good theoretical search space, [8] most of it is not practically useable as users cannot draw some of those patterns. Lack of pattern rotation makes it vulnerable to smudge attack.

Marbles [9] method was found to be secure against smudge attack. But remembering colors is a difficult task for many men. Farnsworth-Munsell 100 Hue Test has found that 1 in 12 men have some form of color vision deficiency. This fact seriously affects the usability of Marbles method. Compass method [9] seems promising except for the fact that the same number cannot be reused. The 3×3 pattern rotation [9] method is effective against smudge attack. But it reduces diagonal length of pattern which makes the dots closer. This will make it less user friendly on phones with smaller screens. Users may dislike it as they may have to physically rotate the phone to enter the pattern.

Biometric and behavior based authentication mechanisms are generally secure from smudge attacks and shoulder surfing. Authentication mechanisms based on how you enter the pattern [10] are promising as they authenticate user based on what he is. Still they are unpopular as they are not accurate enough and require a lot of training. Biometric authentication techniques like fingerprint recognition and retinal scan are not popular on smart phones due to attacks [11, 12]. Touchstroke based method [13, 14] in iOS authenticates users based on four biometric factors namely speed, distance, hold time and inter time. But the fact that its accuracy is heavily dependent on the training at setup time makes it vulnerable and affects its usability. Moreover the possibility of this technique getting misused if the Android device gets compromised cannot be neglected.

Biometric gait recognition [15] which authenticates user based on how he walks was found to be user friendly as it does not require any user interaction or external hardware. But it works only when the user walks and has high error rate. Shoulder surfing is a major threat to smart phone authentication techniques. [7] It can be prevented by asking the user to answer indirect questions about the secret. But this may not be practical as most users may find it difficult to answer the indirect questions.

PIN based authentication schemes are relatively more secure against smudge attack than graphical authentication techniques. It has been found that permitting user selection in any graphical authentication scheme [16] will reduce its security. So it is still not time to write off pin based authentication schemes.

3 Threat Model

People use smart phones for a wide range of tasks and authentication is just the first step. This usage adds smudge noise to the smudge of authentication secret. This reduces success of smudge attack. In the best case user wipes off smudge well with cotton cloth just after unlocking screen and use phone for long enough to add smudge noise. This is the worst case for an attacker. Worst case (for user) is when user wipes screen just before authentication and locks the phone just after unlocking. Here when user wipes screen just before authentication all smudge noise will get lost. The smudge of screen unlock pattern will remain intact as user did not use the phone to add any smudge noise after unlocking. This is the best case for the attacker. The average case is when user rarely wipes the screen.

4 Proposed System

Auto connect was a proposed idea. According to it user should select numbers or points from a 3×3 grid of numbers. When the user selects a second point after selecting the first point, a directed edge is generated automatically connecting the first point to the second. This continues till the user selects OK. The location of numbers is not random. This technique eliminates edges from the smudge of the pattern. The absence of edges in smudge will make it very difficult for an attacker to deduce the order in which the vertices were connected. But since locations of numbers are not randomized, it is equally vulnerable to smudge attack as an ordinary number lock. But it is more user friendly as many people find it easier to remember patterns [1, 2] than numbers. This technique can be combined to randomization of the location of numbers. But then it will become same as the PIN lock proposed below. So this technique was discarded in prototype analysis stage of this project. But this idea is partly used in the split pattern method proposed below.

Overdrawing nine dots was another idea. It is similar to ordinary 3×3 matrix except that the user can select same dots multiple times (not sequentially). Pattern can have a minimum of five points and a maximum of fifteen points. This technique will make the pattern bigger making guessing more difficult. Complexity of pattern combined with small average size of screen will make shoulder surfing difficult. It will confuse a smudge attacker thereby making smudge attack difficult and shoulder surfing difficult. This idea is partly used in the split pattern method proposed below.

We followed a step by step procedure to come up with the proposal of an implementable model. Usability, implementability, smudge resistance and shoulder surfing resistance were considered to choose 3 secure techniques to be implemented from the set of ideas generated during the initial phase. Some techniques were partially or fully discarded. Concepts like auto connect and overdrawing 9 dots were added to selected proposals like split lock.

4.1 Wheel Lock

It uses a group of 4 to 12 digits as secret. Digits of the secret may repeat. The secret is just a group of digits without any order. One scrollable wheel is displayed at a time with axis of rotation horizontally along screen as shown in Figure 2. Digits from 0 to 9 are displayed in order on the wheel. User can scroll up or down to get the required number and long press a number to select it. Screen will vibrate but wheel will not scroll when a number is selected. Then user chooses next digit from the wheel. He may or may not have to scroll the wheel to find the next digit. He need not enter the digits in order.

After entering all digits of the secret, user should swipe the screen below the wheel from left to right. Screen gets unlocked if the secret is correct. Else user should enter the secret again. User can delete the last number entered by swiping the screen below the wheel from right to left. The number of digits entered is not indicated on screen unlike PIN lock. This will make shoulder surfing difficult as attacker will find it difficult to deduce if a digit was entered or not. But user can change this default setting and display the number of digits entered. User can change the default number selection method from log press to touch if he wants to enter quickly.

User should enter secret correctly in three chances. Else he will have to wait for ten minutes to retry. The digit displayed on top of the screen will be the units digit of the clock time in seconds at the instant when phone wakes up to authenticate. Wheel will be as shown in Figure 2 if the user turns on the screen to login at 11:13:03 pm or xx:xx:x3 pm. Wheel will get reset based on clock again only on re login attempt. Smudge left by the wheel lock will look like a vertical line. This will leave no clue for the attacker to deduce the number due to the randomness involved. This will make smudge attack very difficult. Shoulder surfing is difficult as long pressing cannot be detected easily by attackers. Since order of digits is irrelevant user can quickly enter even a ten digit secret.

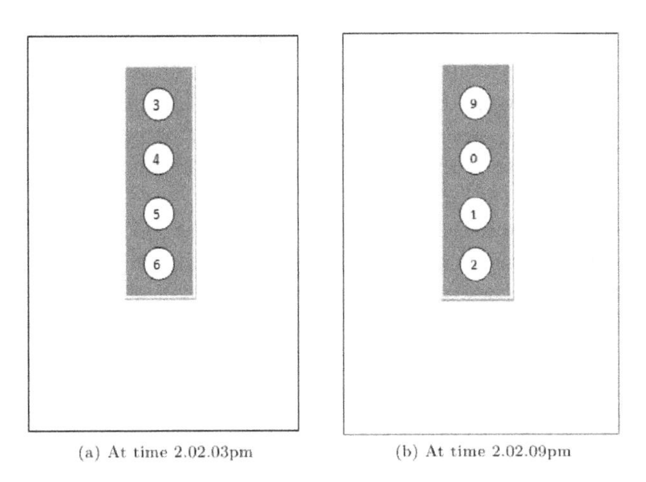

(a) At time 2.02.03pm (b) At time 2.02.09pm

Fig. 2 Prototype of Proposed Wheel lock at different times

4.2 Split Pattern

This technique allows user to enter multiple patterns on same screen as shown in Figure 3 or on different screens as shown in Figure 4. By default the user will have to enter the second pattern along with the first pattern in the same screen.

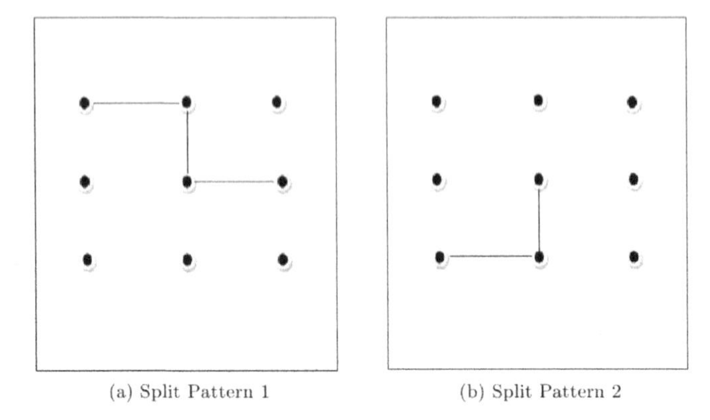
(a) Split Pattern 1 (b) Split Pattern 2

Fig. 3 Prototype of Split Patterns entered on different screens

This technique uses the auto connect feature explained above. According to it user should select points on a 3×3 grid. When the user selects a second point after selecting the first point, a directed edge is generated which automatically connects the first point to the second. This will remove edges and direction of edges from smudge.

To enter multiple patterns on same screen, first user will enter a graphical pattern in an ordinary 3×3 matrix of points by touching vertices in order. User should not draw edges as auto connect is enabled. Then he can split and enter another pattern or complete unlocking by swiping the bottom of screen from left to right. To split he should long press the screen below the pattern. User can enter the next pattern on same screen after the screen vibrates due to long press. He should swipe the bottom of screen from left to right after entering all patterns. Same point can be selected in different patterns due to split option. Splitting can be done atmost once. So user can enter upto two patterns.

During the setup phase user can change the default and choose to enter the second pattern after refreshing the screen as shown in Figure 4. Similar to the above method user will enter the first pattern using auto connect. Then he can split (refresh to remove first pattern) and enter another pattern or complete unlocking by swiping the bottom of screen from left to right . To split he should long press the screen below the pattern. User can enter the next pattern on same screen after the screen vibrates due to long press. He should swipe the bottom of screen from left to right after entering all patterns.

It will make smudge attack and shoulder surfing difficult. Now search space of attacker will become the square of the search space of ordinary pattern on 3×3 matrix. We argue that pattern splitting technique will be as user friendly as ordinary pattern on 3×3 matrix. It can be as simple as ordinary pattern method if user does not use splitting in the authentication pattern. Still it will be more secure from smudge attacks than ordinary patterns due to auto connect feature explained above. Shoulder surfing is made difficult by complexity involved in the pattern.

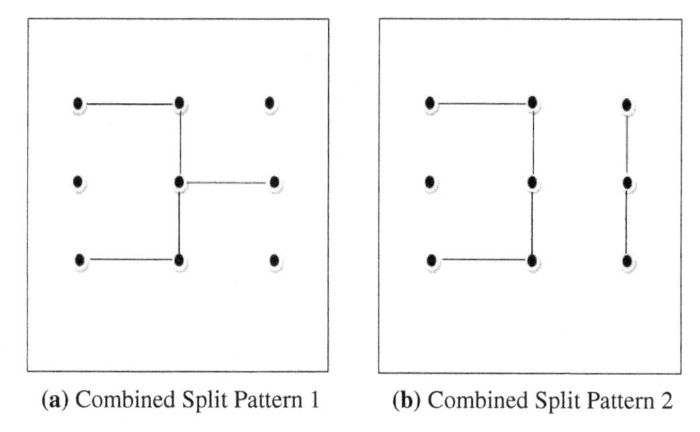

(a) Combined Split Pattern 1 (b) Combined Split Pattern 2

Fig. 4 Prototype of Split Patterns entered in the same screen

4.3 Random PIN Lock

It is similar to Current PIN lock shown in Figure 5 except that numbers will be displayed randomly as shown in proposed Random PIN Lock in Figure 5 . User has to set the PIN which can have 4 to 8 digits. Here the first digit in the lock will be decided based on the units digit of the clock time in seconds similar to the wheel lock explained above. User will have to enter them correctly in three chances. Else user will have to wait for ten minutes to retry.

In ordinary number lock attacker can deduce the number entered from smudge. Random PIN lock adds randomization which makes it almost impossible for attacker to deduce the number entered from smudge. We argue that a shoulder surfer will find it more difficult to deduce selected number based on location as he may not be able to see it clearly from a distance. So we argue that Random PIN lock is more secure than ordinary PIN lock.

4.4 Temporal Lock

Different authentication mechanisms are set for different time intervals. User can choose the authentication technique for each time interval. Length of the interval can be chosen by the user. The mechanisms proposed above can be combined with existing authentication mechanisms. User should set the authentication mechanisms for each time interval when he first uses the phone (during the time when he enters Google account) after flashing the new ROM. The basic objective of this method is to prevent shoulder surfing. It will reduce smudge attacks as it will confuse the attacker. If the user is not able to remember multiple techniques then he can reduce the number of authentication techniques to one. It will support all existing Android authentication techniques.

(a) Current PIN Lock

(b) Random PIN Lock

Fig. 5 Android PIN locking techniques

5 Implementation

Screen unlocking mechanism of Android is not implemented like an Android applications in a sandbox. It is coded in the Android framework. This code can be modified to change the login mechanism. Practically this involves rebuilding Android ROM from new source code. We use Lollipop 5.0 for our implementation.

6 Conclusion

Automated attacks on Android were not considered for this study as expert attackers have better ways to attack it as mentioned in [3].Split pattern lock technique can resist smudge attack effectively while maintaining user friendliness and shoulder surfing resistance. The Temporal lock technique which combines existing authentication techniques with proposed techniques, is very effective against shoulder surfing and smudge attack.

The wheel lock technique based on number grouping logic is very effective against smudge attack. The Random PIN lock method can be implemented by making small changes to the source code of existing PIN lock. It is arguably better than the existing PIN lock mechanism in terms of Shoulder surfing resistance and Smudge attack resistance. Four techniques proposed in this paper were logically analyzed and found to be effective against smudge attack.

References

1. Hockley, W.E.: The picture superiority effect in associative recognition. Memory & Cognition **36**(7), 1351–1359 (2008)
2. Jermyn, I., Mayer, A.J., Monrose, F., Reiter, M.K., Rubin, A.D., et al.: The design and analysis of graphical passwords. In: Usenix Security (1999)

3. Vidas, T., Votipka, D., Christin, N.: All your droid are belong to us: a survey of current android attacks. In: WOOT, pp. 81–90 (2011)
4. Google.com: Google Report: Android Security 2014 Year in Review. Tech. rep., Google.com, April 2015. https://static.googleusercontent.com/media/source.android.com/en/us/devices/tech/security/reports/Google_Android_Security_2014_Report_Final.pdf
5. Brostoff, S., Sasse, M.A.: Are passfaces more usable than passwords? a field trial investigation. In: People and Computers XIVUsability or Else!, pp. 405–424. Springer (2000)
6. De Angeli, A., Coutts, M., Coventry, L., Johnson, G.I., Cameron, D., Fischer, M.H.: Vip: a visual approach to user authentication. In: Proceedings of the Working Conference on Advanced Visual Interfaces, pp. 316–323. ACM (2002)
7. Wiedenbeck, S., Waters, J., Sobrado, L., Birget, J.C.: Design and evaluation of a shoulder-surfing resistant graphical password scheme. In: Proceedings of the Working Conference on Advanced Visual Interfaces, pp. 177–184. ACM (2006)
8. Aviv, A.J., Gibson, K., Mossop, E., Blaze, M., Smith, J.M.: Smudge attacks on smartphone touch screens. WOOT **10**, 1–7 (2010)
9. Von Zezschwitz, E., Koslow, A., De Luca, A., Hussmann, H.: Making graphic-based authentication secure against smudge attacks. In: Proceedings of the 2013 International Conference on Intelligent User Interfaces, pp. 277–286. ACM (2013)
10. De Luca, A., Hang, A., Brudy, F., Lindner, C., Hussmann, H.: Touch me once and i know it's you!: implicit authentication based on touch screen patterns. In: Proceedings of the SIGCHI Conference on Human Factors in Computing Systems, pp. 987–996. ACM (2012)
11. Prabhakar, S., Pankanti, S., Jain, A.K.: Biometric recognition: Security and privacy concerns. IEEE Security & Privacy **2**, 33–42 (2003)
12. Uludag, U., Pankanti, S., Prabhakar, S., Jain, A.K.: Biometric cryptosystems: issues and challenges. Proceedings of the IEEE **92**(6), 948–960 (2004)
13. Damopoulos, D., Kambourakis, G., Gritzalis, S.: From keyloggers to touchloggers: Take the rough with the smooth. Computers & Security **32**, 102–114 (2013)
14. Kambourakis, G., Damopoulos, D., Papamartzivanos, D., Pavlidakis, E.: Introducing touchstroke: keystroke-based authentication system for smartphones. Security and Communication Networks (2014)
15. Derawi, M.O., Nickel, C., Bours, P., Busch, C.: Unobtrusive user-authentication on mobile phones using biometric gait recognition. In: 2010 Sixth International Conference on Intelligent Information Hiding and Multimedia Signal Processing (IIH-MSP), pp. 306–311. IEEE (2010)
16. Davis, D., Monrose, F., Reiter, M.K.: On user choice in graphical password schemes. In: USENIX Security Symposium, vol. 13, pp. 11–11 (2004)

Active and Entire Candidate Sector Channel Utilization Based Close Loop Antenna Array Amplitude Control Technique for UMTS and CDMA Networks to Counter Non Uniform Cell Breathing

Archiman Lahiry, Amlan Datta and Sushanta Tripathy

Abstract The paper introduces a self optimized close loop antenna array amplitude control system to counter the effect of non-uniform cell size breathing in UMTS and CDMA networks by avoiding the overshooting cells in the entire network automatically. In the proposed antenna array amplitude control system the active sector's channel utilization as well as the relative channel utilizations of the entire candidate sectors is considered. We introduce a feedback system to detect physical parameters of the entire cell site sector antennas, like antenna heights above the ground level, antenna azimuths, antenna mechanical tilts, antenna latitudes and longitudes to avoid the chances of overshooting cells when the antenna array amplitudes are controlled remotely from the OMC-R. Remote mechanical antenna tilts are used to eliminate overshooting cells. The contribution of the proposed work is to develop an overshooting resistant antenna array amplitude control system to counter the non-uniform cell size breathing.

Keywords Close loop antenna array amplitude control · Relative channel utilization · Overshooting cells · Non-uniform cell size breathing

1 Introduction

The adjacent cells overlap in CDMA and W-CDMA networks to support soft and softer handoff.

Cell size breathing can lead to the shrinkage of coverage area of a cell and thereby generating the coverage holes in the busy hour in a cellular mobile network.

A. Lahiry(✉) · A. Datta · S. Tripathy
School of Electronics Engineering, KIIT University, Bhubaneswar, India
{archiman.lahiry87,amlandatta01,sushant.tripathy}@gmail.com

S. Berretti et al. (eds.), *Intelligent Systems Technologies and Applications*,
Advances in Intelligent Systems and Computing 385,
DOI: 10.1007/978-3-319-23258-4_4

There are chances of call drops and handoff failures when user equipment (UE) moves into the region with an insufficient signal level which is less than minimum UE receiver sensitivity. The BTS and Node-B sectors have a "non-uniform traffic" [1] utilizations in the real network. In [2,6,9], the specific state of the art problem is illustrated. "Site densification" [2,3], using "smart antennas" [2] and "sectorization" [2,3] are the existing solutions to counter the shrinkage of coverage area due to cell size breathing. Smart and beam-forming antennas are costly and the signal processing of these antenna systems is extremely complex. Beam-forming also cannot increase the antenna footprints uniformly in all the directions like "variable amplitude tapered antenna array" [4,5].

We proposed the method of "OMC-R controlled remotely variable antenna array amplitude tapering" [4, 5] scheme to counter the effect of cell size breathing. We introduced the "variable vertical antenna array amplitude tapering" [4] scheme with hardware design to compensate the cell shrinkage in the soft handoff region. Then we proposed an "improved version of variable antenna array amplitude tapering scheme" [5] in which we stated that if we implement variable horizontal and vertical antenna array amplitude tapering simultaneously, then we can improve and compensate the coverage due to cell shrinkage in the soft as well as in the softer handoff region. In our previous works we also mentioned that "antenna array amplitude tapering is directly proportional to the channel utilization" [4] and the "antenna array amplitude tapering is controlled by an algorithm defined at OMC-R (operation and maintenance center for radio)" [4]. In this paper the channel utilization of the sector will sometimes be called as congestion, traffic utilization or loading.

In our previous works the channel utilization of the active sector was only considered and the channel utilizations of candidate sectors were not considered. In the present work, a close loop two dimensional antenna array amplitude tapering scheme is proposed where the active sector's channel utilization as well as the channel utilizations of entire candidate sectors will be considered in order to improve the overall network performance. Antenna array amplitude tapering can combat cell size breathing, but if the "antenna tilts" [8] and antenna array amplitude tapering are not optimized properly, then it may cause overshooting cells. Therefore, we have developed a hardware system to detect and correct the physical parameters of cell site sector antennas in the entire network. The physical parameters that will be detected from OMC-R are antenna heights above the ground level, antenna mechanical tilts, antenna latitudes, antenna longitudes, inter-site distances and antenna azimuths measured clockwise with respect to true north in degrees. Laser meters will be used for detecting the antenna heights of the entire sector antennas above the ground level. Digital compass will be installed behind the entire sector antennas for detecting the antenna azimuths. Tilt meters will be used for monitoring and correcting the remote mechanical tilts of entire sector antennas. GPS sensors will be used for detecting the latitudes and longitudes of entire sector antennas.

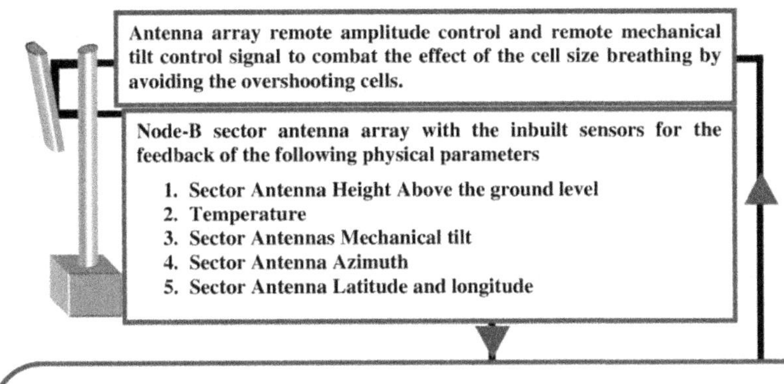

Fig. 1 The close loop remote antenna array amplitude control system to combat the effect of cell size breathing by avoiding the chances of overshooting cells

The Fig. 1 illustrates the overall idea of the proposed system. Antenna array amplitude control and physical self optimization of the entire network will be done automatically by the software installed at the OMC-R. The software installed at OMC-R will evaluate antenna array pattern by calculating array factors for different amplitude distributions. The software will also evaluate and control the footprints of the variable tapered antenna array [4,5] by evaluating the hardware feedbacks and by doing the mathematical calculations. We can also call the proposed method "*a close loop network defined intelligent antenna array amplitude taper control system*"

2 Antenna Footprint Calculations

In this section we will derive all the mathematical expressions for the re-configurable amplitude control antenna array footprint calculations.

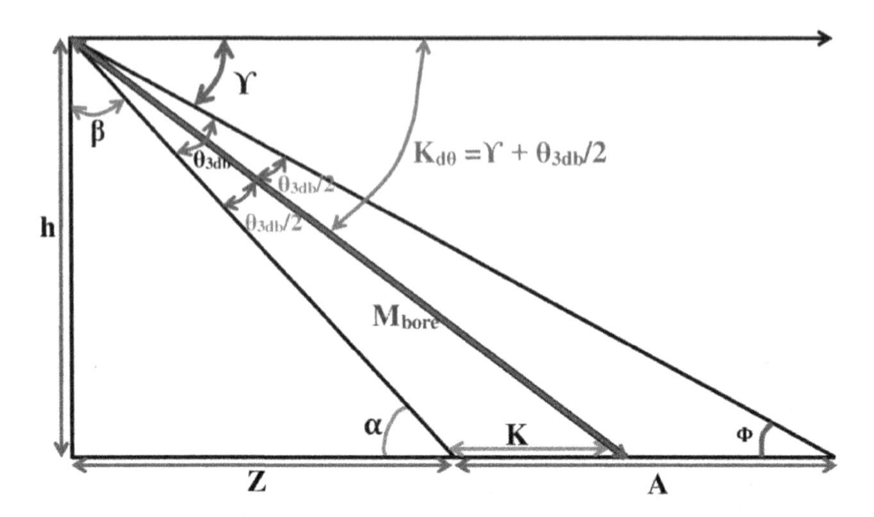

Fig. 2 The footprint A of the antenna array due to the half power ground touchdown points of the main lobe in the vertical direction, antenna height above the ground level and the mechanical down tilt angle

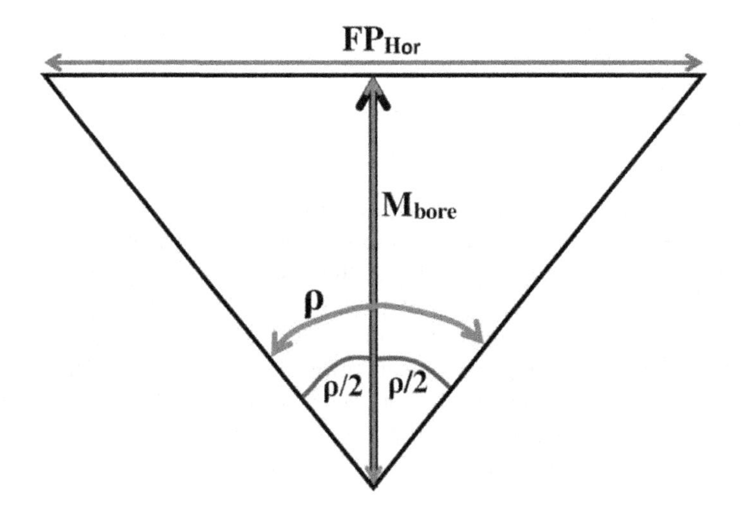

Fig. 3 The footprint FP_{Hor} of the antenna array due to the half power ground touchdown points of the main lobe in horizontal direction, antenna height above the ground level and the mechanical down tilt angle

Antenna parameters like vertical beam widths, horizontal beam widths, mechanical down tilt angles and antenna height above the ground level are expressed mathematically. The footprint A can be controlled by varying θ_{3dB} and the footprint FP_{Hor} can be controlled by varying ρ. Now in the illustrated Fig. 2 we mentioned the tunable optimal mechanical down tilt angle $\Upsilon + \theta_{3db}/2$ which is equal to $K_{d\theta}$. In the illustrated Fig. 2 the re-configurable vertical half power beam width of the variable tapered antenna array is θ_{3db}. In Fig. 3 the re-configurable horizontal half power beam width of the variable tapered antenna array is ρ. The cell site physical parameters can be controlled remotely from OMC-R by retrieving the entire cell site antenna physical parameters accurately. The software will control the footprint by calculating all the mentioned parameters at OMC-R.

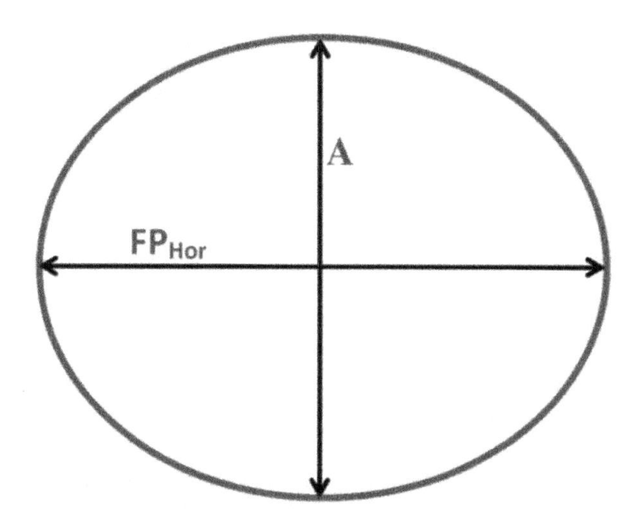

Fig. 4 The overall footprint of the antenna array main lobe half power ground touchdown points in the horizontal and the vertical directions

The mathematical expressions for optimal antenna mechanical down tilt angle is given below

$$K_{d\theta} = (\theta_{3dB}) + \gamma \tag{1}$$

Now relations with other parameters are given below

$$\mathrm{Tan}\,(\theta_{3dB} + \beta) = \mathrm{Cot}\,\phi = (A + Z)/h \tag{2}$$

$$\theta_{3dB} = \mathrm{Tan}^{-1}((A + Z)/h)) - \beta \tag{3}$$

$$\beta = \mathrm{Tan}^{-1}((A + Z)/h) - (\theta_{3dB}) \tag{4}$$

Now the calculation of optimal antenna mechanical down tilt angle

$$K_{d\theta} = ((180 - 2\beta - (\theta_{3dB}))/2) \tag{5}$$

The parameter α is

$$\alpha = \mathrm{Cot}^{-1}(Z/H) = \mathrm{Tan}\text{-}1(h/Z) \tag{6}$$

The parameter ϕ is

$$\phi = \mathrm{Cot}^{-1}((A/h) + (Z/h)) \quad (7)$$

Finally the area of coverage due to the antenna array vertical beam width is given below

$$A = (\mathrm{Sin}(\alpha - \phi)\, h / \mathrm{Sin}\, \alpha \times \mathrm{Sin}\, \phi) \quad (8)$$

In the Fig. 2 M_{bore} is the antenna bore sight ground touchdown distance from the antenna. The expression for calculating the parameter is given below:

$$\mathrm{Cos}(\beta + (\theta_{3dB}/2)) = h/M_{bore} \quad (9)$$

$$M_{bore} = h/\mathrm{Cos}(\beta + (\theta_{dB}/2)) \quad (10)$$

Now the expression for FP_{Hor} calculation is given below:

$$\mathrm{Tan}(\rho/2) = (FP_{Hor}/2)/M_{bore} \quad (11)$$

$$\rho = 2 \times \mathrm{Tan}\text{-}1\,(FP_{Hor}/2.M_{bore}) \quad (12)$$

$$FP_{Hor} = 2 \times M_{bore} \times \mathrm{Tan}(\rho/2) \quad (13)$$

All the above expressions will be calculated and controlled automatically by the software by evaluating the antenna's sensors feedbacks of the entire network. Initially the antenna array physical parameters will be evaluated and then the antenna array amplitude control will be implemented according to the cell site sector antenna's footprint requirements. Software at OMC-R will control FP_{Hor} and A by controlling ρ and θ_{3dB}. Antenna mechanical down tilt angle $K_{d\theta}$ will be controlled remotely from OMC-R by using stepper motors in order to eliminate overshooting cells. The footprint will be controlled mainly by varying the antenna array amplitude taper. The antenna array tilt control will have the second priority.

3 Antenna Inbuilt Sensors for Antenna's Physical Parameter Detection

The sensors inside the Node-B sector antennas will send the accurate data of the antenna array physical parameters. Laser meter will accurately measure the antenna height above the ground level. The formula for measuring the height is given below:

$$h = Ct/2 \quad (14)$$

In the above equation C is the velocity of light, h is the antenna height above the ground level and t is time taken for reflection of the laser beam from ground surface to the laser meter. The antenna sensors will be monitored by OMC-R at regular intervals for retrieving all the physical parameters of the entire Node-B sector antennas for the antenna physical self optimization purpose. The laser meter will simply send the value of t to OMC-R and by using the equation no. 14 the software will calculate the value of h of the entire Node-B sector antennas. The laser beam makes an angle of 90 degrees with respect to the ground surface. The Beam should be reflected from the ground surface and the beam should not be obstructed by any object therefore installation should be carried out properly. The power supply management of the antenna sensors of entire Node-B sector antennas will be handled by OMC-R. The sensors will be switched on by the OMC-R only when the physical parameter data from sites will be required for physical self optimization.

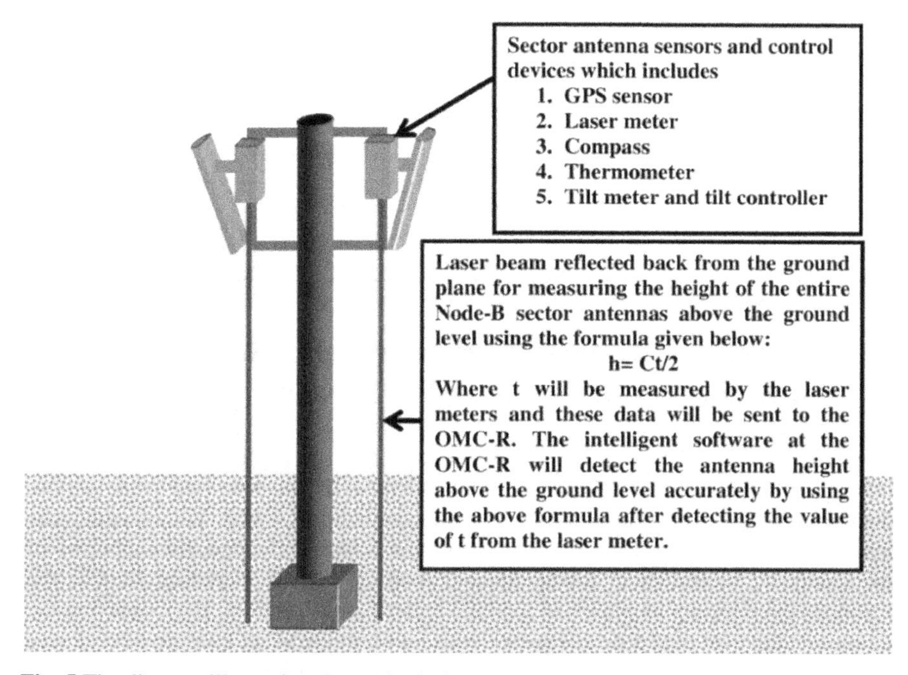

Fig. 5 The diagram illustrating the method of remote antenna array height above the ground level measurement system using the OMC-R controlled laser meter

The GPS sensors in the entire sector antennas will be at the top of all the sensor devices in order to locate the sector antenna's accurate geographical position.

4 Antenna Array Half Power Beam-Widths with the Variation of the Array Amplitude Distributions in Two Dimensions

In this section half power beam-widths of the antenna array factor patterns with "different amplitude distribution" [4,5] will be analyzed. 3×12 element planar array will be used as the base station antenna with remote amplitude control.

Table 1 Antenna array vertical halfpower beam-widths in different amplitude distributions

Vertical inter-element spacing	Amplitude distributions	Side lobe level	Half power beamwidth (θ_{3dB})
0.65λ	Normal Distributions	-13.5 dB	6.53°
	Dolph Tchebyshev taper	-20 dB	7.07°
	Dolph Tchebyshev taper	-23 dB	7.46°
	Dolph Tchebyshev taper	-26 dB	7.82°

Table 2 Antenna array horizontal half-power beam-widths in different amplitude distributions

Horizontal inter-element spacing	Amplitude Distributions	Side lobe level	Half Power beam-width (ρ)
0.32λ	Dolph Tchebyshev taper	-19.5 dB	65.05°
	Binomial taper	No Side lobes	69.33°

5 Methods to Counter the Effect of Non Uniform Cell Size Breathing in UMTS and CDMA Network

We will consider Node-B sector's active or own loading percentage as well as the loading percentage of the entire candidate sectors. In our proposed work we are dealing with non-uniform loading or traffic utilization. We will consider the availability of the channels, therefore if a sector is heavily loaded, then tapering of the candidate sectors with lesser channel loading will be increased so that the footprint overlapping increases in soft and softer handoff region. By increasing the amplitude taper of the Node - B sector with minimum load we can provide service to more no of users. The idea is to utilize the available channels efficiently and therefore if there are more channels available with any candidate sector, then that sector will increase its coverage area by increasing the antenna array amplitude taper and shift its footprint towards its candidate sector with maximum loading. Antenna array amplitude taper variation in our proposed work will be non uniform. Footprint calculations will be done by the software by using all the equations we have derived in the section 2. The database of all these equations will be maintained at OMC-R. Now we will analyze a set of very specific loading conditions thereafter we will further define a set of conditions for our proposed method at OMC-R. The maximum distance up to which A+Z can be varied is less than 80% of the nearby candidate site distance d, which is the maximum distance which is controlled according to the loading conditions. We consider the unequal channel utilization of two sectors 2 of Node-B ID 1 and sector 6 of Node-B ID 2 given in the Fig. 14 with the footprint diagram. We consider that sector 2 with the footprint diagram given in Fig. 14 is having 30 percent channel utilization and its candidate sector 6 with footprint diagram is having 90 percent channel utilization.

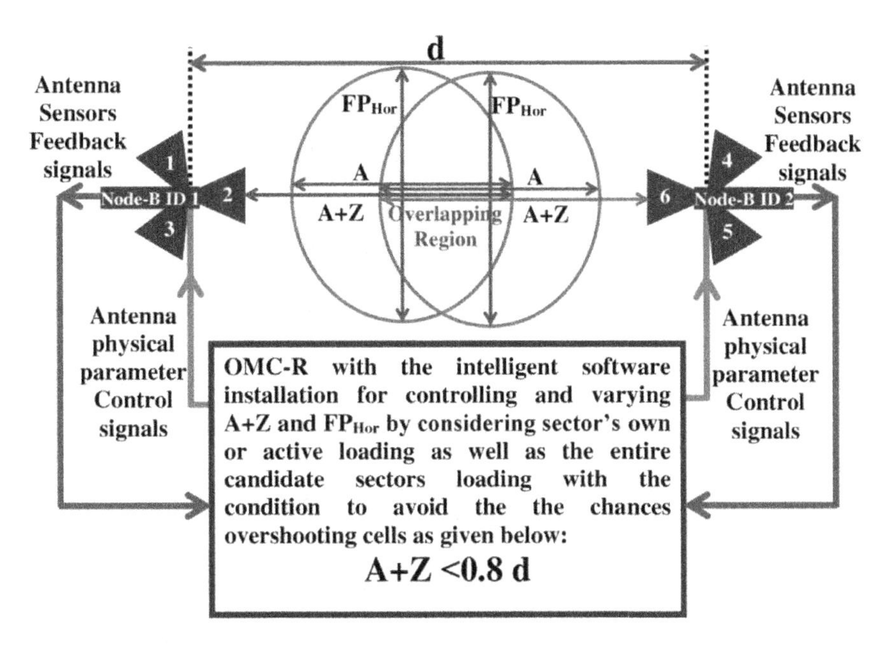

Fig. 6 The figure illustrating the antenna array footprint control systems with the feedbacks and the entire process is controlled from the OMC-R

The value of A+Z will be relatively more for the sector 2 having 30 percent channel utilization and the value of A+Z will be relatively smaller for sector 6 having 90 percent channel utilization. Similarly, FP_{Hor} will also be controlled relatively according to the active and entire candidate sector's channel utilization percentage.

6 Antenna Array Automatic Physical Optimization in the Non-uniform Channel Utilizations or Loading Conditions

The value of A+Z and FP_{Hor} will be controlled by considering the sector's active as well as entire candidate sector loading as given in Fig. 14. Initially, all the physical parameters of entire sectors will be detected and then the conditions will be implemented considering the loading of sector 2 own as well as entire candidate sector loading as given below:

Condition 1: Let us consider relative channel utilization or loading between sector 2 and sector 6
If

$$0\% < \text{loading}\% \text{ (Sector 2)} < 30\%$$
$$80\% < \text{loading}\% \text{ (Sector 6)} < 90\%$$

Then

$$0.7\ d < A+Z \text{ (Sector 2)} < 0.8\ d$$
$$0.65\ d < A+Z \text{ (Sector6)} < 0.7\ d$$

Condition 2: Let us consider relative channel utilization or loading between sector 2 and sector 6

If

$$30\% < \text{loading}\% \text{ (Sector 2)} < 80\%$$
$$80\% < \text{loading}\% \text{ (Sector 6)} < 90\%$$

Then

$$0.7\ d < A+Z \text{ (Sector 2)} < 0.75\ d$$
$$0.65\ d < A+Z \text{ (Sector6)} < 0.7\ d$$

Condition 3: Let us consider relative channel utilization or loading between sector 2 and sector 6

If

$$80\% < \text{loading}\% \text{ (Sector 2)} < 90\%$$
$$80\% < \text{loading}\% \text{ (Sector 6)} < 90\%$$

Then

$$0.65\ d < A+Z \text{ (Sector 2)} < 0.75\ d$$
$$0.65\ d < A+Z \text{ (Sector6)} < 0.75\ d$$

Condition 4: Let us consider relative channel utilization or loading between sector 2 and sector 1

If

$$0\% < \text{loading}\% \text{ (Sector 2)} < 30\%$$
$$80\% < \text{loading}\% \text{ (Sector 1)} < 90\%$$

Then

$$FP_{Hor}\text{(Sector 2) new} = FP_{Hor}\text{(Sector 2)} \times 1.4$$
$$FP_{Hor}\text{(Sector 1) new} = FP_{Hor}\text{(Sector 1)} \times 1.2$$

Condition 5: Let us consider relative channel utilization or loading between sector 2 and sector 1

If

$$30\% < \text{loading}\% \text{ (Sector 2)} < 80\%$$
$$80\% < \text{loading}\% \text{ (Sector 1)} < 90\%$$

Then

$$FP_{Hor}\text{(Sector 2) new} = FP_{Hor}\text{(Sector 2)} \times 1.3$$
$$FP_{Hor}\text{(Sector 1) new} = FP_{Hor}\text{(Sector 1)} \times 1.15$$

Condition 6: Let us consider relative channel utilization or loading between sector 2 and sector 1

If

$$80\% < \text{loading}\% \text{ (Sector 2)} < 90\%$$
$$80\% < \text{loading}\% \text{ (Sector 1)} < 90\%$$

Then

$$FP_{Hor}\text{(Sector 2) new} = FP_{Hor}\text{(Sector 2)} \times 1.2$$
$$FP_{Hor}\text{(Sector 1) new} = FP_{Hor}\text{(Sector 1)} \times 1.2$$

FP_{Hor}(Sector1) and FP_{Hor}(Sector2) are the antenna footprints due to horizontal beam width with -19.5 dB side lobe level. FP_{Hor}(Sector 2) new and FP_{Hor}(Sector 1) new are the footprints after increasing the horizontal antenna array amplitude taper. We considered sector 2 active loading as well as loading of entire candidates sectors 1,3,6. Conditions for sector 2 and sector 3 will be similar to that of sector 2 and sector 1.

7 Numerical Results

Table 3 Footprint variation due to the two dimensional antenna array amplitude control

h	$K_{d\theta}$	d	θ_{3db} and amplitude distributions	ρ and amplitude distributions	A+Z< 0.8d	FP_{Hor}
30 meter	5.7°	1200 meter	6.53° for SLL -13.5 dB	65.05° for Dolph Tchebyshev SLL -19.5 dB	705 meter	385.22 meter
30 meter	5.7°	1200 meter	7.82° for Dolph Tchebyshev SLL -26 dB	69.33° for Binomial	959 meter	417.75 meter

Results show that the main lobe beam-width broadening in horizontal direction should be large for getting the best results as it increases the footprint overlapping in the softer handoff region. It is possible by increasing the antenna array horizontal amplitude taper and reducing the horizontal inter-element spacing.

8 Conclusions

The antenna array designs can be modified according to the main beam broadening requirements. Results reveal that if main lobe beam-width broadening in the horizontal direction is large then we get the best results. The horizontal beam broadening can be increased by increasing the horizontal array amplitude taper and by reducing the horizontal inter-element spacing of the antenna array. The paper introduces an extremely flexible antenna array amplitude control system to counter the effect of non-uniform cell size breathing.

References

1. Anpalagan, A., Sousa, E.: Performance analysis of a CDMA network with fixed overlapping sectors in nonuniform angular traffic. IEEE Transactions on Wireless Communication **5**(8), 2050–2060 (2006)
2. Dupuch, Y.: Alcatel lucent's document on how to minimize the impact of cell breathing on UMTS networks, In: IEEE Workshop on Applications and Services in Wireless Networks (2002). http://www-rst.int-evry.fr/~afifi/index.html/panel2.pdf
3. Kelif, J.M., Coupechoux, M.: Cell breathing, sectorization and densification in cellular networks. In: WiOPT 2009. 7th IEEE International Symposium on Modeling and Optimization in Mobile, Ad Hoc, and Wireless Networks, Seoul, pp. 1–7 (2009)
4. Lahiry, A., Tripathy, S., Datta, A.: W-CDMA busy hour handoff optimization using OMC-R controlled remote electronic variable tapered planar array. In: 3rd IEEE International Conference on Communication and Signal Processing, Melmaruvathur, pp. 31–35 (2014)
5. Lahiry, A., Datta, A., Maiti, S.: Improved self optimized variable antenna array amplitude tapering scheme to combat cell size breathing in UMTS and CDMA networks. In: 2nd IEEE International Conference on Signal Processing and Integrated Networks, Noida, pp. 77–82 (2015)
6. Micallef, G., Mongensen, P., Scheck, H.O.: Cell size breathing and possibilities to introduce cell sleep mode. In: IEEE European Wireless Conference, Lucca, pp. 111–115 (2010)
7. Neimela, J., Lempiamen, J.: Mitigation of pilot pollution through base station antenna configuration in W-CDMA. In: 60th IEEE Vehicular Technology Conference, vol. 6, pp. 4270–4274. IEEE press, Los Angeles (2004)
8. Siomona, I.: P-CPICH power and antenna tilt optimization in UMTS networks. In: Telecommunications, 2005. IEEE Advanced Industrial Conference on Telecommunications/Service Assurance with Partial and Intermittent Resources Conference/E-Learning on Telecommunications Workshop, Lisbon, pp. 268–273 (2005)
9. Thng, K.L., Yeo, B.S., Chew, Y.H.: Performance study on the effects of cell-breathing in WCDMA. In: 2nd IEEE International Symposium on Wireless Communication Systems, pp. 44–49 (2005)

M-SEP: A Variant of SEP for WSN

Tenzin Jinpa and B.V.R. Reddy

Abstract Energy efficiency of the protocol is one of the deciding factor while considering the efficiency of a protocol in WSN (Wireless Sensor Network). Thus, this paper presents an improved version of SEP protocol called M-SEP. M-SEP will inherit some properties of the SEP while introducing the multilevel power transmission in the protocol. Thus extending the lifetime of the network. In a nutshell, the idea is to acknowledge the existence of the different minimum energy requirement while transmitting data packets in WSN i.e. the intra transmission of packets require lower energy than that of the entire transmission or from the cluster head to the base station transmission. By implementing the multilevel power transmission in the SEP protocol we improve the efficiency of the SEP protocol as is shown by the simulation result, we called this protocol M-SEP or Modified Stable election Protocol.

1 Introduction

The world of WSN may be the future of communication for mankind as many eminent technocrats openly advocate about it. Recently, Google boss, Eric Schmidt at the World Economic Forum strongly emphasized on the importance of the sensor nodes in our life. Use of the WSN based technology could be seen in almost every fields like in medicine, oceanography, defense and agriculture, etc. Because of its robust nature and tiny size, we can plant hundreds of nodes in remote places with a limited battery life and a small processing power for information gathering purposes. But for the efficiency of the network we need to rely on the large number of the sensor nodes as nodes will die soon because of its limited battery. Thus, it is advisable to make the nodes as cheap and energy efficient as possible while thought should be given on the data aggregation and the problem of the limited share bandwidth while designing the WSN protocols. Many energy efficient protocols were developed to solve the problems like LEACH, SEP and DEEC etc.

T. Jinpa · B.V.R. Reddy(✉)
USICT, GGSIP University, Dwarka, India
e-mail: Kardol123@gmail.com, bvrreddy64@yahoo.co.in

S. Berretti et al. (eds.), *Intelligent Systems Technologies and Applications*,
Advances in Intelligent Systems and Computing 385,
DOI: 10.1007/978-3-319-23258-4_5

This paper introduces a new variant of SEP [1], we will call it **M-SEP (Modified Stable election protocol)**. The **SEP** protocol stand for "**Stable Election Protocol** " is a heterogeneous two level energy based WSN protocol which divides the nodes into two set of nodes i.e. Higher energy nodes called the Advance Nodes while the Lower energy nodes called the Normal Nodes. It will ensure that the Advance nodes become cluster head more frequently than the Normal nodes, which helps in the extending stability period of the network. The SEP while applying the clustering principle in a heterogeneous network does not consider the different types of transmission in WSN system such as the (i) Intra Cluster Transmission (ii) Inter cluster Transmission (iii) Cluster Head to Base Station transmission. SEP will consider all the transmission same category and allocate the same amount of energy to every transmission, which is practically not true. As the minimum energy requirement for Intra Cluster Transmission is different from that of the Inter cluster transmission or the cluster head to the base station transmission. The **M-SEP** protocol is also a heterogeneous clustering based protocol which considers the existence of different transmission types. Thus we apply the multilevel power transmission, i.e. the Intra cluster energy amplification requirement for transmission will be lower than that of the other two types of transmission. This will also reduce the packet drop ratio, Interference and the collision in the network, which will improve the overall result.

2 Related Work

A considerable number of solutions were proposed on various issues of the WSN technology.Early suggestion like "Direct Transmission" and result in result in loss of efficiency of the WSN as the far away nodes will die out faster than the near one while in "MTE(Minimum Transmission Energy)" nodes near to the Base Station (BS) are over populated, result in early demises of the nodes near to the BS thus both solutions fails to stabilize the WSN.

In Eri.Ent Al [2], asymmetric energy efficient WSN Protocol energy efficient protocol for that deal with the hierarchal clustering mechanism. The Jiang et.at presented a "CBRP (cluster based routing protocol)" [3] hierarchal clustering algorithm, though not very efficient energy saving protocol. The cluster based methods have a proven track record of having an efficient mechanism of energy utilization of energy. Implementing the principle of clustering in WSN will lead to low bandwidth utilization and extension of the network life [4]. In [5] the author gave the idea of the multi-hop transmission, i.e. the packets will be transmitted from one cluster head to another cluster head till it reaches the BS thus enhancing the lifetime of the network. The purpose of all these protocols is to enhance the lifetime by reducing the energy consumption and data aggregation like M. Tahir [6] introduces the "link Quality Metric" which divide the network in three parts that conserve energy, but the author of [7] conserves energy by differentiating idle and the operational mode of a sensor node.

The Mod-Leach protocol introduces the wise head selection and multilevel power transmission in Leach Protocol, which increases the efficiency of the Leach Protocol.The Leach protocol is a homogeneous WSN protocol based on the dynamic head selection in the cluster based protocol. But for the heterogeneous WSN many other protocols were developed such as a SEP (Stable Election protocol) which divides the nodes in two sets, i.e. Advance nodes(High energy nodes) and Normal nodes (Low energy nodes) and ensure that advance nodes become head more frequently than the normal nodes. The DEEC protocol is a "**Distributed Energy Efficient Clustering**"[8] protocol which is an distributed based energy efficient protocol for the heterogeneous wireless sensor network .It is also based on the basic Leach protocol but the Cluster head selection method here is different as the cluster head are selected based probability based ration between the residual energy of each node and the average energy of the system.

In [9] Heinzelman studied the election of cluster head based on the residual energy in each node, but keeping global knowledge of residual energy of the system is difficult to implement. In [10] Duarte-Melo and Liu examined the performance and energy consumption of a wireless sensor network,where there are two types of nodes, i.e. the fewer overlay nodes having more energy and the normal nodes. All the normal nodes will report to the overlay node and the overlay nodes will do the aggregation and sent the data packet to the base station.The drawback of this protocol is that cluster head selection is not dynamic as a result nodes that are far away from the base station will died out fast. Moreover, it is surmised the presence of a set of powerful nodes, which is difficult to use when the system is heterogeneous by its operation. In [11] Mhatre and Rosenberg proposed estimation method for optimal distribution faces the difficulty when heterogeneity is due to its operation. Other energy efficient routing scheme considers that exact location of the nodes is known (using GPS in nodes) and initially all nodes are homogeneous. But the using GPS in nodes will add up to its cost and homogeneity assumption may be difficult to implement. Thus, this paper focuses on reduction of energy wastage during transmission in SEP hence improves the property over the Leach [12] and Mod-leach [13] protocols.

3 First Order Radio Model for M-SEP

The radio model of Fig. 1 shows that to achieve an **SNR** (Signal to Noise ratio) while L bit of data packets is transmitted over the distance "d", the energy expended by the radio model is given by

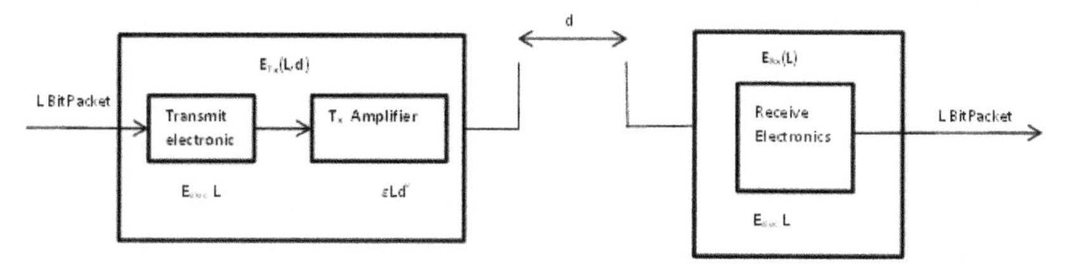

Fig. 1 Radio Energy Dissipation Model [1]

For the inter cluster transmission and for the cluster head to the base station transmission.

$$E_{Tx}(l,d)\begin{cases} L.E_{elec} + L.\varepsilon_{fs}.d^2 & if\ d \le d_o \\ L.E_{elec} + L.\varepsilon_{mp}.d^4 & if\ d > d_o \end{cases}$$

Now E_{elec} is the energy dissipation per bit for running the transmitter and receiver circuit while the ε_{fs} and ε_{mp} depend on the transmitter and amplifier mode.By equating the first equation at $d=d_0$, we have the $d_0=\sqrt{\frac{\varepsilon_{fs}}{\varepsilon_{mp}}}$.

Now for the intra cluster transmission, we will have slightly different model as the energy requirement will be less than the other two types of the transmission.

$$E_{Tx}(l,d)\begin{cases} L.E_{elec} + L.\varepsilon_{fsi}.d^2 & if\ d \le d_{oi} \\ L.E_{elec} + L.\varepsilon_{mpi}.d^4 & if\ d > d_{oi} \end{cases}$$

Now E_{elec} is the energy dissipation per bit for running the transmitter. Now we assume that the energy requirement of the ε_{fsi} and ε_{mpi} is $\varepsilon_{fsi}=\frac{\varepsilon_{fs}}{10}$ and $\varepsilon_{mpi}=\frac{\varepsilon_{mp}}{10}$ thus, by the above logic, we have the $d_{oi}=\sqrt{\frac{\varepsilon_{fsi}}{\varepsilon_{mpi}}}$.

Considering that "n" nodes are uniformly distributed in an area A=M*M square meter and energy dissipated when the node distances from the Base station (BS) is less than d_0 and d_{oi}. As a result the energy dissipation of cluster head is

$$E_{ch}=(\frac{n}{k}-1)\ L.E_{elec}+\frac{n}{k}L.E_{DA}+L.E_{elec}+L.\ \varepsilon_{fs}d^2_{toBS}$$

The database is the average distance between the sink and the cluster head while k is the number of the cluster head. E_{DA} is the processing cost of a bit per round of the sink. The energy used in the non- cluster head nodes is equal to :

$$E_{nonCH}=L.E_{elec}+L..\ \varepsilon_{fsi}.d^2_{toCH}$$

Considering the uniform distribution of nodes than it can be shown that:-

$$d^2_{toCH}=\int_{x=0}^{x=xmax}\int_{y=0}^{y=ymax}(x^2+y^2)\beta(x,y)dxdy=\frac{M^2}{2\pi k}$$

Where $\beta(x,y)$ the node distribution

The total energy dissipated in each network is equal to

$$E_{tot}=L(2nE_{elec}+nE_{DA})+\ \varepsilon_{fs}(\frac{Kd^2_{toBS}}{10}+nd^2_{toCH})$$

Now, differentiating E_{tot} equation by k and equating it to zero ,we get the optimal number of constructed clusters

$$K_{opt}=\sqrt{\frac{n*10}{2\pi}}*\frac{M}{d_{toBS}}=\sqrt{\frac{n*10}{2\pi}}*\frac{2}{0.765}=3.165*\sqrt{\frac{n}{2\pi}}*\frac{2}{0.765}$$

Considering the paper [14], the mean distance between cluster head and Base station or sink is given by:

$$D_{toBS}=\int_A \sqrt{x^2+y^2}\frac{1}{A}dA = 0.765\frac{M}{2}$$

Thus, for each node to become the cluster head the optimal probability P_{opt}, can be computed as follows

$$P_{opt}=\frac{k_{opt}}{n}$$

Now form the [1] paper we found that the P_{SEP} (optimal probability) to be clustered head of any node for the SEP protocol is equal to the

$$P_{SEP}=\frac{k}{n}, \quad K=\sqrt{\frac{n}{2\pi}}*\frac{2}{0.765}$$

Where k is the optimal number of the cluster constructed and n is the number of the nodes. Now comparing the P_{SEP} value to that of the P_{opt}. The overall result of the comparison and replacement is:-

$$P_{opt}=\frac{k_{opt}}{n}=\frac{3.165*K}{n}=3.165*P_{SEP} \quad (1)$$

Thus, in our model we assumed two types of the energy level, one for the entire transmission involving the cluster head and base station while the other intra cluster transmission involving the non-cluster head nodes having lower energy. We also assume that the nodes are uniformly distributed. From the equation (i) it is clear that the new M-SEP has better results than the SEP as the P_{opt} M-SEP is 3.165 times than the SEP.

4 Modified Stable Election Protocol (M-SEP)

The M-SEP is an improvement over the existing SEP protocol. A SEP is an energy efficient heterogeneous cluster based protocol which attempts to maintain the constrain of " well balanced energy consumption". For this SEP divides the nodes into two sets of nodes, i.e. Advance nodes, which got more energy, thus become cluster head more often than the Normal nodes having less energy. The M-SEP implements all the concepts of the SEP but it also adds the concept of the multilevel energy transmission, which say that the energy requirement for intra cluster packet transmission is different from that of inter cluster packet transmission or cluster head to the base station packet transmission. As we see from the above section that the P_{opt} of the M-SEP is 3.165 times than the SEP protocol. We will now see how such a change will affect the overall result of the various parameters in the M-SEP.

Here it needed to understand that the new heterogeneous setting has no effect on the spatial density of the network,so the priori setting of P_{opt} does not change. But the total energy of the system changes as the energy of the advance node E_0 equal to the $E_0(1+\alpha)$ .Thus, initial energy of the new heterogeneous setting equal to the

$$n.(n-1)E_0 + n.m.E_0.(1+\alpha) = n.E_0(1+\alpha.m)$$

Now from the SEP protocol paper [1] we have the P_{nrm} and P_{adv} which is the weighted probability of the advance and normal nodes respectively.

$$P_{nrm} = \frac{p_{opt}}{1+\alpha.m}$$

$$P_{adv} = \frac{p_{opt}}{1+\alpha.m}(1+\alpha)$$

As we know from the equation (i) that P_{opt} (M-SEP) give 3.165 times than the P_{opt}(SEP). Thus the M-SEP still outflank the SEP as far as the weighted probability is concerned. As the P_{nrm} (M-SEP)=3.165*P_{nrm}(SEP) and similarly P_{adv} (M-SEP)=3.165*P_{adv}(SEP) Now the threshold of the normal and advance nodes are given by:-

$$T(S_{nrm}) = \begin{cases} \dfrac{p_{nrm}}{1-p_{nrm}\left(r \bmod \frac{1}{p_{nrm}}\right)} & \text{if } S_{nrm}\ \varepsilon\ G' \\ 0 & \text{otherwise} \end{cases}$$

Where the G' is the set of the normal nodes that has not become a cluster head for the last rounds of the epoch, while the r is the current round, $T(S_{nrm})$ is the threshold applied to a population of n.(1-m) normal nodes. As a result, each normal node will become a cluster head every $\frac{1}{p_{nrm}}(1+\alpha * m)$ round per epoch, and the average number of cluster heads that are normal nodes per round per epoch is equal to n*(1-m)*P_{nrm}.

And for the advance nodes, we have

$$T(S_{adv}) = \begin{cases} \dfrac{p_{adv}}{1-p_{adv}\left(r \bmod \frac{1}{p_{adv}}\right)} & \text{if } S_{adv}\ \varepsilon\ G'' \\ 0 & \text{otherwise} \end{cases}$$

Where $T(S_{adv})$ is the threshold applied to the n*m population of advance nodes, which ensure that the advance nodes will become the cluster head exactly once every $\frac{1}{p_{opt}} * \frac{1+\alpha m}{1+\alpha}$ rounds. G'' is the set of the nodes that have not become the cluster head with last $\frac{1}{p_{adv}}$ rounds per epoch. And the average number of cluster head per round per heterogeneous epoch is equal to

$$n*(1\text{-}m)*P_{nrm}+n*m*P_{adv}=n*P_{opt}$$

This also gives the number of the cluster heads per round per epoch. Now let us discuss on the numerical aspect of the SEP and M-SEP. Let assume that the 20% of the nodes is advance nodes (m=0.2) and has 300%more energy than the normal nodes. We have a total of 100 nodes, which means n=100. We found from table below that the number of the nodes that will become the cluster head has increased from 10 to 32 nodes per rounds while the epoch of both homogeneous nodes and heterogeneous nodes decreases from 10 rounds and 16 rounds to 3 and 5 rounds respectively. And the average number of the normal nodes to become cluster head increases from 5 nodes per round in SEP to 16 nodes per round. Likewise,in advance nodes the number of the nodes increases from 5 nodes per round in SEP to 16 nodes per round in M-SEP.

Parameters	**M-SEP**	**SEP**
P_{opt}	0.32 (avg. 32 nodes become cluster head per round)	0.10(avg.10 nodes become cluster head per round)
Epoch(Homogeneous)	3.03 rounds	10 rounds
Epoch(Heterogeneous)	4.86 rounds=5 rounds	16 rounds
Avg. No of normal nodes become the cluster head	16 normal nodes become cluster head per round	5 normal nodes become cluster head per round
Avg. No of Advance nodes become the cluster head	16 advance nodes become cluster head per round	5 advance nodes become cluster head per round

5 Simulation Result

We will use the "**matlab**" for simulation, we initialize the P_{opt}=0. 5 while having 100*100 region where there will be 100 sensor nodes randomly distributed in the system. We also assumed that the 20% sensor node will be advance nodes (m=0.2) and the advance nodes will have 300% more energy than the normal nodes. Various other parameters are as follows.

Parameter	Value
E_{elec}	50nJ/bits
E_{fs}	10pj/bit/m^2
EDA	5nj/bit/packet
E_o	0.5j
Number of rounds	2500
K_{opt}	3
p_{opt}	0.1
Network size	100*100
Base station location	(50,50)
E_{mp}	0.014pj/bit/m^4

Now we are estimating the performance of the M-SEP protocol with other protocols on various metrics like throughput, network lifetime, stability period and the cluster formation.

1) Network Lifetime:-From figure (2), it's clear that the "number of the alive nodes per round" of the M-SEP is better than the SEP and Mod-Leach protocols, thus enhancing the network lifetime. This enhancement of the performance can be attributed to the multilevel power transmission,which saves energy in the transmission which was not there in SEP. This will in turn improve the performance of the network.

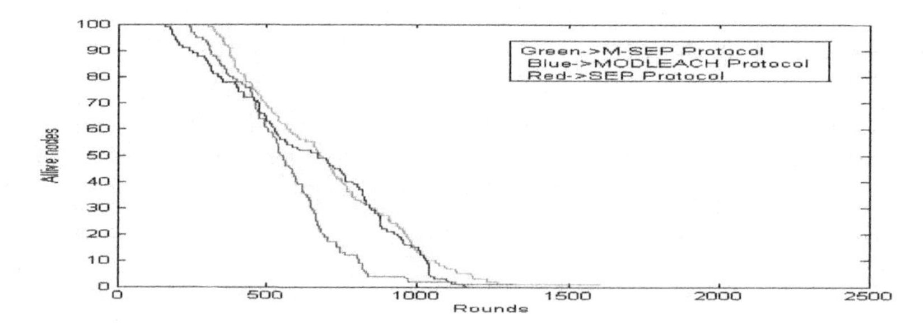

Fig. 2 Number of alive nodes per round

2) Stability Period:-It is the "time interval from the start of the network to the death of the first sensor node". It is also called the "stable region". From the figure (3) we came to know that the stability period of the M-SEP is better than the other two protocols (SEP and Mod-Leach protocols). It will be helpful for many applications where we need a higher stability period. It is apparent from the figure (3) that the lifetime of the network also increases.

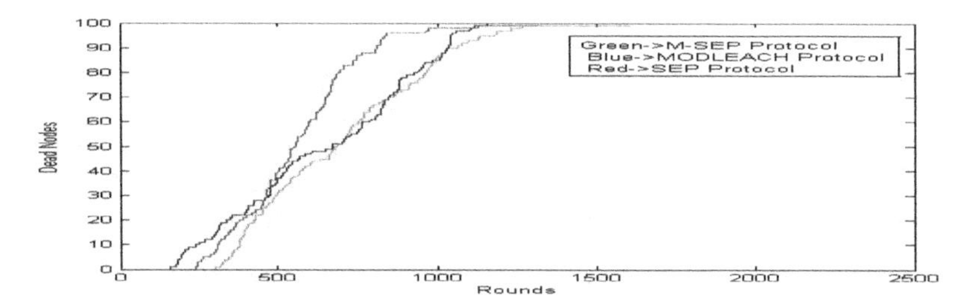

Fig. 3 Number of Dead nodes per Round.

3) Throughput:-It is the "rate of data sends over the network" i.e. the rate of data sent from cluster heads to the base station (sink) or the rate of data send from nodes to the cluster head. Above two figures clearly show the network lifetime has been increased, thus we expect the throughput of the system to be increased.

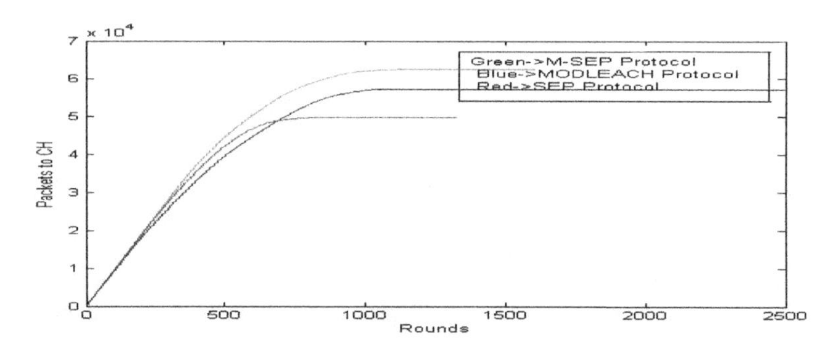

Fig. 4(a) rate of data send from noes to cluster head(CH)

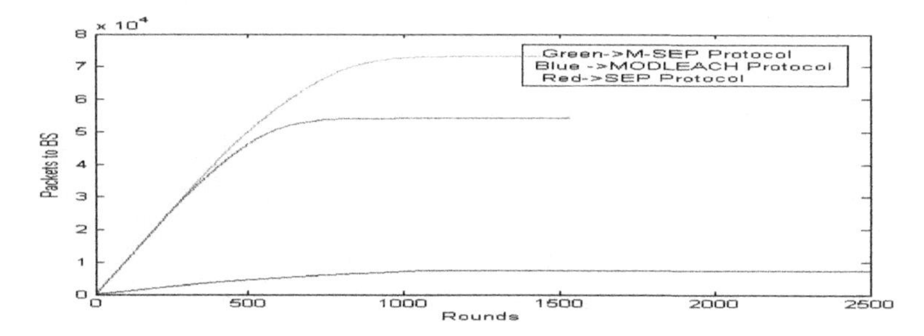

Fig. 4(b) Rate of data send from the cluster head to Base station

Fig. 4(a) show that the M-SEP's rate of data send from the nodes to the cluster head has been increased significantly as compare to the SEP and Mod-Leach. Fig. 4(b) shows M-SEP's rate of data transmission from the cluster head to the base stations compared to other two protocols. The M-SEP gives much better result than the other two protocols. The reason of these enhancements is that M-SEP implements the multilevel energy transmission, which saves the energy of the network, hence better throughput than the SEP or Mod-Leach protocols.

4) Cluster Formation:-The cluster formation of the three protocols per round is to be considered for idle situation, the cluster formation should be optimal and the number of the cluster head should be same throughout the network life. The Fig. (5)'s cluster formation of the M-SEP shows that the M-SEP has the similar cluster formation pattern of SEP and Mod-LEACH [11]protocols.

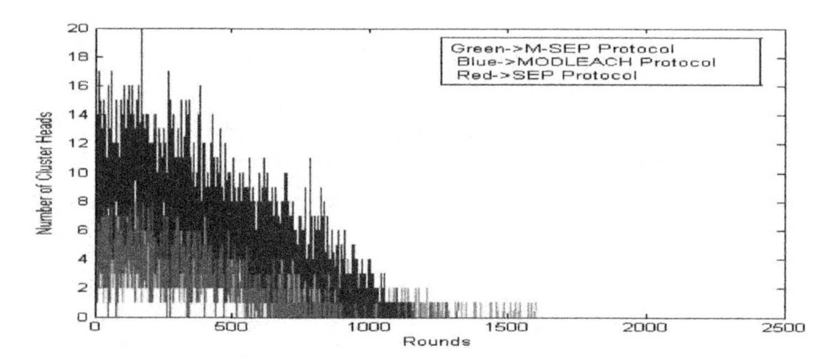

Fig. 5 Number of cluster Head formation per rounds

We will now summarize our observation from the simulation result and the theoretical calculation we made in the above sections.

i) Our M-SEP protocol extends the stable region by employing the multilevel power transmission as compared to SEP protocol, which has same working except that the SEP will assume same energy for all kinds of transmission.

ii) Due to the higher network lifetime and extended stable region, we have higher throughput than the SEP and Mod-leach protocol.

iii) The implementation of multi-level power transmission in M-SEP made it more effective than other energy efficient protocols by employing two levels of energy consumption in which energy requirement of the intra transmission and that of the inter transmission or that of the base to cluster head transmission is different.

6 Conclusion

We proposed M-SEP(Modified Stable election Protocols) protocol that is an energy efficient Wireless sensor network, which elect its cluster head independently based on its initial energy relative. We are able to show the positive impact of the multilevel power transmission through M-SEP which implement it i.e. the energy requirement of the intra cluster transmission is lower than the energy requirement of the inter cluster transmission or cluster head to the base station transmission .Which greatly improves the M-SEP performance compares to other SEP or Mod-Leach protocols.M-SEP needs no global knowledge of energy at every election round as assume in the SEP. In future studies, we will try to improve the head selection method of the M-SEP which is same as that of the SEP and improve the M-SEP further and also we would like to see how the multilevel power transmission affects the others WSN protocols. The table (1) shows the comparison of three WSN protocols under consideration under various performance metrics.

Table 1 comparison of the Mod-leach, SEP and DEEC protocols.

Criteria	**Mod-Leach**	**SEP**	**M-SEP**
Type of system	Homogeneous WSN	Two levels of heterogeneous WSN	Two levels of heterogeneous WSN
Energy Efficiency	Moderate	Moderate	Higher than SEP
Stability period	Moderate	Moderate	Higher than SEP
Rate of packet transfer	Low	Moderate	Higher than SEP
Cluster stability	Moderate	Moderate	Moderate
Network life time	Batter than Leach and similar to SEP	Moderate	Higher network lifetime than the Mod-leach and SEP.

References

1. Smaragdakis, G., Matta, I., Bestavros, A.: SEP: a stable election protocol for clustered heterogeneous wireless sensor networks. In: Second International Workshop on Sensor and Actor Network Protocols and Applications (SANPA 2004), vol. 3 (2004)
2. Estrin, D., Govindan, R., Heidemann, J., Kumar, S.: Next century challenges: scalable coordination in wireless networks. In: Proceedings of the 5th Annual ACM/IEEE International Conference on Mobile Computing and Networking (MOBICOM), pp. 263–270 (1999)
3. Jiang, M., Li, J., Tay, Y.C.: Cluster Based Routing Protocol. Internet Draft (1999)
4. Arboleda, L.M.C., Nasser, N.: Comparison of clustering algorithms and protocols for wireless sensor networks. In: Canadian Conference on Electrical and Computer Engineering, pp. 1787–1792, May 2006
5. Mhatre, V., Rosenberg, C.: Design Guidelines for Wireless Sensor Networks: Communication, Clustering and Aggregation. Ad Hoc Networks **2**(1), 45–63 (2004)
6. Tahir, M., Javaid, N., Khan, Z.A., Qasim, U., Ishfaq, M.: EAST: energy-efficient adaptive scheme for transmission in wireless sensor networks. In: 26th IEEE Canadian Conference on Electrical and Computer Engineering (CCECE 2013), Regina, Saskatchewan, Canada (2013)
7. Shah, T., Javaid, N., Qureshi, T.N.: Energy efficient sleep awake aware (EESAA) intelligent sensor network routing protocol. In: 15th IEEE International Multi Topic Conference (INMIC 2012) (2012)
8. Qing, L., Zhu, Q., Wang, M.: Design of a distributed energy-efficient clustering algorithm for heterogeneous wireless sensor networks. ELSEVIER, Computer Communications **29**, 2230–2237 (2006)
9. Heinzelman, W.R.: Application-Specific Protocol Architectures for Wireless Networks. Ph.D. thesis, Massachusetts Institute of Technology (2000)
10. Duarte-Melo, E.J., Liu, M.: Analysis of energy consumption and lifetime of heterogeneous wireless sensor networks. In: Proceedings of Global Telecommunications Conference (GLOBECOM 2002), pp. 21–25. IEEE, November 2002
11. Mhatre, V., Rosenberg, C.: Homogeneous vs. heterogeneous clustered sensor networks: a comparative study. In: Proceedings of 2004 IEEE International Conference on Communications (ICC 2004), June 2004
12. Heinzelman, W., Chandrakasan, A., Balakrishnan, H.: Energy-efficient communication protocols for wireless microsensor networks. In: Proceedings of Hawaiian International Conference on Systems Science, January 2000
13. Mahmood, D., Javaid, N., Mahmood, S., Qureshi, S., Memon, A.M., Zaman, T.: MODLEACH: a variant of LEACH for WSNs. In: Proceeding of IEEE 8th International Conference on Broadband and Wireless Computing, Communication and Applications (BWCCA 2013), Compiegne, France, July 2013
14. Minimizing communication costs in hierarchically-clustered networks of wireless sensors. Computer Networks **44**(1), 1–16, January 2004

Network Monitoring and Internet Traffic Surveillance System: Issues and Challenges in India

Rajan Gupta, Saibal K. Pal and Sunil K. Muttoo

Abstract NETRA or Network Traffic Analysis is an Internet traffic surveillance system developed by the Indian Government. It has been designed by various intelligence agencies of India. The conceptualization, monitoring & development have been done by a group consisting of the Centre for Artificial Intelligence and Robotics, Defense Research & Development Organization and National Technical Research Organization. The prime purpose of this application is to analyze the internet traffic and draw inferences for various suspected activities in the country. But the design scheme and implementation level analysis of the system shows few weaknesses like limited memory options, limited channels for monitoring, pre-set filters, ignoring big data demands, security concerns, social values breach and ignoring ethical issues. These can be covered through alternate options which can improve the existing system. The paper reviews the architectural framework and existing scheme of NETRA system and suggests improvements for the weak areas. The existing framework of NETRA system has been compared to similar international level surveillance systems like Dish Fire, Prism and Echelon. The similarity and differences amongst these systems are identified and recommendations are provided based on them. The analysis of surveillance system will help in developing several other mini spy-cum-monitoring models which can be further customized for various applications and communication channels in India.

R. Gupta(✉) · S.K. Muttoo
Department of Computer Science, Faculty of Mathematical Sciences,
University of Delhi, Delhi 110007, India
e-mail: guptarajan2000@gamil.com, skmuttoo@cs.du.ac.in

S.K. Pal
SAG Lab, Metcalfe House, DRDO, Delhi 110054, India
e-mail: skptech@yahoo.com

S. Berretti et al. (eds.), *Intelligent Systems Technologies and Applications*,
Advances in Intelligent Systems and Computing 385,
DOI: 10.1007/978-3-319-23258-4_6

1 Introduction

The extensive usage and penetration of technology, specifically internet, is one of the most crucial and pivotal factors that have possibly led to an increase in the criminal activities. Terrorist activities have also started to gain a wide range and spectrum of supporters due to the easy availability and access of internet and differed platforms of social media. Terrorist groups have been reported to use internet for a wide range of activities such as recruitment, funding, training and instigation for committing heinous acts of violence and destruction, and the gathering and dissemination of information for terrorist purposes. This strengthens the fact that a proper surveillance is essential and mandatory in order to be in a position to combat such terrorist acts. Lots of traffic monitoring and surveillance applications have been started due to this [1].

There are large numbers of different data sets which are currently maintained by the world net data warehouse. One of them is router configuration which includes information related to security, access list, topology, and the likes. A 'triple A' system covering authentication along with authorization and accounting systems makes up as the components of the registration records. Similarly, the data under call record comprises of summary related to customer's dial-up session on per-session basis. Email server logs' include SMTP and POP3 transaction summaries. Router statistics is obtained by SNMP polling. It includes link/router utilization, access and gateway routers. Apart from them, numerous packet scopes exist for IP packet headers' collection and they are designed for high performance systems. They act as passive link access in which the modification of the device driver was done for all the 'read' commands but not 'write' commands for the network interface which was under monitoring. This monitoring can be T3 which was, for a case, terminated at router modelled 7505 by Cisco and is designed for the forwarding of the packets towards monitor for the capture. Now there captured packets are utilized for the collection of header which contains vital information. Similarly, data apart from textual form like multimedia data is monitored by passing the traffic through various protocols like RTSP, SIP and other protocols related to session-control for set-up and packet filters tear down for capturing multimedia sessions [2].

NETRA (NEtwork TRaffic Analysis) is a networked software system developed by Centre for Artificial Intelligence and Robotics (CAIR), a departmental laboratory under Defense Research and Development Organization (DRDO) in response to the ever increasing terrorist and criminal threats received via the internet and other means of data communication [3]. This paper analyzes the architecture and scheme of NETRA and discusses various issues related to it. This study is a working paper currently at analysis and design phase. The aim of this paper is to review an existing system which has been designed in Indian conditions for monitoring and surveillance, so that the shortcomings can be utilized for existing system's improvement and new system's features selection.

2 NETRA: Evolution and Challenges

As a part of the Government of India's efforts to develop an internet monitoring system, two different agencies, CAIR by DRDO and NTRO were given the responsibility of developing technical systems that would concentrate on scanning through internet data and detecting suspicious words. *Vishwarupal*, NTRO's monitoring and surveillance system, faced several problems such as the involvement of some external private company in issues concerning security, and NTRO's ability and competence in operating such a system independently without the help of Paladion. NETRA performed better and got preference over *Vishwarupal* as the Government of India's official Internet Monitoring System.

In present context, the system can analyze the data flow through internet even when the messages being transmitted are encrypted in a manner that is difficult to decrypt. NETRA is modeled to process and filter out the content that is being generated by the *"Netizens"* (Internet Citizens) these days [4] and report any such word usage and their related IPs to the associated authorities and agencies which can take appropriate actions on immediate basis.

NETRA, though an advanced national level security measure, is doomed not to be a 100% success as per the analysis because of the large number of challenges that it may face. NETRA, at the moment is prone to a partial success only due to the multiple issues that hinder its expected success. They are discussed below.

Inefficiency Due to Use of Pre-set Filters
NETRA as a system operates by using pre-defined filters that match keywords like *'attack' , 'bomb' , 'blast', 'terrorist' and 'jehad'* among others [3]. But this could lead to an inefficient system as the conversation or data that is being intercepted may end up having negligible relevance to national security. The system should be more adaptable towards words analysis.

Restricted Memory for Storage
The specifications of the NETRA system reveal that three national security agencies, the Cabinet Secretariat, Intelligence Bureau and the Research & Analysis Wing will be given only limited combined storage space of 300 GB with law enforcement agency getting an additional 100 GB. Statistics reveal there are 1.6 Billion Social Media users, 1.28 Billion Facebook users (102 Million in India), 294 Billions emails are sent every day and there about 500 million tweets in a day [5]. Though these statistics are on a global level, but even on a national level of India, NETRA will not be able to manage and intercept a large amount of data that is being transmitted and transferred on a daily basis. It is highly probable for the system to fail because a storage space of 300 GB for three Security agencies will exhaust in a very quick succession of time [4].

Ever Growing Internet and Big Data
Internet is a vast network that is growing at an exponential speed. Its monitoring and surveillance can be a much more difficult and handy task than was initially predicted. There are millions of social network users who transmit and transfer

billions of messages and voice talks on a daily basis. Scrutiny and inspection of such a large number of users will require resources and technology that is much more advanced than what India presently owns. Thus burgeoning internet usage and ever growing big data poses a serious threat to the NETRA's working.

Important Missing Channels

Popular file sharing websites and channels such as *Dropbox*, *Rapid share* and *Fileshare* [1] have not been included in the affecting channels by CAIR. Well in demand messenger services like *Whatsapp* and other mobile based Applications have also not been included. Such omissions leave huge gap in the proper implementation of the system and may hinder complete interception of online activities of suspicious individuals and groups. Moreover, constantly evolving such platforms and channels also make it difficult for the monitoring system to be more dynamic in its approach.

Improvement in Specification and Efficient Mechanism

During a small scale demonstration in January of 2012, NETRA was able to pass only 3 GB traffic out of a total of 28 GB through its probes. It is the only high point, which worked as an advantage at that time. It was the only system out of those tested, that was able to capture the Internet data traffic without any stalling. Additionally, for a system such as NETRA that is aimed as a measure for providing security, quick response is a must and probably the most important feature. Based only on key-words filtration won't solve the efficiency part and improved big data mining techniques will have to be supported for future activities.

Security Concerns

CAIR needs to protect and safeguard the NETRA system against external hacking. An external hacking of the NETRA system can lead to chaos in internet traffic and probable threats being transmitted without any interception rendering the main function of using NETRA system useless. Also flooding attack of the keywords may create panic for the system. Thus, there must be a pattern recognizer that should be able to keep control over the system's filters and keep a track of abnormal behavior.

Ethical Issues

Surveillance of one's private talks and messages on platforms such as Google Talk and Skype is considered to be highly unethical. Even largely advanced economies like United States and China have not been completely successful in implementing their Internet monitoring systems due to the same reason. Internet surveillance can work and be successful only till an extent beyond which its success rate and effect becomes stagnant. Even one of the founding fathers of World Wide Web, Dr. Vint Cerf believes that Internet surveillance can never be successful as it goes against the basics of Internet, which in the first place is based on the freedom of expression and action [4].

3 Surveillance Systems Outside India

Every nation is trying to develop a surveillance & monitoring system that can collect and analyses the data which can help them to detect terrorism activities well in advance. This section discusses about the various leaked international monitoring systems that officially didn't exist. They are Echelon, Prism and Dishfire.

Echelon
ECHELON is not just a system but a code name which was used to describe collection of any signal's intelligence and analysis of its network. Due to a secret treaty around late 1940s, the five nations (USA, UK, Canada, Australia, New Zealand) formed Echelon under an Anglo-Saxon Club without any commercial implications. ECHELON was directly linked to the headquarters of the US National Security Agency (NSA) at Fort Mead in Maryland [6]. The system has been designed to tackle multiple channels of communications for the message transfer. It can intercept and inspect the data transfer through facsimile, telephone conversations, messages through teletext, internet usage, e-mail communications and other digital communication forms taking place in different regions of the world. SILKWORTH and SIRE are the functional programs that make up the core DNA of the ECHELON system and the interception was done through a satellite named VORTEX [6]. The system works in 3 stages, viz., data capturing, data processing & data matching [7].

Dish Fire
Dish Fire is a surveillance system, developed by US & UK, working at global level monitoring the traffic along with maintaining the database for the communication. A related analytical tool known as PREFER is used to analyze the data collected on daily basis through text messages. This tool is used for processing the SMS sent through cellular network and extract vital information from the phone like call alerts, location of the phone, financial details of the payments done and electronic card processing. Dish Fire works by collecting and analyzing automated text messages such as missed call alerts or texts sent to inform users about international roaming charges [8].

Prism
PRISM, launched in 2007, is also used for surveillance through monitoring and data mining by NSA in USA. It collects the stored communication from internet and compares it against the court-approved terms. It is more inclined towards the assessment of the encrypted data sets. NSA programs collect two kinds of data: metadata and content. Metadata, a sensitive byproduct of communications, such time, number of calls, phone records and contents, which the NSA PRISM Program includes i.e. emails, chats, VoIP calls, cloud-stored files, and more [9]. The system also has the similar comparative style of searched content and approved content within the legal boundaries.

Comparison of Various Systems

The similarity between all the systems is the common approach used that is data capture and data mining/analysis but the techniques are different. ECHELON and NETRA both uses the concept of matching the words from their dictionaries respectively. ECHELON intercepts transmissions from satellite and also from PSTN whereas NETRA only aims at observing the internet activities of the citizens. Also, Dish Fire is one system that extracts anything from text messages while prism program stores the contents of emails, chats, and more. NETRA only focuses on data from the internet services.

Some of the features of the international systems can be extracted which can help make NETRA – a better system. One of those features can be taken from Dish Fire that is to use a tool like PREFER that can extract information from miss call alerts, travel alerts and automated messages and more. The other feature that could be extracted is from ECHELON network i.e. the capability to intercept transmission from PSTN that can listen to and analyze telephone conversations, fax transmission and more. Like ECHELON the 3 agencies can be provided 3 different locations to cover and then integrate as one to analyze the data and then send to its required department.

4 Recommendations

There are certain proposals that can effectively enhance the working, operation and probability of the success of the NETRA system in India.

Use of Efficient Algorithms

NETRA system uses pre-set filters to match the intercepted conversations with a set of pre-defined keywords. Instead of using only filters, the NETRA system should switch to sentiment analysis in order to understand the intent behind usage of the keywords. And also the system will have to use real time text mining rather than a static implementation of the algorithms. Specific lightweight algorithms are available [10] which can cover both text data and multimedia data.

Use of Expandable Memory Techniques

The use of expandable memory techniques is a must for the NETRA system as the current allocated memory does not look to be sufficient enough. Cloud Storage services can resolve the issue of restricted memory allocation to the Security Agencies for the storage of intercepted data and will be better used for the protection of data stored on them [11]. Government has already launched *MeghRaj* as its official cloud platform for various applications.

Inclusion of Other Popular Channels

The NETRA system will have to expand its targeted channels. The whole system will have to be built around the dynamic environment so that every new application developed in the market could fit into the surveillance system. Currently NETRA will need to include various popular file sharing and messenger services like *Fileshare*, *Dropbox* and *WhatsApp* among others to widen its range of monitoring that it seeks to intercept data transmission from.

Appropriate Security Measures
The system can start adding alternate keywords to broaden its range of relevant intercepted conversations. The system should prevent hacking and spamming through the pre-defined keywords. Also an exhaustive IP Traffic Monitoring scheme will be useful in enhancing the security [12], along with Network Monitoring tools integration like NetFlow [13].

Type of Data Expansion
Currently NETRA only uses the surveillance system for textual type of data only. Rising multimedia data will be a concern for the monitoring agencies and mostly text based data is converted into images or videos and is then spread across the various channels. NETRA will have to take care of both types of data in order to be more efficient and effective in its implementation. Also deep packet inspection [14] can also help in much exhaustive analysis for the Government Surveillance System.

Network Breach Techniques
Accessing traffic in each zone can be accomplished by hubs, span ports, taps, inline devices and more. Some of the most common ways for breaching a network is Social Hacking, Cracking, Network Sniffing, and Packet Sniffing. **Quantum Insert** is a leaked NSA technique where fake servers are created to mimic real websites between traffic on a targeted network and the real website. **Serendipity** was a NSA tool developed to capture data from Google network. **S.N.A.C.K.S.,** Social Networking Analysis Knowledge Collaboration Services, is a NSA application that looks at SMS text messages, to derive organizational structures. **EDGEHILL,** UK's anti encryption technique aimed at cracking all the major internet company's encryption technique [15].

Inclusion of Self Similarity Models
Self-similarity models, as studied by various researchers in past, can be utilized within NETRA system for better efficiency. According to statistics, the idea of self-similarity can be used to describe traffic i.e. bursts on wide range of scales. Self-similar process can exhibit long range dependence as it has noticeable bursts at wide range of time scales. "*Self-similarity is a property which links with one type of fractal; where an object's appearance is unchanged despite of the scale at which it is viewed*" [16]. Self-similarity can be explained mathematically as,

$$Xt=m^\wedge\text{-}H \sum (i= (t-1)\ m+1 \text{ to } tm)\ Xi, \text{for all } m \in N \tag{1}$$

If X is H-self-similar, it has the same autocorrelation function

$$r(k) = E\ [(\ Xt - \mu)\ (Xt+k - \mu)]/\ \sigma^\wedge 2 \text{ as the series } X^\wedge m \text{ for all } m \tag{2}$$

Various traffic models and conditions of its self-similarity have been described in the past which can be very well utilized while strengthening NETRA system as a whole. Another model has the buffer content distribution in a fluid queuing system which receives input from N independent on/off sources. "*It considers a traffic which is a superposition of N sources. A source j constantly transmits at rate Rj> 1 when active, contributing RjAij volume to the traffic during its ith active period*" [17] [18]. The results on self-similarity of considered traffics are not provided by this model. Web traffic similarity model [19] appears to be more relevant in context to the NETRA which can improve the monitoring capabilities of the system through the web mode.

SCNM Implementation
The SCNM provides a function to detect network problems, by allowing users to monitor their individual data. It also allows users to monitor specific data streams without providing special access privileges or intervention by network administrators. The SCNM monitoring host has a network interface which is used for sending output data and managing the SCNM host. This model allows user to only monitor their individual data [20]. With the implementation of SCNM, the three agencies involved in NETRA would be able to manage and monitor the traffic at their respective ends and thus a more improved monitoring environment can be created.

5 Conclusion

NETRA seems to be an innovative approach towards internet surveillance by the Government of India. But in order to ensure its efficient and effective implementation, the key concerns pointed out need to be addressed. The ethicality of the system will always remain a controversial issue, but system's fair use and positive results can overshadow the doubts concerning the citizen's breach of privacy. In such tough situation where India is facing terrorist and criminal activities on high instances, NETRA may prove to be a constructive measure towards a pragmatic change.

For monitoring huge amount of internet traffic flowing at high speed, trade-off between different parameters has to be practically addressed. Theoretically, many things look promising but these may not be practically feasible. In this regard, capabilities of this application are limited right now but scope for improvements looks promising. If high channel and bandwidth issues can be monitored along with the concerned problems identified in the study, NETRA could become a powerful internet monitoring tool for government of India.

References

1. Brown, I., Korff, D.: Terrorism and the proportionality of internet surveillance. European Journal of Criminology **6**(2), 119–134 (2009)
2. Caceres, R., Duffield, N., Feldmann, A., Friedmann, J.D., Greenberg, A., Greer, R., Johnson, T., Kalmanek, C.R., Krishnamurthy, B., Lavelle, D., Mishra, P.P., Rexford, J., Ramakrishnan, K.K., True, F.D., van der Memle, J.E.: Measurement and analysis of IP network usage and behavior. IEEE of the Communications Magazine **38**(5), 144–151 (2000)

3. Wadhwani, D.: NETRA – You will be under surveillance by Indian Internet Spy System (June 9, 2014). http://newtecharticles.com/netra-indian-government-internet-spy-system/
4. Ghosh, M.: Beware, Government Plans To Spy On Your Internet Activity Using Netra (2013). http://trak.in/tags/business/2013/12/17/govt-spy-internet-netra/ (June 9, 2014)
5. Smith, C.: 20 social media statistics you probably didn't know (2014). Social media user stats. social/digital stats. http://expandedramblings.com/index.php/social-media-statistics/#.U5avXHKSzl8 (June 10, 2014)
6. Chandler, P.: ECHELON – The Spy System That Knows Everything, philipcfromnyc (2013). http://philipcfromnyc.hubpages.com/hub/ECHELON—The-Spy-System-That-Knows-Everything/ (March 4, 2015)
7. Bomford, A.: The Echelon spy network-Wednesday, November 3, 1999, 11:35 GMT,Echelon spy network revealed (1999). http://news.bbc.co.uk/2/hi/503224.stm
8. Hahn, D.J.: DISHFIRE: The Program That Lets the NSA Capture Almost 200 Million Texts a Day (2014). http://www.complex.com/pop-culture/2014/01/dishfire-nsa-collects-textson (March 4, 2015)
9. Greenwald, G., MacAskill, E.: NSA Prism program taps in to user data of Apple, Google and others (2013). http://www.theguardian.com/world/2013/jun/06/us-tech-giants-nsa-dataon (March 4, 2015)
10. Gupta, R., Aggarwal, A., Pal, S.K.: Design and Analysis of New Shuffle Encryption Schemes for Multimedia. Defence Science Journal **62**(3), 159–166 (2012)
11. WildPackets: Network Forensics in a 10G World, Wild Packets, White Paper (2013). http://www.wildpackets.com/elements/whitepapers/network_forensics_in_a_10_g_world.pdf
12. Wei, D., Ansari, N.: IP traffic monitoring: an overview and future considerations. In: Shum, H.-Y., Liao, M., Chang, S.-F. (eds.) PCM 2001. LNCS, vol. 2195, pp. 335–342. Springer, Heidelberg (2001)
13. Cecil, A.: A Summary of Network Traffic Monitoring and Analysis Techniques. cit, 10–25 (2012)
14. Bendrath, R., Mueller, M.: The end of the net as we know it? Deep packet inspection and internet governance. New Media & Society **13**(7), 1142–1160 (2011)
15. Takhar, S., et al.: If I Was … Analyzing Edward Snowden. National Security Agency, USA (2014)
16. Crovella, M.E., Bestavros, A.: Self-similarity in World Wide Web traffic: evidence and possible causes. IEEE/ACM Transactions on Networking **5**(6), 835–846 (1997)
17. Boxma, O.J.: Fluid queues and regular variation. Performance Evaluation **27**, 699–712 (1996)
18. Tsybakov, B., Georganas, N.D.: Self-similar processes in communications networks. IEEE Transactions on Information Theory **44**(5), 1713–1725 (1998)
19. Roberts, J.W.: Traffic theory and the Internet. IEEE Communications Magazine **39**(1), 94–99 (2001)
20. Agarwal, D., González, J. M., Jin, G., Tierney, B.: An infrastructure for passive network monitoring of application data streams. Lawrence Berkeley National Laboratory (2003)

An Empirical Study of OSER Evaluation with SNR for Two-Way Relaying Scheme

Akshay Pratap Singh, Anjana Jain and Shekhar Sharma

Abstract The two-way relaying scheme is a network coding scheme which provides efficient utilization of spectrum. The performance of a two-way relaying scheme based on analog network coding (ANC) can be evaluated in context of many parameters such as overall symbol error rate (OSER), overall outage probability (OOP) and ergodic sum rate (ESR). In this paper we analyse the OSER over Nakagami-m fading channels for analog network coding scheme. The variation of OSER with respect to signal to noise ratio (SNR) has been observed for various factors such as fading parameter, modulation scheme used and number of antennas at the source and destination terminal. The threshold range of SNR for a tolerable value of OSER of 0.001 is computed as 20 to 25 dB. The results demonstrate the necessity of high SNR in two way relaying scheme for tolerable OSER. Further the analysis may be useful for selection of various physical layer parameters in design of communication networks.

Keywords Two-way relaying · Analog network coding (ANC) · Overall symbol error rate (OSER) · Threshold range and signal to noise ratio (SNR)

1 Introduction

The need of efficient use of spectrum is a key factor in wireless communication system. The two-way relaying scheme [1] based on analog network coding (ANC) [2] is one of the various ways which can be used for efficient utilization of spectrum. In this paper analysis of network coding has been done towards improving the spectral efficiency [3]. The approach for designing the wireless communication networks has been changed fundamentally with the network coding schemes [4], in which the intermediate node is allowed to process the packet.

A.P. Singh(✉) · A. Jain · S. Sharma
S.G.S.I.T.S., Indore, India
e-mail: {akshay44444p,jain.anjana,shekhar.sgsits}@gmail.com

S. Berretti et al. (eds.), *Intelligent Systems Technologies and Applications*,
Advances in Intelligent Systems and Computing 385,
DOI: 10.1007/978-3-319-23258-4_7

Some practical techniques are used to exploit the spectrum by allowing user terminals to share their time-slots co-operatively named as relaying schemes. This can be categorised in two ways, namely full- duplex relaying and half duplex relaying. In full-duplex relaying (FDR) transmission and reception is done in same time slot whereas in half-duplex relaying (HDR) transmission and reception is done in orthogonal time slots [5] hence there is no co-channel interference in half-duplex relaying as compared to full-duplex relaying [6]. The spectral efficiency achieved with FDR is much greater than HDR. The full-duplex relaying protocols have larger difference in receive and transmit signal's power levels as compared to that of half-duplex relaying hence half-duplex relaying protocols are easier to implement as compared to full-duplex relaying protocols [7].

The motivation to use two way relaying scheme is the efficient use of radio spectrum. Analog network coding is a latest emerging scheme, which gives an opportunity to achieve high spectral efficiency and a solution to the hidden-node problem.

The assumption taken for this paper is that all the terminals in a relaying network are operating in a half-duplex manner. Traditional communication method results with very less spectral efficiency, further advancements have been done to increase spectral efficiency by using digital network coding (DNC) [8] and analog network coding (ANC) with decode and forward (DF) relaying protocol and amplify and forward (AF) relaying protocol respectively. Further orthogonalize and forward (OF) protocol can be used for multiple source and multiple destination terminals by using space division duplexing (SDD) relaying scheme [9].

In analog network coding, the interference of source and destination terminals is exploited instead of avoiding it to increase the network capacity [10] and as a solution of hidden node problem.

The paper presents the overall symbol error rate (OSER) [11], [12] for the two way relaying scheme based on analog network coding. The variation of OSER with respect to SNR for different parameters such as fading parameter, different modulation schemes used and different number of antennas used at the source and destination terminal is analyzed. A threshold range of SNR for tolerable value of OSER of 0.001 based on mathematical results is calculated.

Section (2) of the paper presents the analysis of network coding, section (3) shows an expression of OSER for two-way relaying scheme based on analog network coding, and numerical results with discussions are presented in section (4). Finally the paper concluded in section (5).

2 Analysis of Network Coding

The design and analysis approach for wireless networks has recently become very interesting. Analysis of network coding schemes has been done in this section.

The assumption taken in this section is that there is no direct link between source and destination terminal and both terminals are communicating in half-duplex scenario. First of all the discussion about the traditional co-operative communication method is done.

Consider the Fig. 1(a) in which there are four time-slots required to exchange the information between source terminal (A) and destination terminal (C) using a relay terminal (B). For doing the same process in less time-slots network coding schemes is used, which results in a better spectral efficiency.

The digital network coding scheme uses decode and forward protocol. Consider Fig. 1(b) in which both the source and destination terminals (A and C) send their information sequentially in first two time-slots and relay terminal (B) mixes their information by a basic X-OR operation and then relay broadcast the mixed packets to both the terminals and then the information of other terminal is extracted by again X-ORing the broadcast information with itself.

The same process can be done only in two time-slots to result a much better spectral efficiency using amplify and forward protocol with analog network coding scheme. Consider Fig. 1(c) in which both the source and destination terminals (A and C) send their information simultaneously in first time-slot and relay terminal (B) mixes naturally their signals and then relay amplify and broadcast the mixed signals to both the terminals and then the information of other terminal is extracted by again subtracting the broadcast information with itself. Hence the spectral efficiency is just double in case of analog network coding as compare to traditional communication method. The analog network coding is exploiting the interference instead of avoiding it and this can be also used as a remedy to hidden node problem. *"The Fig. 1 is taken from the reference paper [13]"*.

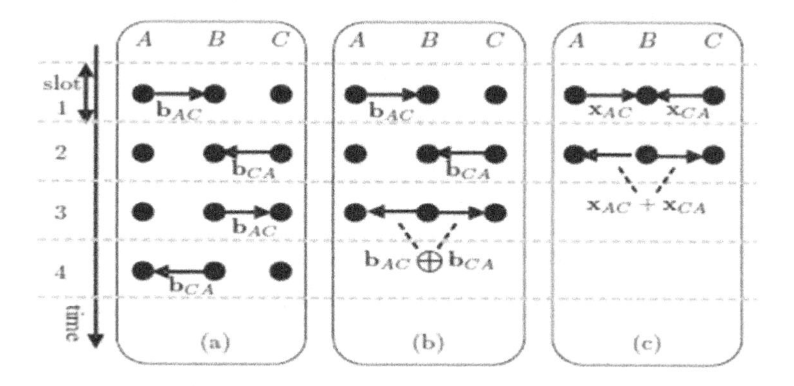

Fig. 1 ((a) Traditional communication; (b) digital network coding; (c) analog network coding) [13]

3 Oser for Two-Way Relaying Scheme Based on ANC

Overall symbol error rate (OSER) comes out as a crucial metric aspect for the performance of system in two-way relaying scheme. Mathematical expression of OSER can be expressed for different coherent modulation schemes as [14, eq. (6.33)],

$$P_s = \alpha Q\sqrt{\beta\gamma_0} \tag{1}$$

Where α and β are modulation scheme dependent parameters such as for binary phase-shift keying (α = 1, β = 1) and for quaternary phase-shift keying (α = 1, β = 0.5).

Let us assume the source terminal S and destination terminal D having Ns and Nd number of antennas respectively and a relay terminal R with single antenna and the channel for both hops are reciprocal having Nakagami-m fading distribution.

The end to end signal to noise ratio resulting for two way relaying scheme from terminal x to y can be given as [11]

$$\gamma_{xy} = \frac{P_x P_r}{\sigma^2}\left[\frac{\|h_y\|^2\|h_x\|^2}{(P_y + P_r)\|h_y\|^2 + P_x\|h_x\|^2}\right] \tag{2}$$

Where

$$h_p = \left[h_p^1 h_p^2 h_p^3 \ldots\ldots\ldots\ldots. h_p^{Np}\right]^T, p \in \{s,d\} \tag{3}$$

the equation (3) denotes a $N_p \times 1$ channel vector from S to R having a fading parameter as m_p and a fading power as Ω_p. P_i is the power allocated to the ith terminal and $\|\bullet\|$ denotes frobenius norm.

The OSER in terms of CDF can be represented as

$$P_s = \frac{\alpha}{2}\sqrt{\frac{\beta}{\Pi}}\int_0^{\infty}\frac{e^{-\beta z}}{\sqrt{z}}F_{\gamma_0}(z)dz \tag{4}$$

And the CDF, $F_{\gamma_0}(z)$ for a threshold value 'z' can be given as

$$F_{\gamma_0}(z) = P_r[\gamma_{sd} < z, \gamma_{sd} < \gamma_{ds}] + P_r[\gamma_{ds} < z, \gamma_{ds} < \gamma_{sd}] \tag{5}$$

An exact expression of OSER can be determined by considering only the first few terms of the infinite series involved.

Let us assume $\frac{P_s}{\sigma^2}$ as SNR with $\frac{P_d}{P_s}$ and $\frac{P_r}{P_s}$ as finite constants.

The equation (5) can be modified for high SNR as

$$F_{\gamma_0}(z) = \left(\frac{1}{\Gamma m_s N_s}\right)\Phi\left(m_s N_s, z\frac{\lambda_1 m_s}{\gamma_{1s}}\right) + \left(\frac{1}{\Gamma m_d N_d}\right)\Phi\left(m_d N_d, z\frac{\lambda_2 m_d}{\gamma_{2d}}\right) \tag{6}$$

Where $\lambda_1 = \frac{P_b + P_r}{P_r}, \lambda_2 = \frac{P_a + P_r}{P_r}$ and $\Gamma(\bullet)$= gamma function.

Now by using equation (6) in equation (4) and after further calculation in equation (4) a close form representation to OSER for high SNR can be given as

$$P_s = \frac{\alpha}{2}\sqrt{\frac{\beta}{\Pi}}\left[\left(\frac{\lambda_1 m_s}{\gamma_{1s}}\right)^{m_s N_s} \frac{\Gamma\left(m_s N_s + \frac{1}{2}\right)}{m_s N_s \left(\beta + \frac{\lambda_1 m_s}{\gamma_{1s}}\right)^{m_s N_s + \frac{1}{2}}} \times {}_2F_1\left(1, m_s N_s + \frac{1}{2}; m_s N_s + 1; \frac{\frac{\lambda_1 m_s}{\gamma_{1s}}}{\beta + \frac{\lambda_1 m_s}{\gamma_{1s}}}\right)\right] +$$

$$\frac{\alpha}{2}\sqrt{\frac{\beta}{\Pi}}\left[\left(\frac{\lambda_2 m_d}{\gamma_{1d}}\right)^{m_d N_d} \frac{\Gamma\left(m_d N_d + \frac{1}{2}\right)}{m_s N_s \left(\beta + \frac{\lambda_2 m_d}{\gamma_{1d}}\right)^{m_d N_d + \frac{1}{2}}} \times {}_2F_1\left(1, m_d N_d + \frac{1}{2}; m_d N_d + 1; \frac{\frac{\lambda_2 m_d}{\gamma_{1d}}}{\beta + \frac{\lambda_2 m_d}{\gamma_{1d}}}\right)\right] \quad (7)$$

By keeping $\Phi(\alpha, p) \underset{p \to 0}{\approx} \frac{p^{\alpha}}{\alpha}$ in eq. (4) and after doing further calculation the asymptotic behaviour of OSER for high SNR can be determined as

$$P_s = \frac{\alpha}{2\sqrt{\Pi}}\left[\frac{\Gamma\left(m_s N_S + \frac{1}{2}\right)}{\Gamma(m_s N_S + 1)}\left(\frac{(P_d + P_r) m_s \sigma^2}{P_s P_r \Omega_s \beta}\right)^{m_s N_S} + \frac{\Gamma\left(m_d N_d + \frac{1}{2}\right)}{\Gamma(m_d N_d + 1)}\left(\frac{(P_s + P_r) m_d \sigma^2}{P_d P_r \Omega_d \beta}\right)^{m_d N_d}\right] \quad (8)$$

4 Numerical Results and Discussion

In this section by considering the path loss exponent (v) = 4, and assuming the uniform power allocation to all the three terminals (S, R, D) i.e. (Ps= Pr = Pd = Pt /3) and the location of relay is exactly at the centre of source and destination terminals i.e. (d=0.5) with normalized distance between source and destination terminal is unity. All the following results are calculated mathematically. Further if we assume the tolerable OSER value as equal to 0.001 then for this the threshold value of SNR for different parameters are calculated.

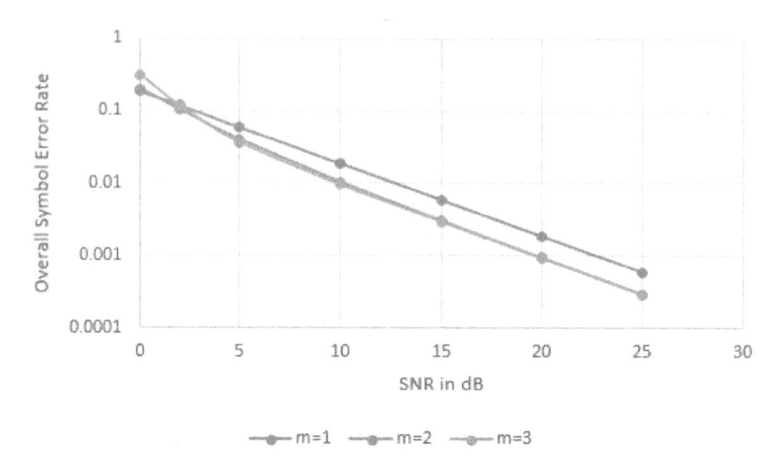

Fig. 2 Variation of OSER with SNR for different values of fading parameter (m)

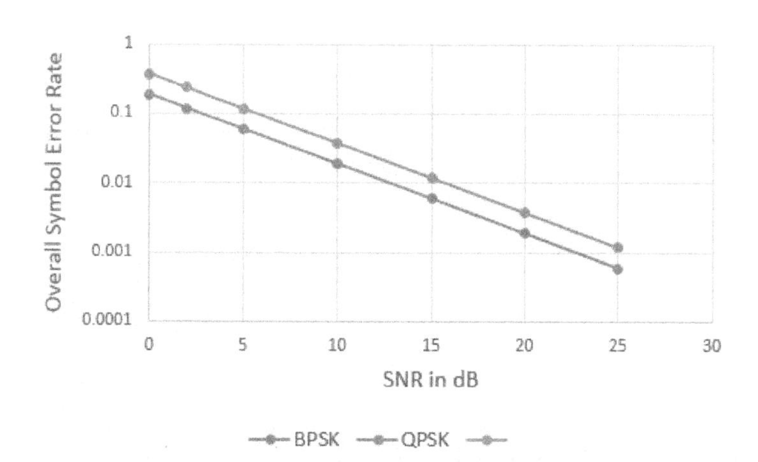

Fig. 3 Variation of OSER with SNR for different modulation schemes

The Fig. 2 shows the variation of OSER with respect to SNR for different values of fading parameter 'm'. With this figure the conclusion can be drawn that with increase in fading parameter value, the value of OSER is decreasing and for the tolerable OSER of 0.001 the value of SNR threshold is (23 dB for m=1) and (20dB for m=2 and m=3)

Consider Fig. 3, It shows the variation of OSER with respect to SNR for different modulation schemes for which the value of modulation dependent parameters α and β. For binary phase-shift keying ($\alpha = 1$, $\beta = 1$) and for quaternary phase-shift keying ($\alpha = 1$, $\beta = 0.5$). Here the conclusion can be made as with the increase in order of modulation scheme, the value of OSER is increasing and for the tolerable OSER of 0.001 the value of SNR threshold is (23 dB for BPSK) and (25dB for QPSK).

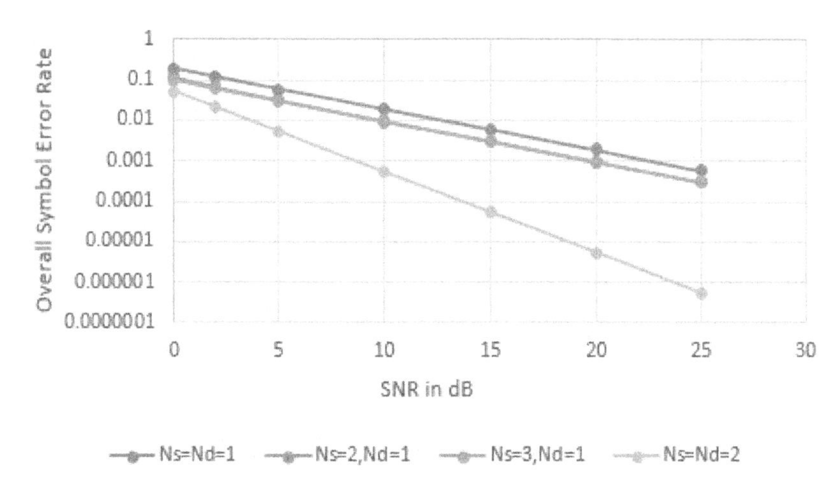

Fig. 4 Variation of OSER with SNR for different values of number of antennas (N_s, N_d)

Consider Fig. 4, It shows the variation of OSER with respect to SNR for different values of number of antennas used at source and destination terminal (N_s, N_d). This shows that with increase in number of antennas, the value of OSER is decreasing and for the tolerable OSER of 0.001 the value of SNR threshold is (23dB for $N_s = N_d = 1$), (20dB for $N_s = 2$, $N_d = 1$), (20dB for $N_s = 3$, $N_d = 1$) and (9dB for $N_s = N_d = 2$).

From the above results we can conclude that the range of SNR value for tolerable OSER may vary from 20dB to 25dB. Hence a statement can be drawn that the two-way relaying scheme based on ANC can work effectively for the region having SNR value greater than 20 dB which is not practical in wireless communication so there are many other constraints which are available to decrease the threshold value of SNR such as by using more number of antennas at source and destination terminals and by increasing the fading parameter value.

5 Conclusion

In this paper the derivation of the expression for overall symbol error rate (OSER) has been done also the variation of OSER with SNR for different parameters has been observed and the threshold range of SNR has been calculated for tolerable OSER of 0.001. Which comes out as 20dB to 23dB for different values of fading parameter 'm', 23dB to 25dB for different modulation schemes used and 20dB to 23dB for different number of antennas used at both terminals, which shows that the threshold range of SNR is 20dB to 25dB for a tolerable OSER of 0.001. The paper also discusses the need of two-way relaying scheme and two half-duplex relaying protocols decode and forward (DF) protocol for digital network coding and amplify and forward (AF) protocol for analog network coding to achieve high spectral efficiency. Further work may be done for implementing the relaying protocol and calculating the threshold SNR range for multiple input and multiple output (MIMO) system.

References

1. Fu, Y.: Single relay selection schemes in two-way relaying networks. In: Asia Pacific Conference on Postgraduate Research in Microelectronics and Electronics (PrimeAsia), September 2010, pp. 304–307
2. Maric, A.G., Medard, M.: Analog Network Coding in the High-SNR Regime. IEEE Wireless Network Coding Conference (WiNC) 1–6 (June 2010)
3. Kwon, J.W., Ko, Y.C., Yang, H.C.: Maximum Spectral Efficiency of Amplify-and-Forward Cooperative Transmission with Multiple Relays. IEEE Trans. on wireless communications **10**(1), 49–54 (2011)
4. Routi, R.R., Ghosh, S.K., Chakrabarti, S.: A network coding based probabilistic routing scheme for wireless sensor network. In: Sixth International Conference on Wireless Communication and Sensor Networks (WCSN), December 2010, pp. 1–6
5. Han, Y., Ting, S.H., Ho, C.K., Chin, W.H.: High rate two-way amplify-and-forward half-duplex relaying with OSTBC. In: IEEE Vehicular Technology Conference, May 2008, pp. 2426–2430
6. Yamamoto, K., Haneda, K., Murata, H., Yoshida, S.: Optimal Transmission Scheduling for a Hybrid of Full- and Half-Duplex Relaying. IEEE Communications Letters **15**(3), 305–307 (2011)
7. Han, Y., Ting, S.H., Ho, C.K., Chin, W.H.: Performance bounds for two-way amplify-and-forward relaying. IEEE Trans. on wireless communications **8**(1), 432–439 (2009)
8. Managayarkarasi, P., Jayanthi, E., Jayashri, S.: Analysis of Digital Network Coding in Dual Hop Networks. IEEE International Conference on Communications and Signal Processing (ICCSP), April 2013, pp. 1078–1081
9. Unger, T., Klein, A.: Duplex schemes in multiple antenna two-hop relaying. EURASIP J. Adv. Signal Process. **2008**, 128592-1–128592-13 (2008)
10. Katti, S., Gollakota, S., Katabi, D.: Embracing wireless interference: analog network coding. In: Proc. ACM SIGCOMM, August 2007, pp. 397–408
11. Yadav, S., Upadhyay, P.K., Prakriya, S.: Performance Evaluation and Optimization for Two-Way Relaying With Multi-antenna Sources. IEEE Transactions on Vehicular Tech. **63**(6), 2982–2989 (2014)
12. Ferdinand, N.S., Rajatheva, N., Latva-aho, M.: Performance analysis of two-way relay system with antenna correlation. In: Proc. IEEE GLOBECOM, December 2011, pp. 1–5
13. Popovski, P., Yomo, H.: Wireless network coding by amplify-andforward for bi-directional traffic flows. IEEE Commun. Lett. **11**(1), 16–18 (2007)
14. Goldsmith, A.: Wireless Communications. Cambridge University Press, Cambridge (2005)
15. Yang, N., Yeoh, P.L., Elkashlan, M., Collings, I.B., Chen, Z.: Twoway relaying with multi-antenna sources: Beamforming and antenna selection. IEEE Trans. Veh. Technol. **61**(9), 3996–4008 (2012)
16. Ikki, S.S., Aissa, S.: Performance evaluation and optimization of dualhop communication over Nakagami-m fading channels in the presence of co-channel interferences. IEEE Commun. Lett. **16**(8), 1149–1152 (2012)

An Optimization to Routing Approach Under WBAN Architectural Constraints

Aarti Sangwan and Partha Pratim Bhattacharya

Abstract WBAN is having its significance in various application areas but the most critical dedicated application is to monitor the medical patient. WBAN is a specialized network with data level, node level criticalities. In this work, an optimization to the routing approach under architectural constraints is defined. The work has analyzed various node level, data level and communication level criticalities. The paper has defined a route optimization algorithm under these critical vectors. The work is here analyzed in terms of network life and energy consumption over the network. The experimentation is provided with real time constraints and the obtained results shows that the presented work has improved the network life.

Keywords WBAN · Architecture · Real time · Medical patient

1 Introduction

Wireless Body Area Network is one critical real time network designed for specialized applications. The specialty of the network is respective to the network architecture, application area and node capabilities. The architectural specifications represent the placement of nodes in the network. These specifications depend on multiple associated vectors. The first and foremost constraint for WBAN architecture is its localization. The localization is about to decide whether the nodes are placed outside or inside human body. It is required to decide, which human organs are required to monitor. The decision about this architectural localization is based on the application specification. Such as for medical monitoring, the nodes are placed on each critical organ such as heart, veins etc. WBAN provides the online and real time monitoring to the patient

A. Sangwan(✉) · P.P. Bhattacharya
Mody University of Science and Technology, Laxmangarh, India
e-mail: {aarti.sangwan1,hereispartha}@gmail.com

S. Berretti et al. (eds.), *Intelligent Systems Technologies and Applications*,
Advances in Intelligent Systems and Computing 385,
DOI: 10.1007/978-3-319-23258-4_8

health that provides two main advantages. First advantage of this network is the mobility of patients so that the monitoring can be done respective to specific activities. This is helpful to identify the effect of some activity operation on patient or healthy person. Such as for athletes, the stamina observing applications can be designed using activity specific BAN network. In such network, the inter-BAN communication can also be established to identify the cooperative and relational activity monitoring. This kind of analysis can be done when some teamwork is going on such as monitoring the players' stamina during a cricket match. In such application, the comparative analysis can be obtained for batsman, fielder and bowler. In same way, according to the field position, the analysis of players can be done. This stamina analysis can be done based on number of activities and calories burn during each activity. This is represented as the location independent monitoring over the network.

1.1 Scenario Constraints for Patient

The evolution of WBAN network was specifically defined for a health care monitoring for a patient. For this, the monitoring devices can be attached on human body or they can be implanted in body itself. This kind of body area network is heterogeneous specific to the node constraints. These constraints are listed in fig. 1.

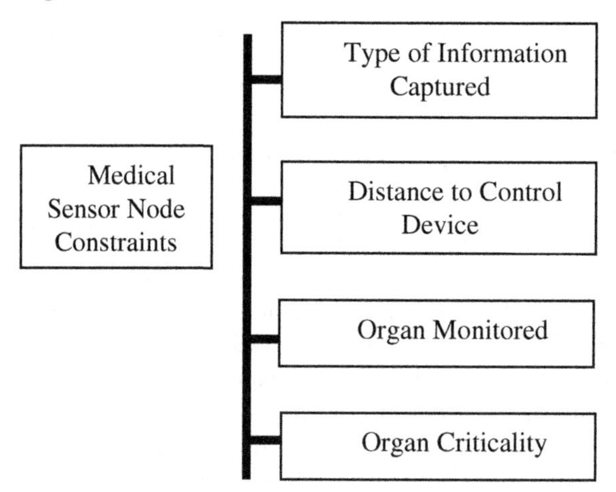

Fig. 1 Sensor Node Constraints

The major constraint to organ monitoring is the type of information communicated. This electronic signal form includes ECG (Electrocardiogram) signal transmission, photoplethysmogram (PPG), pulse rate analysis, blood pressure, and body temperature analysis. Each of these statistics transfers different kind of information at different frequencies and after different extraction time interval. These all parameters are explained in section 3. Another criticality in the formation of architecture is the specification of nodes and their localization respective to the centralized controller so that regular and time effective information capturing and updation can be performed. It is also required to prioritize the hardware component so that critical information will be extracted uninterruptedly. The criticality of the organ is another vector that different from individual as well as disease of the person. The type of data communicated over the component also represents the criticality vector. These all vectors are based on the physical specification or the hardware constraints. But there also exist some criticality parameters specific to the communication architecture. These parameters includes the energy specification of nodes, energy dispersion respective to the communication type, distance and operation, load sensitivity, fault tolerance etc. In section 3 and 4 all these environmental and application specific parameters are explained in detail.

In this present work, a routing architecture for WBAN network is defined under hardware level criticalities, application level criticalities and environment level criticalities. The WBAN network is here defined for medical patient. In this section, a standard specification for WBAN architecture is defined along with relative constraints, applications and challenges. Section 2 covers the work defined by different researchers on optimization of WBAN under architectural and routing specifications. In section 3, the architectural specification of proposed WBAN architecture is done. Section 4 covers the proposed algorithm along with its description. In section 5, the results obtained from the work are presented and discussed under different parameters. In section 6, the conclusion obtained from the work is presented.

2 Literature Survey

WBAN is the most critical wireless architecture with different associated challenges under environmental and application specific challenges. Lot of work is defined to handle these challenges. In this section, the work defined by the earlier researchers is discussed. Samaneh Movassaghi [1] defined a study based work on body area network. Author identified the technological changes in network that includes the architectural constraints, optimization constraints and technological constraints. Author defined the significance of BAN for health monitoring system. Author defined under various standards and provided the simplifications so that the improvement to the network QoS can be achieved in terms of speed, accuracy and reliability. Author also discussed various challenges and issues of WBAN and relative inspirations. Aung Aung Phyo Wai [2] has presented a work on reliability analysis on virtualization of WBAN network. Author analyzed the challenges associated with real time implementation of network and tests them in an integrated virtual environment. Author designed the network with real time hardware and architectural constraints and provided the network modeling under various parameters including the energy specification, routing scheme, motion awareness etc. Author applied the routing communication in WBAN network and analyzes the communication under reliability constraint.

Zahoor Khan [3] has defined an improvement to the Body area network under QoS parameter. Author defined a peered communication over the network under reliability and efficiency constraints. Author provided the optimization to the network in terms of energy effectiveness, network life and communication throughput with specification real time constraints over network nodes. Author analyzed the network under delay sensitivity and provided mechanism to reduce the communication delay. Author implemented the work in OMNeT++ environment and provided the simulation on realistic patient network. Jocelyne Elias [4] provided a topological improvement to the network so that energy effectiveness over the network will be improved. Author designed the work for health monitoring for patient so that effective routing over the network will be achieved. Author provided the joint communication based on relay positioning to the network so that network problems will be reduced and the improvement to the network effectiveness will be achieved. The observations shows that the optimized network designed not only reduced the design cost but also increased the effectiveness in terms of communication statistics. Author demonstrated the characteristics of planned network and provided its significance respective to network life and energy consumption.

Christian H. W. Oey [5] provided a work to the WBAN network under various energy effective protocols. Author analyzed the network issues in health care system and medical patient monitoring. Author analyzed the constraints relative to the biomedical environment and provided the design time analysis over it. Author also analyze the issues associated with WBAN and achieved the reliability and energy effectiveness over the network communication. N. Javaid [6] defined a specialized WBAN architecture to measure the fatigue of soldiers. Author defined an event driven activity monitoring architecture to define a comparative analysis over various soldiers. The analysis is here performed in terms of communication throughput, network life and energy level. Author defined an improved protocol under specialized constraints to perform effective comparative analysis on network. Md. Tanvir Ishtaique ul [7] defined a work on energy adaptive WBAN network. Author presented the architectural formation for different integrated applications including the medical, sports, entertainment, military etc. Author identified the architectural and application specific challenges and provides the changes required to optimize the network under energy vector. Author presented the clustered routing approach to provide multi hop communication over the network. Author obtained the spatial information over the network nodes and provided the opportunistic routing over the network. Author defined the centralized operations to reduced the computational efforts and optimized the network communication. Muhannad Quwaider [8] presented an HMM based probabilistic framework to identify the physical context to the network architecture and body posters. The framework was here designed in two main stages. In first stage, the statistical information pattern over the network was obtained and in second stage, the physical context based extensive analysis is obtained so that the architectural model modeling over the network will be obtained. Author provided the significant improvement to the model based on various activities. The activities considered in this work include walking, running, sitting standing, lying down etc. Author defined a control probabilistic experimentation under HMM model to identify the effect of poster on network and provide the optimized solution in terms of effective communication and network life.

Lu Shi performed [9] has defined an authentication network architecture based on channel characteristics of network. Author provided the physiological signal architecture and its integrated promising technology to perform real time signal modeling for medical application. Author defined a trust adaptive framework to obtain the information to decide network reliability and provide the solution against physical and experimental challenges. Author also provided the secure cryptographic authentication with light weight key exchange so that trust adaptive communication will be performed. The work effectiveness is verified under different attacks.

Benoit Latre [10] presented a study based work on body area network. The work includes the architectural analysis and network formation under different specifications to the node and the environmental constraints. Author provided the descriptive overview to different constraints associated with network and provided the technological aspects associated with work. Author also identify the crosslayer protocols and service formation while performing the secure communication in network. JeongGil Ko [11] defined an architectural view of sensor network for health care system. Author provided the analytical description of network under physical, behavioural and physiological aspects. Author also discussed the application and environment based challenges.

3 Proposed WBAN Architecture

In this section, the WBAN architecture is defined to monitor the patient. The architecture is here defined with node level constraints described specific to the environmental parameters and criticality constraints. According to this presented architecture, the nodes are placed on patient body at fix position and with specification to the organ. The centralized controller is defined so that it can capture the information from all nodes effectively as well as placed at optimized distance from the external controller. This centralized controller is responsible to aggregate the node information captured from different medical sensors. This coordinator will forward this collected aggregative information to the external controller node so that the information will be maintained in medical database. The architectural view of this model is shown in fig. 2.

Here the figure is shown with four WBAN networks. Each network is here specified with 9 specialized nodes and one coordinator. The coordinator node is here defined on patient body to capture the information from different sensor nodes and collect this information in an aggregative form. This aggregative information will be then forwarded to the centralized external controller. This controller will update this information to the database as well as analyze the deficiencies or errors in input data. If some such error identified, the relative message will be generated and process will be performed to reduce such error. This centralized controller can also connect to web to publish the information globally or to pass it to remote system.

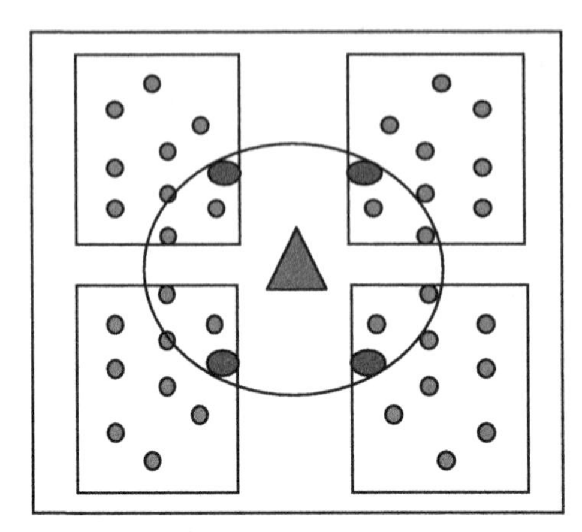

Fig. 2 WBAN Architecture

Each of the sensor node attach to the human body are defined with specialized features and constraints. These constraints are defined in detail in table 1. The table has explained the description respective to the node position, type of information captured, required frequency and the criticality vector. For each kind of information, the specialized sensor nodes are defined.

Table 1 Sensor Node Constraints

Node Number	Placement To	Properties	
1	Heart	Parameter	ECG Signal
		Form	Wave Form
		Parameter Range	0.5-4mV
		Frequency	.01 to 250 Hz
		Criticality	High
		Type	Continuous
2	Brain	Parameter	EEG Signal
		Form	Wave Form
		Parameter Range	3 MicroV-300MicroV
		Frequency	.5 to 60 Hz
		Criticality	Low
		Type	Continuous
3	Brain	Parameter	EMG Signal
		Form	Wave Form
		Parameter Range	10 MicroV-15mV
		Frequency	10 to 5000 Hz
		Criticality	Low
		Type	Continuous

Table 2 *(continued)*

4	Heart	Parameter	Respiratory Rate
		Form	Discrete
		Parameter Range	2-50 breath/min
		Frequency	.01 to 10 Hz
		Criticality	Low
		Type	Time Bound
5	Arm	Parameter	Body Temperature
		Form	Discrete
		Parameter Range	32 to 40 C
		Frequency	0 to 0.1 Hz
		Criticality	Low
6	Arm	Parameter	Blood Pressure
		Form	Discrete
		Parameter Range	10 to 400 mm Hg
		Frequency	0 to 50 Hz
		Criticality	Low
		Type	Time Bound
7	Skin	Parameter	GSR
		Form	Wave Form
		Parameter Range	30 Micro V to 3 m V
		Frequency	.03 to 20 Hz
		Criticality	Low
		Type	Continuous
8	Nerve	Parameter	Nerve Potential
		Form	Discrete
		Parameter Range	.01 to 3mV
		Frequency	10000 Hz
		Criticality	High
		Type	Continuous
9	Finger	Parameter	Sugar
		Form	Discrete
		Parameter Range	100 to 500
		Frequency	10 to 50 Hz
		Criticality	Low
		Type	Time Bound
10	Controller	Position	Centralized
		Type	Receiver, Forwarder and Aggregator
		Processor	High Processing Capability
		Connection	Connected to Real World
		Criticality	High
		Memory	High

Once the information is captured from various sensors, the centralized control collects this information and transforms it to the required form. This intelligent sensor is able to take the early decision on critical vector. If the criticality is identified, the relative decision is taken respective to the identified problem and specific to the associated organ. This centralized controller is also responsible to identify the problem in capturing process. If the information is collected without any interruption, the aggregation to this all information is obtained. This aggregative information is finally forwarded to the external controller. If some instruction passed by the external controller, the body network controller will accept this information and perform the instructed operation. The table has explained the relative parameters for this centralized controller including the position vector, memory, processing capability and its connectivity in real world to perform the communication. The controller is having the decision adaptive capability for route selection and communication.

4 Research Methodology

In this paper, the route optimization to the WBAN network is defined under different constraints defined specific to the network, application and environment. These architecture level constraints are described in previous section. In this section, the consideration of these constraints is defined in terms of communication parameters and based on which the routing decision is taken. Some of architectural constraints transformed to the communication restrictions are shown in table 2.

Table 2 Transformation of Architectural Constraints to Communication Decision

Serial	Architectural Constraint	Communication Decision
1	Criticality : High	Node will not participate in forwarding Data
2	Criticality : Low	Node can forward Information
3	Form : Waveform	Packet Size 32K
4	Form : Discrete	Packet Size : 4K
5	Type : Continuous	Load Restriction < 3
6	Type : Time Bound	No Load Restriction
7	Node Strength	Energy Level
8	Node Life	Energy Remaining
9	Communication Cost	Energy Consumption (Transmission, Forwarding, Receiving)

Table 3 Routing Algorithm

```
RoutingAlgorithm(SNodes, N, Controller)
/*SNodes is the list of N sensor nodes localized to different position on patient
body. Controller is the centralized controller defined as destination node*/
{
1.      Set the Decision Vector for Hop Selection as limit thresholds named:
        EnergyThreshold, SensingRange, LoadThreshold and
        CriticalityThreshold
2.      For Source=1 to N
        {
3.      Set Cur=Source
        Set Dst=Controller
4.      while Cur<>Dst
        {
5.      Neigh=GetNeighbors(Cur)
        NextHop=1
6.      For i=1 to Neigh.Length
        {
7.      if (Dist(Neigh(i),Cur)<SensingRange And
        Dist(Neigh(i),Dst)<Dist(Cur,dst))
        {
8.      If(Neigh(i).Energy>EnergyThreshold And  Neigh(i).Criticality=Low)
        {
9.      If (Source.Type=Continuous and   Neigh(i).Load>LoadThresold)
        {
10.     Neigh(j).ForwardBlock=True;
        }
11.     if(Neigh(i).Energy>Neigh(NextHop).Energy And
        Neigh(i).Criticality<Neighbor(NextHop).Criticaltiy And
        Neigh(i).FailureProb< Neighbor(NextHop).          FailureProb)
        [Identify the Next Hop with High Energy , Low Criticality and
        Low Failure Rate]
        {
12.     Set NextHop=i
        }
13      Elseif(Neigh(i).Criticality<Neighbor(NextHop).      Criticaltiy
        And Neigh(i).FailureProb<          Neighbor(NextHop). FailureProb)
        {
```

```
14.     Set NextHop=i
        }
15      Elseif(Neigh(i).FailureProb<Neighbor(NextHop).   FailureProb)
        [Find Next Hop with low Failure Rate]
        {
16.     Set NextHop=i
        }
17.     Else
        {
16.     Set NextHop=i
        }
17.     Set Cur=NextHop
        }}}}
```

After specification of these constraints and restriction, the algorithmic approach is defined to perform the multi-hop communication over the BAN network. As of the architectural specification, each of the sensor node work as source node and the centralized controller act as destination node. The communication to the destination end also includes aggregation. During this stage, the information encoding and aggregation is performed to forward the information to external controller. The algorithmic approach adapted in this work is shown in table 3.

4.1 Explanation

In this sub section, the algorithmic approach applied for the research is discussed. The work is here defined to optimized the WBAN network with architectural specification and route optimization. The route optimization is here provided to achieve the data adaptive and node criticality based route routing. As the WBAN provides the transmission of critical information including the discrete and continuous data forms, because of this there is the requirement to provide the adaptive route selection. The work is defined to consider this criticality while forming the route in BAN network.

At the early stage of architecture, the node criticalities are described respective to the data type and organ association as per table 2. Once the nodes are described, the limits on the node participation vector in route selection are described. These parameters include energy threshold, sensing threshold, load threshold etc. EnergyThreshold is the minimum energy required for communication, sensing range is the communication range, loadthreshold will limit number of simultaneous communication and criticalitythreshold is the limit rate applied on failure rate. Now as the communication begins, each node act as the source node and the controller node is the receiver. The current node, cur and destination node, dst are set for the current communication. The individual communication route is build to provide effective communication. The first level analysis in route formation is done based on distance analysis. This distance adaptive analysis is defined to generate the neighbor node list. The neighbors list respective to the current node under distance parameter is generated and the adaptive possible next hop is set at node 1. All the neighbors are processed for reliable communication and if the Neighbor Node is in sensing Range and in same direction of the coordinator (destination) node, it will be considered as expected next hop. In second stage, the energy and node criticality vectors are analyzed to identify the feasibility of node as next hop selection. Only a High Energy and Low Criticality Node is considered as Next Hop. In final stage, the load and communication data is verified to discard the heavy load under critical information analysis. If communication type is continuous only restricted communication can be performed over the network. A Neighbor Node with heavy load will not be elected for continuous data communication. "ForwardBlock" is a dynamic parameter which is here used to identify the node's feasibility for election as next hop.

Once the adaptive nodes are identified, the best node is elected under energy, acceptability and directional vectors. This identified best hop is set as the current node. This process of hop selection is repeated till the destination node not occurs. After this stage, the effective and reliable route is identified over the network. These routes are taken by the WBAN units to perform the information transmission.

Once the information is transmitted through the generated effective route, the coordinator accept this information and combine this node information using some aggregative operator shown as under

AgreeInfo= $\sum_{i=1}^{N} Nodes(i).Info \{ where\ Node(i).Type = Discrete\}$

AgreeInfo= $\sum_{i=1}^{N} Nodes\ (i).SignalSegment(j)\{ where\ \ Node(i).Type = Continuous\}$

For the Discrete information, the information contents are aggregated in the actual form whereas in case of continuous signal form, the signal information will be collected for the specific segment interval. This information will be then combined in an aggregative form with specification of instance value at particular interval. Finally, both kind of discrete and continuous signals will be transformed to numerical form and the information will be forwarded to the external controller.

5 Results

In this paper, an experiment to the body network is been performed under the specification of real time parameters and their transformation to the communication parameters. The work is implemented in Matlab environment. The parameters considered for the experimentation of network are shown in table 4.

The work is implemented to generate the route from each source node to the destination node. The strength of the network is here represented in terms of network life and the dead nodes over the network. The network life estimation respective to number of communication rounds is described in terms of alive nodes and dead nodes.

Table 4 Network Parameters

Parameter	Value
Number of Nodes	9
Type of Nodes	Heterogeneous
Type of Communication	Discrete and Wave Communication
Initial Energy	Random
Packet Size	4 K for Discrete 32 K for Wave
Sensing Range	.5 M
Transmission Energy Loss	.5 nJ/Byte/100cm
Receive Energy Loss	.5 nJ/Byte/100cm
Forwarding Energy Loss	.1 nJ/Byte/100cm
Position	Architecture Specific
Number of Communications	5000 Rounds
Forward Block	0 => Valid Node 1 => Block Node

To provide the analysis, the presented work is compared with energy and distance adaptive route generation approach. The comparative analysis on the route selection parameters considered in this work given here in table 5.

Table 5 Parameter Adaptive Comparison (Existing Vs. Proposed)

	Existing	Proposed
Energy Adaptive	Yes	Yes
Distance Adaptive	Yes	Yes
Node Criticality Analysis	No	Yes
Communicating Data Consideration	No	Yes
Destination Directed	Yes	Yes
Load Consideration	No	Yes

The table has shown the comparison in the algorithmic approach applied in the existing and proposed work. The table provided the clear view that the existing work is based on mainly distance and energy parameters whereas in this present work, architectural and communication constraints are considered. The parameters added in this work include node criticality analysis, communicating data criticality analysis and the load criticality analysis.

The network level constraints for energy consumption and the energy vectors are taken same in both existing and proposed approach. Once the parameters are defined, the analysis is here done under energy and network life. The comparative results obtained from the work are shown here under

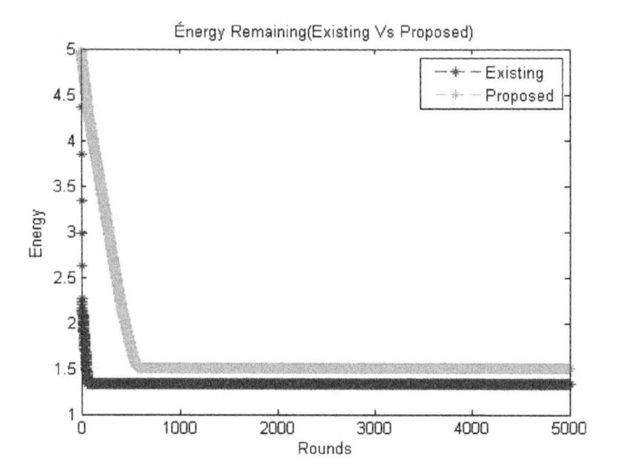

Fig. 3 Network Energy Analysis (Existing Vs. Proposed)

Here fig. 3 is showing the network energy analysis for existing and proposed work. Here x axis represents the communication rounds and y axis represents the remaining network energy. Initially, the nodes are defined with random energy specification and with each communication round some amount of energy are lost. The figure shows the energy loss in case of existing work is performed at early stage so that the overall network energy remained in proposed work is higher. It shows that the proposed work is more reliable under energy parameter.

Here fig. 4 is showing the dead node analysis over the network. Here x axis represents the communication rounds and y axis represents the nodes. Initially, the nodes are defined with random energy specification and with each communication round some amount of energy are lost. The nodes with energy 0 are represented as dead node. Initially no nodes are dead and later on about 7 nodes are dead over the network because of high communication. In existing work, the nodes are dead around 10 rounds where as in this proposed work, the network resides for about 900 rounds. It shows that the overall network communication in this proposed work is improved.

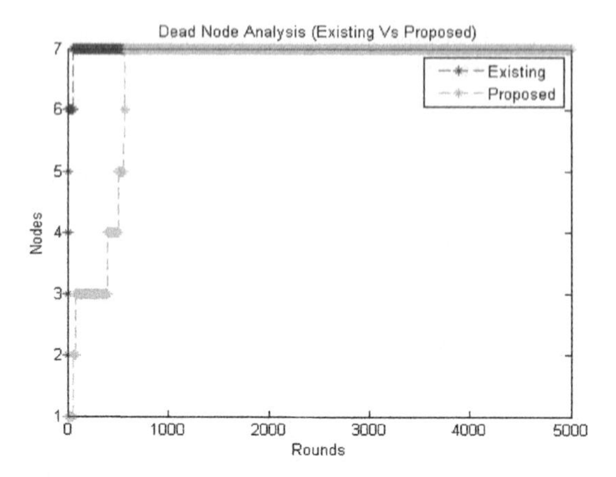

Fig. 4 Dead Node Analysis (Existing Vs. Proposed)

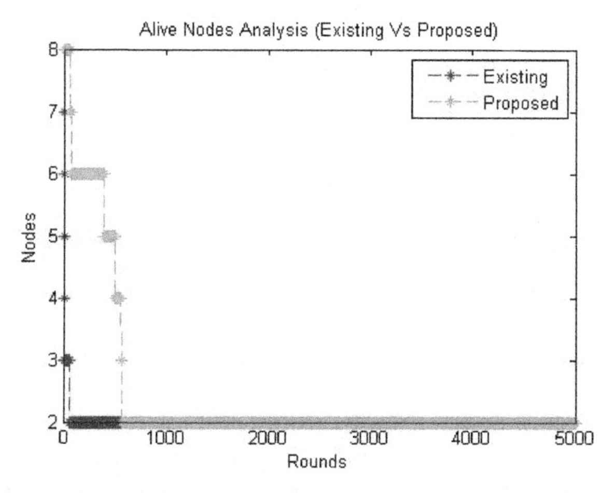

Fig. 5 Alive Node Analysis (Existing Vs. Proposed)

Here fig. 5 is showing the alive node analysis over the network as the comparison vector. The results show that the network resides for more communication rounds in proposed work so that the overall network life in proposed work is improved.

6 Conclusion

In this paper, an optimization to WBAN network is provided in terms of network route under architectural constraints and communication constraints. The work has first transformed the architectural parameters to communication parameter and real time setting to energy specific constraints. Based on these parametric adjustments the network is simulated for route generation from all nodes to centralized controller. The work shows the estimation of work in terms of network life and energy consumption over the network. The analytical results show that the work has improved the network life and the energy resides in the network for maximum time.

References

1. Movassaghi, S., Abolhasan, M., et al.: Wireless Body Area Networks: A Survey. IEEE communications surveys & tutorial **16**(3), 1658–1686 (2013)
2. Wai, A.A.P., Ge, Y., et al.: Reliability enhancement and performance visualization system for wireless body area network. In: Proceedings of PERCOM Workshops, Lugano, pp. 498–500 (2012)
3. Khan, Z, Sivakumar, S, Phillips, W, Robertson, B.: QPRD: QoS-aware peering routing protocol for delay sensitive data in hospital body area network communication. In: Proceedings of 7th International IEEE Conference on Broadband, Wireless Computing, Communication and Applications (BWCCA), Victoria, BC, Canada, pp. 178–185 (2012)
4. Elias, J., Mehaoua, A.: Energy-aware Topology Design for Wireless Body Area Networks. In: Proceedings of IEEE ICC 2012 - Selected Areas in Communications Symposium, Ottawa, ON, pp. 3409–3410 (2012)
5. Oey, C.H.W., Salim, S, Moh, S.: Energy-aware routing protocols in wireless body area networks. In: Proceedings of 8th International Conference on Computing and Network Technology (ICCNT-2012), Geuongju, pp. 159–162 (2012)
6. Javaid, N., et al.: Measuring fatigue of soldiers in wireless body area sensor networks. In: Proceedings of Eighth International Conference on Broadband, Wireless Computing Communication and Applications, pp. 227–231 (2013)
7. Ishtaique ul Huque, M.T., et al.: EAR-BAN: energy efficient adaptive routing in wireless body area networks. In: Proceedings of 7th International Conference on Signal Processing and Communication Systems (ICSPCS), Carrara, VIC, pp. 1–10 (2013)
8. Quwaider, M., Biswas, S.: Physical Context Detection using Wearable Wireless Sensor Networks. Journal of Communications Software and Systems. Special Issue on Medical Applications for Wireless Sensor Networks **4**, 191–201 (2008)
9. Shi, L., Li, M.: BANA: Body Area Network Authentication Exploiting Channel Characteristics. IEEE Journal on selected Areas in Communications **31**, 1803–1816 (2013)
10. Latre, B., Braem, B., et al.: A Survey on Wireless Body Area Networks. Wireless Netw **17**, 1–18 (2011)
11. Ko, J., et al.: Wireless Sensor Networks for Healthcare. Proceedings of IEEE **98**, 1947–1960 (2010)

Real Time CO_2 Monitoring and Alert System Based on Wireless Sensor Networks

Parvathy Pillai and M. Supriya

Abstract Carbon Dioxide (CO_2) is an inevitable part in atmosphere for the existence of life on the Earth. But if increased beyond a certain limit it is harmful as it is the major cause for global warming. The only effective way to fight this is to store CO_2 away from the atmosphere for a long time. This is achieved by using Carbon Dioxide Capture and Storage (CCS). But one major concern that arises when CO_2 is stored in large quantities is it's leakage from reservoir. An online CO_2 emission monitoring and alert system is developed in order to monitor the leakage of CO_2 in real time in CCS, try to reduce the leakage up to a controllable level and to give alert to the consent person, if the CO_2 leakage goes beyond an uncontrollable limit. The data is stored in real time for further analysis. Advantages of wireless sensor networks are exploited in order to precisely sense the CO_2 concentration.

Keywords Wireless sensor networks (WSN) · CO_2 emission · Online monitoring · Global system for mobile communication (GSM)

1 Introduction

Carbon dioxide in the atmosphere is directly responsible for the oxygen for breathing and the food and hence it is inevitable in the atmosphere for the existence of life. But if it is increased beyond a certain limit it is turns out to be very harmful as it is the major cause of global warming and ocean acidification. The major cause of CO_2 emission is combustion, usually from thermo electric

P. Pillai(✉) · M. Supriya
Department of Computer Science and Engineering,
Amrita School of Engineering, Amrita Vishwa Vidyapeetham, Bangalore, India
e-mail: pillaiparvathy87@gmail.com, m_supriya@blr.amrita.edu

S. Berretti et al. (eds.), *Intelligent Systems Technologies and Applications*,
Advances in Intelligent Systems and Computing 385,
DOI: 10.1007/978-3-319-23258-4_9

plants, Vehicles, industrial processes, extraction wells of fossil fuels etc. CO_2 Capture and Storage (CCS) is identified as an effective means to accomplish successful storage of greenhouse gas, and moreover to enhance the production of oil and gas. CCS includes separating emitted CO_2 from other gases, transport to a storage location and long term isolation from the atmosphere. All the attempts humans have made to reduce global warming would go waste in case if the CO_2 leaks from the container. For that reason, in order to make sure that CO_2 is stored away safely from the atmosphere, keeping an eye on the leakage of green house gas is of supreme importance. Thus, after storing the CO_2 isolated from the atmosphere for a long time, monitoring of the CO_2 leakage at that region is doubtlessly essential. The rise of CO_2 concentration in a region above some uncontrollable limit can be fatal and has to be dealt with extreme caution. So an attempt must be made to keep the CO_2 concentration always in an acceptable level. This can be accomplished by running a fan, in case if CO_2 concentration goes beyond an acceptable limit, to keep the CO_2 concentration always in a normal level. But in case the CO_2 concentration goes beyond uncontrollable limits which cannot be cured by using a Fan, developing a monitoring and alert system for the surveillance of the region and providing essential warning to the consent personals for immediate action is of great importance. This system can be implemented, with a little modification, in various scenarios like CO_2 monitoring at traffic signals, CO_2 level monitoring at various industries where combustion of fuel or wood is a main process etc. For the monitoring purpose the advantages of WSNs can be exploited.

Wireless sensor networks are the most developed technology of present generation. A link between the real and the virtual world is provided by WSNs. Most of the previously unobservable things can be closely observed with a fine resolution using WSNs, with minimum sensing cost. They have wide range of industrial, science, and transportation and security applications. WSNs are group of spatially dispersed and dedicated sensors used for monitoring and recording the physical conditions of the environment such as temperature, humidity, wind speed, pressure, and direction etc, which organize collected data at a central location. WSN consists of a few hundreds to thousands of sensor nodes with radio transceiver along with an antenna, microcontroller, interfacing electronic circuit, energy source (battery). Advantages of wireless sensor networks over wired networks are increased mobility, better installation speed and simplicity as wires through walls and ceilings need not be pulled and reduced cost.

2 Background

Global warming is one of the biggest problem world is facing now and this could be very dangerous for the human life. Atmospheric concentration of CO_2 above pre-industrial levels causes rise in average temperature of Earth and ocean acidification [1]. The problem of global warming could be solved using many ways out of which stopping the production of CO_2 is one solution. This can be achieved by replacing oil, coal and gas by renewable energy and by planting more trees. But CCS is identified as the one hand solution to decrease the amount of CO_2 in the environment substantially. Processes of CCS separate the emitted CO_2 from other gases, transport to a storage location and long term isolation from the atmosphere. But if this gas leaks out from its reservoir all the efforts that have been used in storing away CO_2 from atmosphere will go waste. So leakage of this gas has to be dealt with utmost concern.

WSNs are the most emerging technology of present scenario and can be used exclusively for monitoring natural environments and providing vital hazard warnings [2]. The most important feature which makes WSN extremely suitable for monitoring purposes is its accessibility, that it can be deployed in most dangerous or remote locations which could be unreachable by humans. Other advantages which make WSN suitable for such applications are low cost, low maintenance and small size. In that case nodes will remain unattended for long time and they will be self organizing on tasks. Geosensor networks are the sensor networks which monitors phenomena's in geographic spaces such as confined environments for example a room to ever changing environments for example an habitat of a living being, is used to gather, interpret and consolidate the geospatial information [3]. These networks should be designed in such a way that they should blend in to the environment which they are monitoring and should avoid any component that pollutes the environment at which it is deployed [4]. They are very helpful in real-time detection of events, wireless communication, data aggregation and processing. CO_2 monitoring equipment that can be kept inside the boots of emergency personals to improve safety and efficiency is developed in [5]. It exploits the possibilities of WSNs such as high sensitivity, selectivity, high robustness and low power demand. This paper also explains certain outcomes of CO2 sensor tests and the integration of sensors with wireless data transmission.

One of the most extensively used applications in the world is SMS or Text messaging. Along with a proper system with an appropriate sensing system the vast scopes of SMS can be exploited as it is considered as world's most reliable source of information. Using a GSM technology, a reliable, real time monitoring and alert system can be established. A remote water level alarm system is developed in [6] by applying liquid sensors and GSM technology as discussed above. GSM uses a variation of Time division multiple access (TDMA). Data is transferred from sensors to the corresponding users through their mobile phone by using GSM and SMS and thus a reliable and real time monitoring system is established. The two main advantages of GSM exploited in this paper are they work on different frequencies and even at the times of line congestion they can be used.

GSM modem for SMS Alert is used in [7] and a real-time application of Home Surveillance system is established. This exploits the advantages of GSM such as fast response, low cost, security of data, high data transfer speed and support of international roaming. A SIM card is required for the proper working of a GSM module and can be configured using AT commands. One thing which has to be taken into careful attention is that GSM module should be compatible with the receiving mobile phone.

Basic idea of our work is derived from [8] in which a remote real time CO2 concentration monitoring system is developed by exploiting the advantages of wireless sensor networks, and which can be used for monitoring the gas leakage in CCS. It consists of a bunch of sensors which collects information about CO_2 concentration, temperature, humidity and light intensity along with a Central processing unit (CPU) which controls all the modules, global positioning system (GPS) for getting information about time and to locate the leakage, a protected digital storage system (SD card) and a liquid crystal display (LCD) module to display the data in real time. A GPRS module is used to aggregate the data to a data server. Later on the data is monitored in real time using a WebGIS application. The block diagram of this system is given in fig. 1.

The conclusion of literature survey enables us to propose a model of remote CO_2 monitoring and alert equipment which tries to minimize the amount of CO_2 in a particular region up to certain level by running a fan and also GSM module starts sending the PPM value of CO_2 concentration to the consent person's mobile phone. When the CO_2 concentration exceeds beyond a controllable limit, along with the fan an alarm rings and alert message for immediate action is sent to the supervisor along with the CO_2 concentration in PPM.

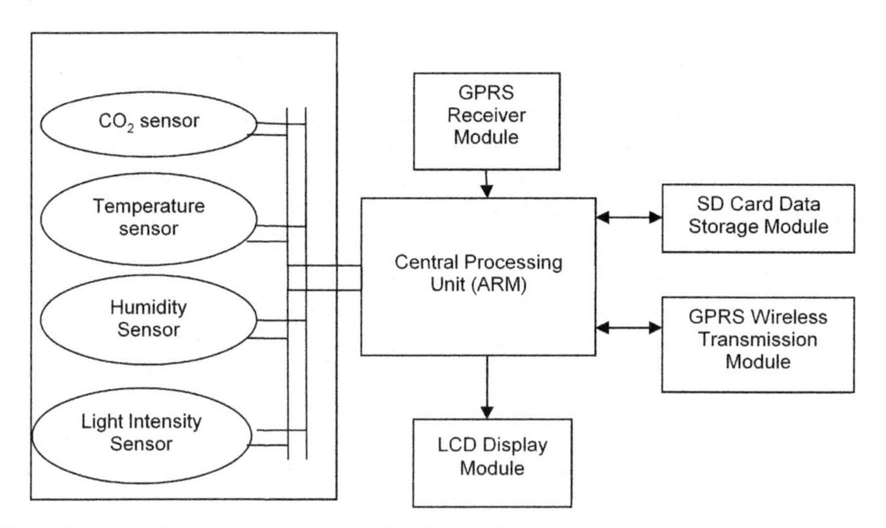

Fig. 1 Hardware Infrastructure diagram for CO2 Leakage Monitoring

Deploying WSNs allows inaccessible areas to be covered by minimizing the sensing costs. Using suitable sensors and GSM technology, a reliable & real time monitoring & alert system can be developed.

3 Proposed Model for CCS

A real time monitoring and alert system is developed which will monitor CO_2 concentration in real time and rings an alarm and gives alert message to the consent person when the concentration goes beyond an uncontrollable limit, using the technology of WSNs. This is developed by considering CCS as a platform. This system senses the CO_2 concentration and displays the value in parts per million (PPM) on an LCD module. CO_2 concentration in PPM is stored in real time in a SD card for further analysis. The effect of leakage up to a certain limit is minimized by running a fan. But when the leakage goes beyond uncontrollable levels, along with the fan, vital warnings using an Alarm and SMS Alert is developed. When the CO_2 concentration crosses the given threshold PPM value of concentration along with the alert message is sent to the operator's mobile. As WSN is used as the base technology to develop this, one main advantage is that inaccessible areas can be covered with a minimal sensing cost.

Real time CO_2 monitoring and alert system consists of CO_2 sensor to collect data about the CO_2 concentration in the atmosphere. The operation of each module is managed by CPU & it automatically stores collected data in the data card. CPU further displays the data in real-time on the LCD display module. GSM module is used to send SMS alert and PPM value of CO_2 concentration into the operator's mobile phone. The digital memory card Storage Module is used to store data in real time for further analysis. A fan runs when the CO_2 concentration is increased above to a certain threshold, considered as 3000 PPM in this work. An alarm and fan and a SMS alert is used to give hazard warning when the CO_2 concentration goes beyond an uncontrolled limit, which is considered as 8000PPM, such that running the fan will not be sufficient to reduce it. The detailed block diagram is given in fig. 2. Operation of each module is controlled by Microcontroller. ARM7TDMI LPC2148 microcontroller is used as CPU. It has 512 Kbyte program flash with 32+8 Kbyte SRAM. It consists of 12.0000 MHz crystal and a Phase-locked loop (PLL) which helps to multiply the frequency with five; 5 x 12 MHz = 60 MHz 32.768 kHz RTC crystal is available. SPI interface with buffering and variable data length capabilities is available.

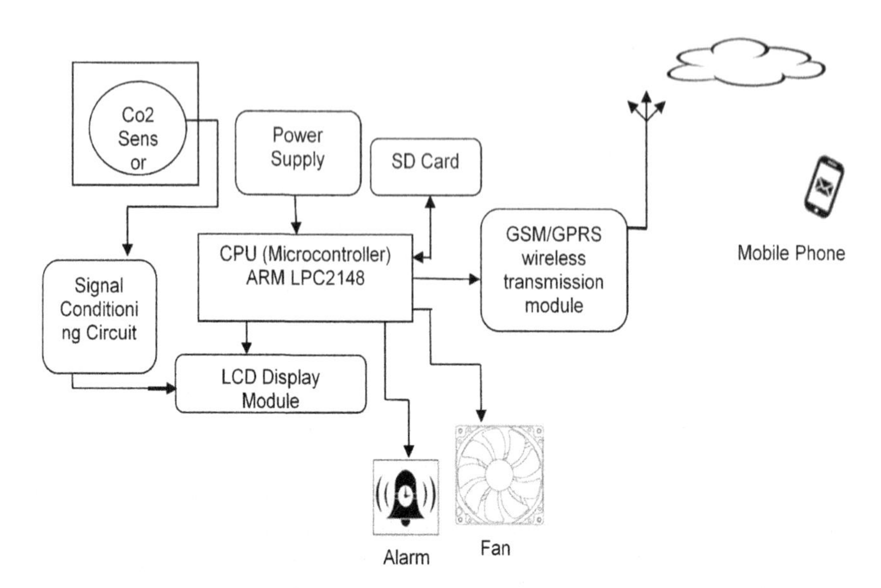

Fig. 2 Block diagram of CO2 Monitoring and Alert system

Onboard Peripherals, 2x16 characters LCD with background light and two UARTs are incorporated in the board. Two (14 channel) ADCs and 1 DAC is available to convert analog to digital and vice versa. Temperature range is from -40°C to +85°C. Interrupt input is given by pushbutton on P0.14. 9-15 VDC, ≥200 mA from 2.1 mm power connector is used as power supply. Mini-B USB connectors can also power it directly. CO2 Sensor Module used is MG 811. It measures from 400 PPM up to 10,000 PPM. It consists of analog output voltage from 200- 600mV. Higher concentration will give lower voltage. Working voltage for the sensor is 6v. It has good sensitivity and selectivity to CO2. Its operating temperature is from -20 to +50 °C.

3.1 Firmware Flow

Firmware Process includes three main parts: Real-time collecting of data where CO_2 Sensors are used to collect data. Hazard handling where a Fan will run when the CO_2 concentration is increased up to a certain threshold and an alarm and fan will run to give hazard warning when the CO_2 concentration goes beyond an uncontrolled limit such that running the fan will not be sufficient to reduce it. An alert message is also sent via text SMS. Wireless transmission through GSM wireless transmission module is used to access data from anywhere real timely in a mobile phone. The firmware flow is shown in fig. 3. The major steps includes
Step 1: Initialize the entire CO_2 remote real-time monitoring and alert system

Step 2: The central processing unit achieves the connection to remote mobile network using GSM
Step 3: Point to point communication will be established
Step 4: Wait for the data from CO_2 sensor
Step 5: If data collection is completed, the CPU will automatically store the collected data into SD card, otherwise go to Step4
Step 6: CPU displays the collected data on the LCD display module real-timely
Step 7: If the collected data is greater than 8000 PPM, Alarm and Fan runs and an Alert message is send along with PPM value of CO_2 concentration to the operators mobile
Step 8: If the collected data is greater than 3000 PPM, Fan runs and PPM value of CO_2 concentration send to the operators mobile
Step 9: When the transmission time interval is reached pack the stored data and access the mobile network using GSM

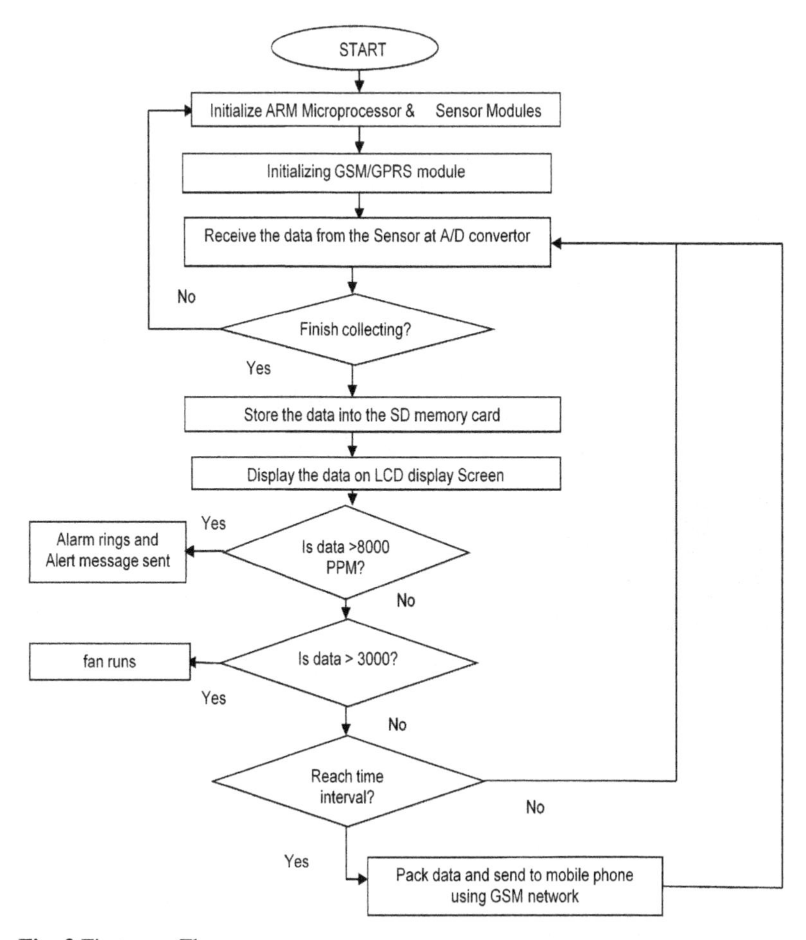

Fig. 3 Firmware Flow

4 Model Implementation

CO_2 sensor MG811 is employed to sense the CO_2 concentration from the atmosphere. It measures from 400 PPM up to 10,000 PPM and gives a corresponding analog output voltage from 200- 600mV. It doesn't need any extra power supply, but takes the supply from the ARM 7 kit. Higher concentration gives lower voltage. This analog voltage is amplified using an amplifier circuit. The entire system's latency is approximately estimated as 2 seconds for one PPM value to be sensed, processed and corresponding alert is given.

4.1 Software Simulation

The code for ARM processor is developed using software called Keil µVision. In Keil editor and debugger are integrated in a single application and this provides a seamless embedded project development environment. Simulation environment for the simulation of the system is set up using Proteus 8 which is an application with many service modules offering different functionality (schematic capture, PCB layout, etc)

In fig. 4 simulation of LCD display showing PPM value of CO_2 is shown in the extreme top right and GSM module simulation in virtual window is shown in the left side. LEDs shown at down left, D1 and D2, represent Alarm and fan respectively. Here it can be seen that since the PPM value is 3520 PPM, according to the conditions only D2 is glowing.

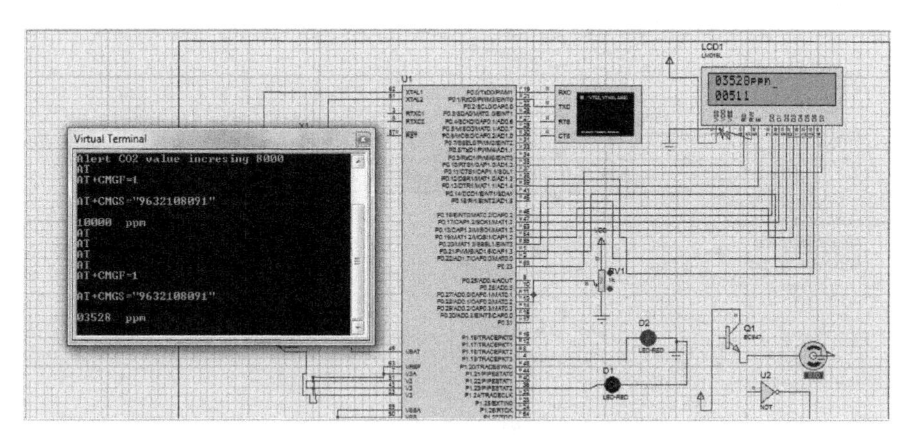

Fig. 4 Simulation of LCD display, fan, alarm and GSM using Proteus

4.2 Hardware Implementation

CO_2 sensor MG811 is employed to sense the CO_2 concentration from the atmosphere. It measures from 400 PPM up to 10,000 PPM and gives a corresponding analog output voltage from 200- 600mV. The main modules used are an amplifier, analog to digital converter, an alarm and a fan as discussed below.

Amplifier. The voltage output of the sensor is in the rage of mV. This voltage has to be amplified into a voltage compatible with ARM processor. As 3.3V is used as the reference voltage for the successive approximation ADC a voltage below 3.3V needs to be given as the input to the ARM processor. The amplified voltage from the Amplifier circuit is given to P0.28 pin of the ARM processor. LM358 IC is used to realize the amplifier circuit as shown in fig. 5. The LM358 series consists of two independent, high gains, internally frequency compensated operational amplifiers.

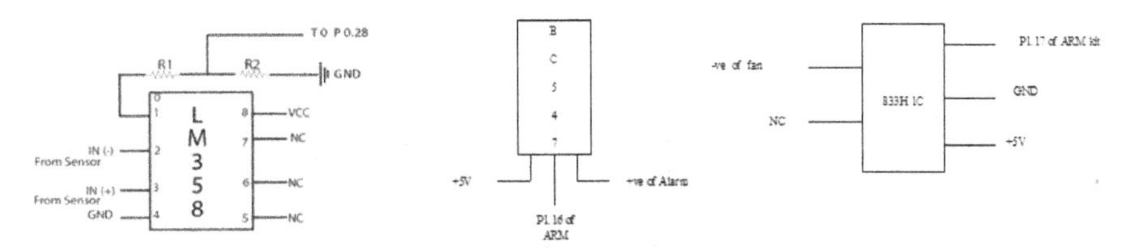

Fig. 5 Amplifier Circuit **Fig. 6** Amplifier using BC547 **Fig. 7** Relay Circuit

The voltage obtained out of pin 1 (output pin) of LM358 is approximately 4.5 volt. But this voltage is not compatible with ARM processor as maximum voltage that it can convert to digital is 3.3V. For this purpose a voltage divider circuit is employed in the amplifier. For a voltage divider

$$\mathrm{Vout} = \mathrm{Vin} * \left(\frac{\mathrm{R1}}{\mathrm{R1} + \mathrm{R2}}\right)$$

$$3\mathrm{V} = 4.5\mathrm{V} * \left(\frac{\mathrm{R1}}{\mathrm{R1} + \mathrm{R2}}\right)$$

$$\mathrm{R2} = .5 * \mathrm{R2}$$

As per the above calculated result R2=1KΩ and R1=2KΩ.

Analog to Digital Converter. The voltage obtained from the amplifier is an analog voltage. In order to make it compatible with ARM LPC2148 it has to be converted into digital. In LPC2148 successive approximation ADC is used for this purpose. The reference voltage which is used for successive approximation is 3.3V. It cannot convert any voltage greater than 3.3V to digital value. The voltage is then converted into parts per billion values and displayed in the LCD module of LPC2148 kit.

Alarm. In order to interface the alarm with LPC2148 BC547 PNP transistor is used as shown in fig. 6. A BC547 transistor is a small device having three terminals. With the help of this transistor it's possible to convert small input signals into amplified outputs. Port1 16th pin of the Arm will become high if the condition for Alarm becomes true.

Fan. A 12V fan is used and in order to interface it with the ARM processor, 833H 1C relay is used as shown in fig. 7. It is a single pole double throw relay. It is an electromagnetic switch consisting of a coil, 1 common terminal, 1 normally closed terminal, and one normally open terminal. When the coil of relay is not energized, the common terminal and the normally closed terminal have continuity. The common terminal and the normally open terminal have continuity, when the coil is energized. Port 1 17th pin will become high when conditions for running fan are satisfied.

Fig. 8 Normal CO_2 Concentration

When the equipment is kept in the normal environment and the CO_2 concentration detected is approximately 300 PPM. At this stage both the fan and alarm is off. The snap shot of the LCD display of the sensor reading in the normal environment of laboratory in which the experiment is set up is shown in fig. 8.

When CO_2 concentration went slightly above the normal level, i.e. above 3000 PPM, only fan works and GSM module starts sending the PPM value to consent people's mobile for monitoring purposes. If the CO_2 concentration goes above 8000 PPM it is considered to be beyond the controllable limit and in that case both fan runs and alarm rings and GSM module sends an alert message to the consent people's mobile phone. Alarm and alert message provides vital hazard warning to consent people. Fig. 9 shows a real time scenario where CO_2 concentration is sensed by the sensor as 8784 PPM, displayed in LCD module, fan runs and alarm rings.

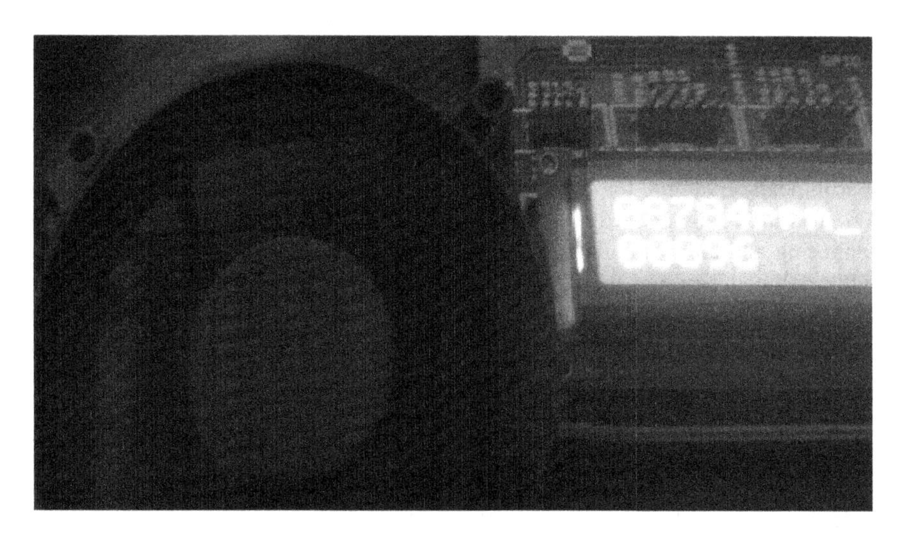

Fig. 9 Real time LCD display of 8784 PPM along with running fan

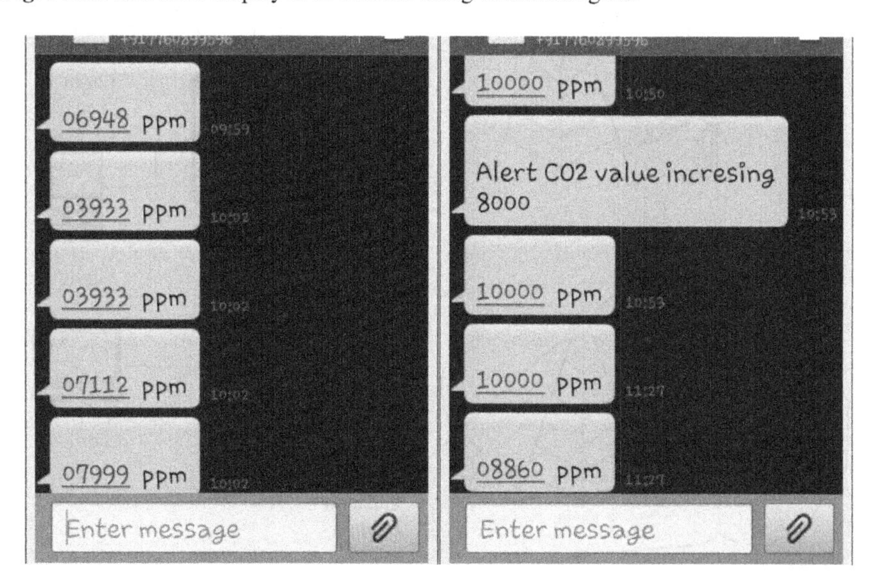

Fig. 10 Real time messages in the mobile phone showing CO_2 concentration monitoring and alert message

Fig. 10 shows the message sent by the GSM module to the registered mobile number in real time, during the various iterations of experiment. 6948 PPM is the first iteration sensor reading which is captured by the sensor. The subsequent values show the PPM values obtained in the succeeding iterations of the experiment. It can be noted that if the CO2 concentration exceeds 8000 PPM an alert message “Alert CO2 value increasing 8000” is sent along with a beep sound.

5 Proposed Model for Traffic Signal

The above developed CO_2 monitoring and alert system can be implemented in various scenarios with a little modification that will suit the corresponding environments. Example for one of such scenario is monitoring and alert system for CO_2 at traffic signals. Fig. 11 represents the system model in which CO_2 monitoring and alert module can be implemented on different traffic signals of a highly polluted city and the data about CO_2 emission in different signals can be aggregated and sent to a data center server, through GSM/GPRS, from which it can be accessed by different clients. Data can be stored in the memory card and can be used for further analysis.

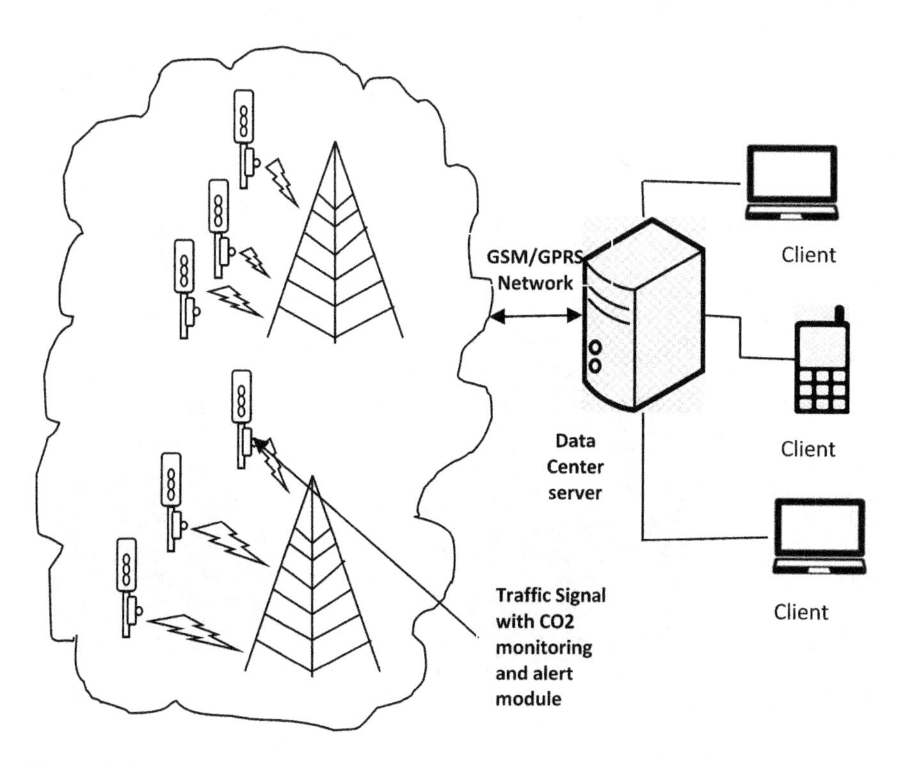

Fig. 11 System Structure for the Implementation of CO_2 Monitoring and Alert Equipment on Traffic Signals

6 Conclusions and Future Scope

A real time CO_2 monitoring and alert system for CCS has been developed in this paper. Though it works in a small set up, the results show that the CO_2 leakage can be monitored and regulated to a certain level. Considering the range of the sensor this work is limited to measure from 400 PPM up to 10000 PPM in an enclosed space such as a laboratory. This system can also be extended in future to a closed loop system with a Fuzzy PD controller in feedback loop to make the system more stable. Implementing such a fuzzy controller will have the advantages of self tuning, capability of handling non-linear systems and reasonably small gain. A file system can be implemented along with SD card to analyze the data that is captured during the leakage. General Packet Radio Service (GPRS) network can be used to send the data to data server. This equipment is not only useful in industrial applications like monitoring of CO_2 leakage in CCS but also can be enhanced and used to analyze CO_2 concentration in environments like traffic junctions, sewage treatment plants, chemical plants etc. This system can be completely automated and can be used as an App for monitoring.

Acknowledgement This work is funded by Vision Group of Science and Technology (VGST), Department of Information Technology, Biotechnology and Science & Technology, Government of Karnataka, India, for SPICE project. Authors are extremely thankful to VGST for their support and motivation.

References

1. Zwaan, B.V.D., Gerlagh, R.: Economics of geological CO2 storage and leakage. Climate Change **93**, 285–309 (2009)
2. Hart, J.K., Martinez, K.: Environmental sensor networks: a revolution in the earth system science. Earth – science Reviews **78**(3), 177–191 (2006)
3. Nittel, S., Labrinidis, A., Stefanidis, A.: Introduction to advances in geosensor networks. In: Nittel, S., Labrinidis, A., Stefanidis, A. (eds.) GSN 2006. LNCS, vol. 4540, pp. 1–6. Springer, Heidelberg (2008)
4. Jung, Y.J., Lee, Y.K., Lee, D.G., Ryu, K.H., Nittel, S.: Air pollution monitoring system based on geosensor network. IEEE Geoscience and Remote Sensing Symposium **3**, 1370–1373 (2008)
5. Radu, T., Fay, C., Lau, K.T., Rhys, W., Diamond, D.: Wearable sensing application-carbon dioxide monitoring for emergency personnel using wearable sensors. In: ICESE (2009)
6. Asmara, W.A.H.W.M.: SMS flood alert system. In: IEEE Control and System Graduate Research Colloquium (2011)
7. May, Z.B.: Real-time alert system for home surveillance. In: IEEE International Conference on Control System, Computing and Engineering, pp. 501–505, November 2012
8. Yang, H., Qin, Y., Feng, G., Ci, H.: Online Monitoring of Geological CO2 Storage and Leakage Based on Wireless Sensor Networks. IEEE Sensors Journal **13**(2), 556–562 (2013)
9. Malki, H.A., Li, H., Chen, G.: New design and stability analysis of fuzzy proportional-derivative control systems. IEEE Transactions on Fuzzy Systems **2** (1994)
10. Mills, B., Heidel, K., Chung, C.-F.: Alternative earth-based and space-based techniques for mitigating global climate change: what can we learn by examining them? In: IEEE EIC Climate Change Technology, pp. 1–10 (2006)
11. Yordanova, S., Merazchiev, D., Mladenov, V.: Fuzzy control of carbon dioxide concentration in premises. In: Proceedings of the 17th International Conference on Systems, Rhode (2013)

Wind Farm Potential Assessment Using GIS

Bukka Bhavya, P. Geetha and K.P. Soman

Abstract Wind energy harvest mainly resides on the place where the trapping of wind energy is efficient along with a self-sustained transmission grid. It becomes a compete source of renewable energy and a proper Geographic Information System (GIS) is required for mapping. This paper employs such techniques for the identification of wind potential area along with the pavement of the transmission grid in Avinashi taluk of Tamil Nadu. This helps in the potential identification of the wind farm installation place, there by solving the energy crisis in Avinashi.

1 Introduction

The wind energy plays a most significant role in the nations economy. Due to more and more usage of indigenous sources of energy they are gradually getting depleted. And especially in country like India renewable energy sources are emerging as an alternative, of these conventional sources, the most potential is the wind energy. India has very huge coast line and hills which helps in wind formation and speed. It covers the coastal region of 7,517 km and it has the land cover of 15,200 km. Dincer (1999) The imperative renewable energy(wind energy) can be achieved in terms of number of wind mills.Building of wind turbine is mainly affected by site selection, availing permits, environmental concerns and socio-economic factors. For building a wind farm the construction industry must see the multiple sites. And for the site selection of wind farm should be based on long term based, which require dependable and real time data of the natural resources and its distribution over space. Hence geospatial information technology forms an important element in decision making process of wind energy Iyappan and Pandian (2012). A Geographical Information System (GIS)

B. Bhavya(✉) · P. Geetha · K.P. Soman
Center for Excellence in Computational Engineering and Networking,
Amrita Vishwa Vidyapeetham University, Coimbatore, India
e-mail: bbhavya.1602@gmail.com, p_geetha@cb.amrita.edu, kp_soman@amrita.edu

S. Berretti et al. (eds.), *Intelligent Systems Technologies and Applications*,
Advances in Intelligent Systems and Computing 385,
DOI: 10.1007/978-3-319-23258-4_10

is a computer system proficient to capture, store, analyze, and display. It is used for general site selection, Lejeune et al. (2010) and also for the complete wind farm installation, GIS can be useful for technical skill and creativity.

An integrated environment for the analysis of for above mentioned problem is accomplished on employing GIS. It seems to be the geographical location, closeness and spatial distribution cr. This paper address the problem that is pretained in Avinashi taluk of Tiruppur. As Tiruppur is facing huge shortage of electricity that makes the need to find a better substitute for energy generation.

2 Study Area

Avinashi is a taluk in tirupur district having an area of 11.65square kilometer in the industrial hub of tiruppur. It has the co-ordinates of 11°11′43″N 77°16′7″E. The boundary of Avinashi have been prepared from the map provided by the Geological survey of India. Figure.1 shows the study area of Avinashi taluk. Power consumption for the industries is also growing exponentially with the increasing demand and household consumption of floating population of migrants from all over the country.

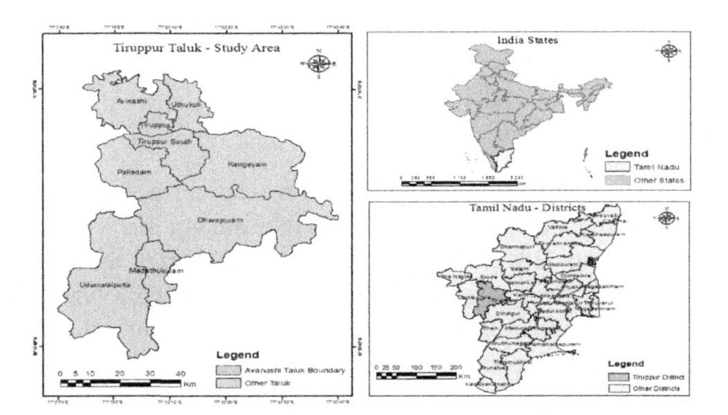

Fig. 1 Study area

3 Methods and Materials

For the preparation and analyzing of the thematic maps the QGIS and the GRASS softwares are used for the purpose of the geographic and the geo spatial data analysis. The functions of the GIS are included into the system to improve the capability and the appropriate locations for the wind farm installations. The functions of the GIS are depicted in the figure 2. The process is divided into three main process .The first is the digitization and collection of data, Second is that data is incorporated as a single system. Last is the geo spatial analysis.

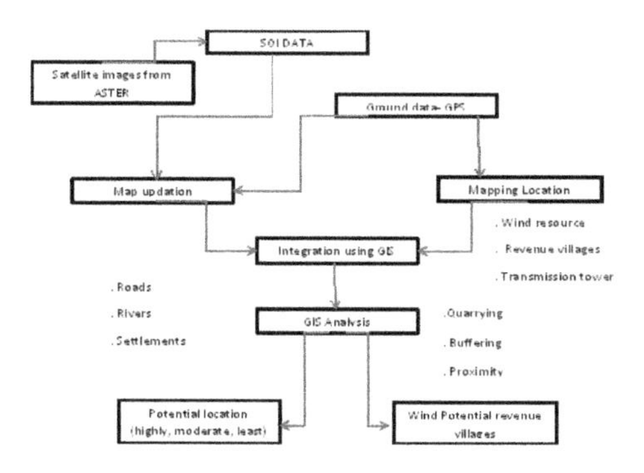

Fig. 2 Process flow

3.1 Data Collection

The first important stage in the GIS enviroment is the data collection.And the second step is the Identification of the ecological objectives and the economic possibility criteria it requires different geographic datas. And it also includes the collection of the data and digitization of the study area. The depicted figure shows the collected spatial data. Survey of India Toposheet 58E\ 3, 58E\4, 58E\ were collected. Geo-

Table 1 spatial data collected

Name of the Data	Resolution/Scale	Source
Topographic Map	1:50000	Survey of India
Wind Power Density Map	5km	CWET CHENNAI
Taluk boundary	1:75000	Open street
ASTER imagery	15m	USGS
Transmission network		GPS surveying

referenced top sheets are used to prepare land use pattern map of Avinashi. The Satellite Imagery of Avinashi is taken from the TERRA/ASTER ASTER (2009) which has a great potential to track the conversation which is due to the various characteristics like spatial and spectral resolution. Avinashi taluk covers replicated colors which are given in the image shows the difference between both the images which are geo-referenced to follow the geographical coordinate system in (WGS84),which is clipped from ESRI-Shape file using SAGA GIS(System for Automated Geoscientific Analysis) software which is used for editing spatial data and the cropped image was re-projected and merged with QGIS software. The spatial resolution of Avinashi taluk ASTER image is 15m as shown in figure

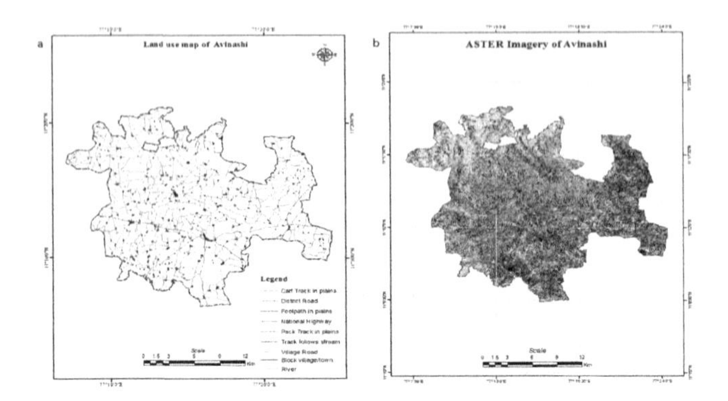

Fig. 3 (a)Land use Map of Avinashi Taluk (b) ASTER image of Avinashi Taluk

3.2 Digitization

By using QGIS and GRASS software the data is digitized with the help of ASTER imagery. Wind power density, slope, proximity to roads settlements, water bodies associated with wind turbines are identified. All these data are correlated and standardized on the basis of World Geodetic System.

4 Results and Discussions

4.1 Slope

Graphical interface of the GIS software system is employed in creating slope map of Avinashi. In setting up an wind farm potential slope is an important. Activities in the wind farm are easier when the slope is flatter. The economic variability of the wind farm installation Clarke (1991) depends on the terrain chosen (the terrain also decides the variability of the energy, it generates). According to the figure6 the suitable area which is recommended for the wind farm development is the flat area.

4.2 Wind Power Density Map

The map obtained from Center for Wind energy Technology(50m) resolution is used during the course of processing.And the map is geo-referenced by assigning the wind speed data, the density is divided into three layers such as the High (250 w/sq.m), Moderate (200 w/sq.m) and Least l(100 w/sq.m) as shown in figure 7. Analysis on wind power density with its terrain feature is shown in table2. The field calculator is estimated in finding the length of the road. The entire length of the road is 473km. The figure 8 shows the power transmission network of the Avinashi taluk which has been obtained through GPS surveying. Transmission network map gives the information about power stations that are available.

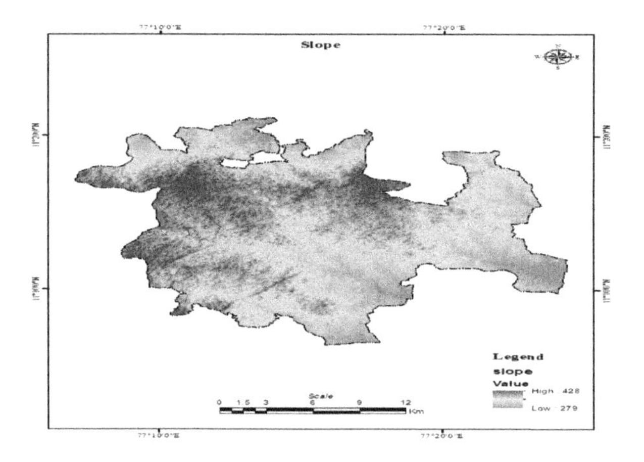

Fig. 4 Slope area map of Avinashi Taluk

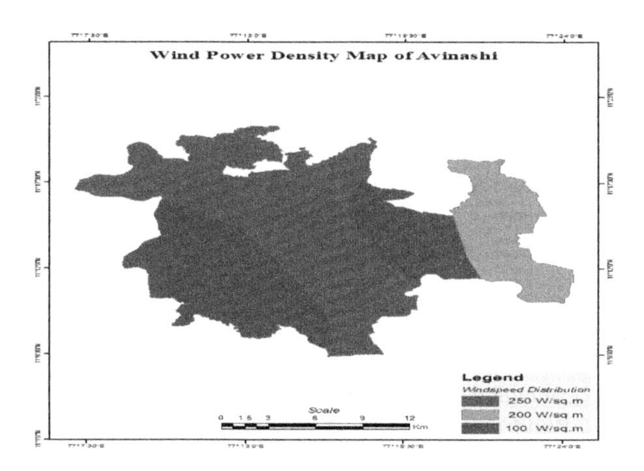

Fig. 5 Wind power density map of Avinashi

Table 2 Analysis of Wind power density

classes	power generated from Wind density (speed m/s)	Available area in sq.km
Highly potential	250	213.2969km^2
Moderate potential	200	69.11932km^2
Least potential	100	56.17956km^2

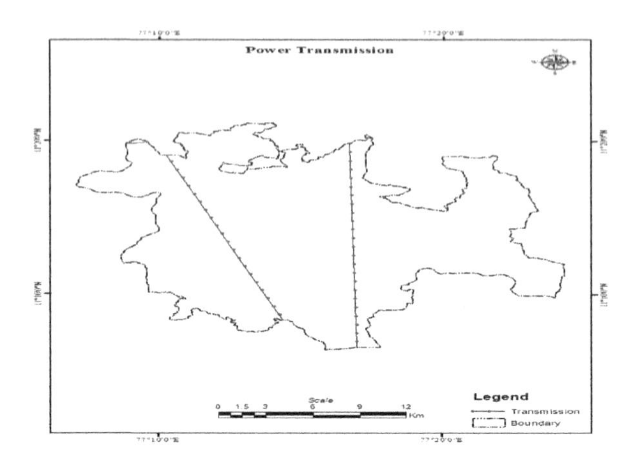

Fig. 6 Power transmission network of Avinashi

4.3 Hydrographic Features

The images obtained by satellite are collected during the winter which helps in identifying the hydro graphic features. The digitization of Avinashi is done using satellite images and topographic sheet. With the help of GRASS soft ware the base map of Avinashi is prepared. As shown in the figure 1the base map is prepared by using the India map and Tamil Nadu map.

4.4 GIS Integration

The statistical data obtained from different functional units of research (land use, Revenue Villages, Wind power density, Transmission network) are collected and are added into desirable software unit. Topological representation of vector data aids us to make and sustain Gupta (2011) vector maps and topological errors are corrected accordingly. The corrected layers are used for analysis GIS and is used in the various operations like questioning etc. By using the method of QGIS software various structural attributes are buffered.

5 Investigation and Validation

Spatial analytics provides a logical solution for the installation of wind mill. The target areas are tagged and the remaining areas which is having at least one negative facts consists of the site that is suitable for wind farm development.

5.1 GIS Analysis

The purpose of the site selection wind speed, area, proper ground condition and the settlement to access the electrical grid are the required important necessities . In Avinashi area the suitability of wind farms is found through the open source . The effect of wind farm installation has the direct influence on the nearby area and the population. And the different effects of the wind energy are also accepted. In this study the ecological impacts linked with the wind energy is studied to develop the acceptability for the other locations for the installation of the wind turbines. Gray et al. (2013)

5.2 Exculsion of Sites Using Buffer Area

The surrounding areas are identified using buffer analysis. Olędzki (2004) the buffer and the buffer distance is taken from the different articles are crawled. According to

Table 3 Details of layer name, distance from the buffer

S.NO	Layer name	Distance from the buffer
1	Stream lines	100m[3]
2	Water bodies	100m[3]
3	Roads	250m[3]
4	Villages	500m[4]

the omission the buffers are created around the each of the feature which is as shown in the table[3] Aydin (2009) and depicted in the following figure 8. The individual buffer layers are over layered into a single layer.

5.2.1 Settlements

With the help of ASTER satellite Imagery and in reference to the topographic sheet the settlement areas has been interrupt. Village areas and town area boundaries are captured and stored as a polygon feature. The attributes of the settlement is obtained from topographic sheet. The clipped portion of the buffers around the features which exceed the boundary limit is clipped and sent for overlay analysis. The total area is that which is not included is taken from the clipped area.

5.3 Overlay Analysis

Spatial overlay provides the solution of spatial relationships among the attributes. And it is done by combining the various data sets that splits or only the part of the similar area. The combination of the original data set is that which identifies the spatial interaction as the result. Figure 9 and figure 10 shows the total available area and the suitability for the wind farm and is done by investigating the variation aimed between the boundary layer and the total area of prohibiting of the Avinashi.

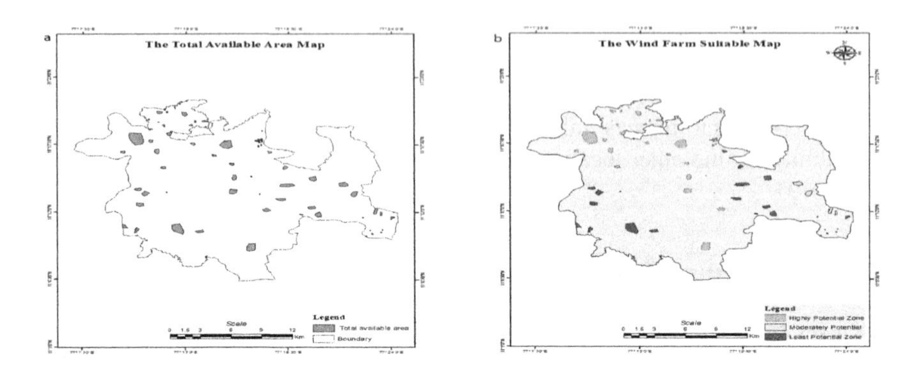

Fig. 7 (a) Total Available Area map (b) Wind farm suitability Area map

6 Conclusion

The wind power assessment and installation has become a major point of interest in this scenario. Wind power is a source of electricity, and also used for agricultural uses. This study gives the importance of Geospatial Information Technology which is used in identifying the potential wind farm locations. The Geospatial softwares like QGIS and GRASS is designed and used for the purpose of combining the thematic layers such as wind, land use etc. Remote sensing features and information along with a proper attribute data elevates the level of accuracy while the process is done. The aim of the study is to present the majority of the accurate and overall analysis of the available data. In future research the analysis of the study may be overcome by considering two additional professionals. First step is that the selected location accuracy can be enhanced by preparing high resolution satellite image which includes a land use map. Next step includes creation of wind suitability map based on the revenue data.

References

GDEM Validation Team Aster. Aster global dem validation summary report (2009)

Aydin, N.Y.: GIS-based site selection approach for wind and solar energy systems: a case study from Western Turkey. PhD thesis, Middle East Technical University (2009)

Baban, S.M.J.: Attaining a balance between environmental protection and sustainable development in the caribbean using

Clarke, A.: Wind energy progress and potential. Energy Policy **19**(8), 742–755 (1991)

Dincer, I.: Environmental impacts of energy. Energy policy **27**(14), 845–854 (1999)

Gray, P., Horan, T.A., Pick, J.B.: Geographic information systems. In: Encyclopedia of Operations Research and Management Science, pp. 635–642. Springer (2013)

Gupta, K.S.: Sustainable wind energy system: Role of energy policy and security-a case study from india. Journal of Economics and Sustainable Development **2**(5), 97–107 (2011)

Iyappan, L., Pandian, P.K.: Identification of potential zone for wind farm in ambasamudram taluk-a geospatial approach (2012)

Lejeune, P., Rondeux, J., Ducenne, Q., Gheysen, T.: Development of an Open Source GIS Based Decision Support System for Locating Wind Farms in Wallonia (Southern Belgium). INTECH Open Access Publisher (2010)

Oledzki, J.R.: Geoinformatics-an integrated spatial research tool (2004)

Real Time Water Utility Model Using GIS: A Case Study in Coimbatore District

G. Praveen Kumar, P. Geetha and G.A. Shanmugasundaram

Abstract Water has become the eternal wonder in 21st century with rapid increase of population and expansion of city limits. The demand for the water has grown up exponentially. Water distribution network needs an efficient modeling for the operation and maintenance with minimal errors in catering to the needs of people with the equitable amount of water through out the year. Creating a simulation model of a real time water distribution network with the account of the pressure and elevation to analyze the flow distribution between the nodes and demand in the network. Geographical information system is an effective tool for decision support using ArcGIS and Water-gems software. Here we tried to characterize the size of pipes with the different diameters of pipes used in the network. The results of the simulation model shows drastic change in the demands resulting in consequences like back-flow, high pressure zone and negative pressure leading to the leakage of pipes making more investment towards the installation and maintenance cost. Main aim of this research is to carry out hydraulic modelling of water distribution network using GIS and reducing the leakage in the pressure zones in saving the time and to minimize the expenditure towards the maintenance. Thus by creating an equity model for the water distribution network in fulfilling minimal demand required across the city.

Keywords Water distribution · Smart city · GIS · Watergems · Coimbatore city

G.P. Kumar(✉) · P. Geetha
Center for Excellence in Computational Engineering and Networking,
Amrita Vishwa Vidyapeetham University, Coimbatore, India
e-mail: praveen16003@gmail.com, p_geetha@cb.amrita.edu

G.A. Shanmugasundaram
Department of Electronics and Communication Engineering,
Amrita Vishwa Vidyapeetham University, Coimbatore, India
e-mail: ga_ssundaram@cb.amrita.edu

S. Berretti et al. (eds.), *Intelligent Systems Technologies and Applications*,
Advances in Intelligent Systems and Computing 385,
DOI: 10.1007/978-3-319-23258-4_11

1 Introduction

Main objective of the study is spatial modelling of water utility modelling using GIS. In developing countries like India, availability of water at source is limited due to resource constraints and hence water shortage is a regular problem and supply is intermittent. Chandapillai et al. (2012) In summer season water consumption rate will grow higher where end user tries to collect maximum creating a demand through out the city and it continues every year. Creaco and Franchini (2013) Chandapillai et al. (2012) So we try to analyze and visualize the impact on the household demand in water distribution in accordance with the population distribution to fulfill the need. Watergems which is being integrated with the ArcGIS for the spatial modelling of the water utility network using GIS. Pressure dependent demand is being analyzed in the network model. Maintaining appropriate pressure levels leads to equivalent water distribution in the network. This paper describes the hydraulic simulation and describes the flow distribution in the water distribution network monitoring the different levels of pressure.

2 Methodology

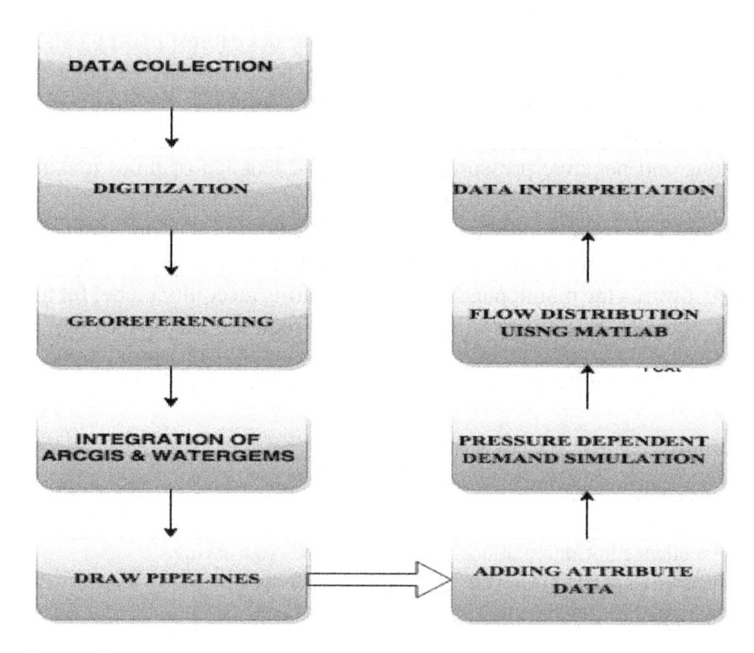

Fig. 1 Process flow

3 Data Interpretation and Analysis

Water utility model is modelled by creating a geo database using ArcGIS and Watergems software by adding necessary attribute data. Then the water utility analyst is being simulated to find out the demand accumulated in the different locations in the entire network. Spatial modelling of water utility network is being designed based on the flow distribution model and demand. Xu et al. (2014) Several features of water pipeline networks have been modelled with help of GIS. Different diameter of pipes which is being used is being modelled with the different color annotations calculating the flow distribution between nodes using a MATLAB program.

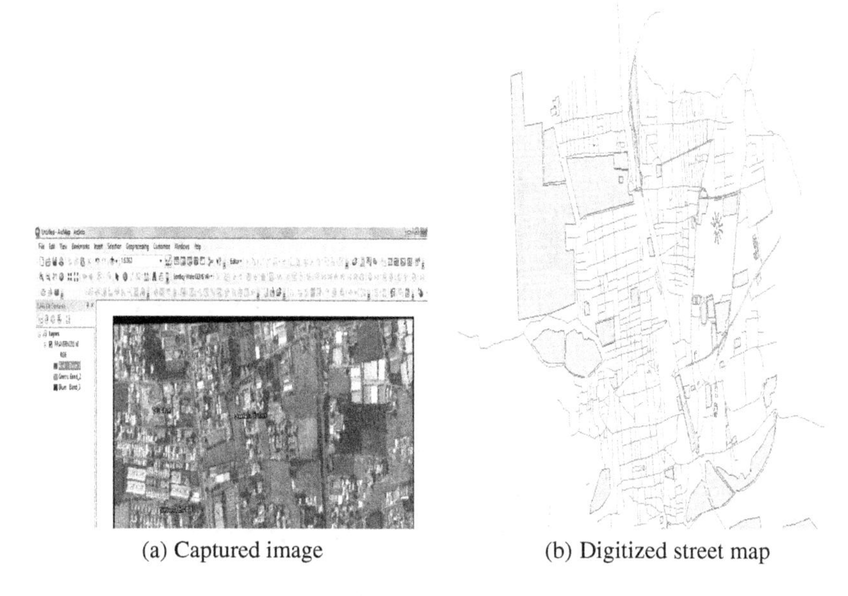

(a) Captured image (b) Digitized street map

Fig. 2 Digitization process

Predominant factors like pipeline capacity, Flow velocity, Pressure head, roughness coefficient and elevation details is being maintained constant among the networkAraujo et al. (2006). Since the values are made on the assumption which is being used worldwide for hydraulic modelling. Demand factor is calculated with the block wise census data of population along with the storage capacity in the overhead tank in the block. Pipeline network is being classified into Main supply line, primary supply networks and distribution among the community for the household purpose. Chandapillai et al. (2012)

4 Results and Discussion

4.1 Flow Distribution

Our goal was to compute the quantity of outflow and inflow of water between the nodes fulfilling the demand in the node, considering the socioeconomic factors of the zone like movement of the population in the zone. This enhanced approach will help even distribution of water through the city fulfilling minimal demand. Flow and velocity between the nodes is computed using the MATLAB program with the help of simulated result from Watergems. The result illustrated in fig. 4 shows that there has been uneven flow distribution between the nodes plotted with the help of table listed in appendix section.

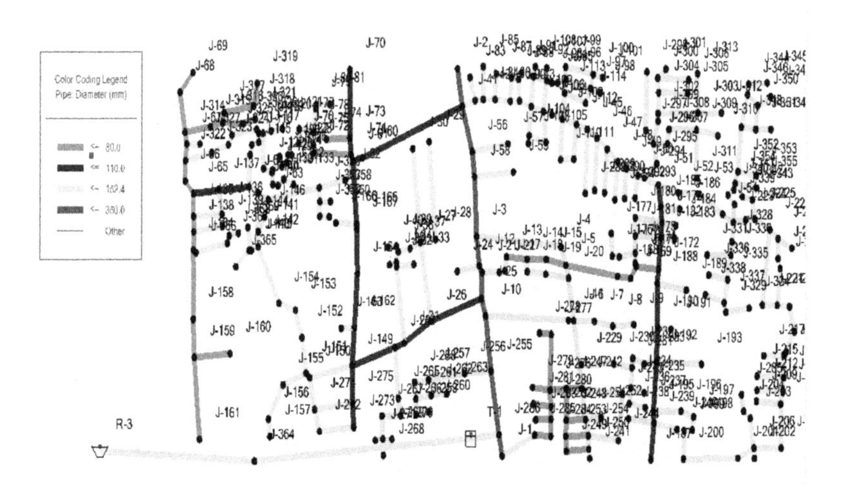

Fig. 3 Water distribution network modelling using Watergems

4.2 Pressure Dependent Demand

Nodal pressure gets affected due to the outage of water and fall below reference pressure in certain locations. Reference pressure should be maintained for supplying the assumed reference demand throughout network. Wu et al. (2006)When pressure level falls below reference level, then the nodal demand and water availability will depend on the pressure in the junction. Wu et al. (2006) In the conventional methodology pressure is the function for the calculation of PDD from demand driven demand. This helps in such a way that quantity of water that needs to be supplied in deficit period can be predicted with the simuation run using Watergems. In pressure driven demand

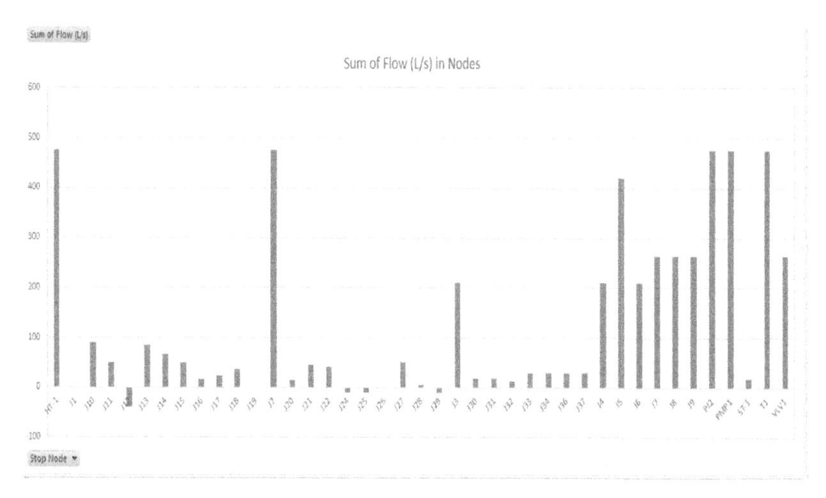

Fig. 4 Flow distribution between nodes

the nodal demand and pressure levels remains unknown. The reference pressure level is maintained at 60 psi and power PDD exponent is set to 0.5 for the extended period simulation gives a insight to the pressure level in network. Wu et al. (2006)

$$\frac{D_i^{\ S}}{D_{ri}} = \begin{cases} 0 & P_i \leq 0 \\ \left(\frac{P_i}{P_{ri}}\right)^{\alpha} & 0 < P_i < P_t \\ \left(\frac{P_t}{P_{ri}}\right)^{\alpha} & P_i \geq P_t \end{cases}$$

Where

$P_i =$ calculated pressure at node i;
$D_{ri} =$ requested demand or reference demand at node i;
$D_i^{\ S} =$ calculated demand at node i;
P_{ri} = reference pressure leve lto fulfill demand;
$P_t =$ pressure threshold above which the demand is independent of nodal pressure
α = exponent of pressure demand relationship

From the result computed throughout the PDD in watergems in appendix table 1, we try to characterise the pressure level which may vary with the reference to the elevation and pipe diameters thus creating critical zones in the network.

4.3 Critical Zone

As a result of the extended period simulation using Watergems showed that there has been a fact of negative pressure event in the distribution system. Pressure head

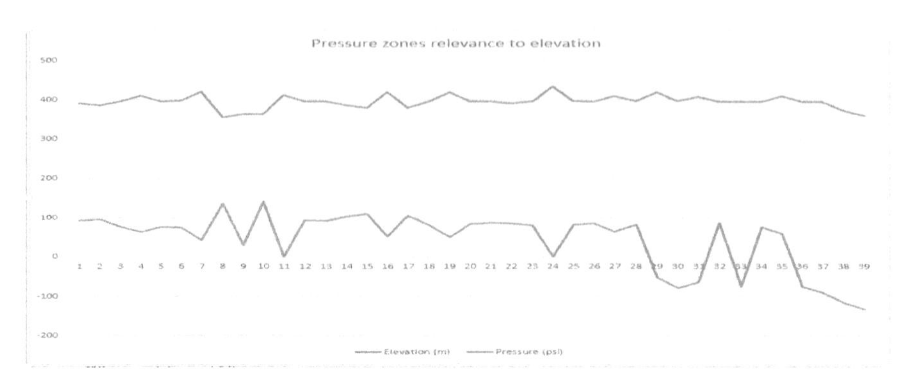

Fig. 5 Pressure level with reference to elevation

is maintained same through out the network analyses the volume of water that could be intruded. Besner et al. (2010)When a transient pressure event takes place in a real system pressure gets unloaded through the pipes which travels along the pipeline finally reaching the tank when the valve is open. But when the valve is closed water comes to rest in the pipeline causing the back flow between the pipes and another unloading of pressure waves happens simultaneously through out the day causing the negative pressure. Thus the pressure level varies with the elevation illustrated in fig 5 with the results simulated shown pressure variations through out the network creating critical zone in the city.

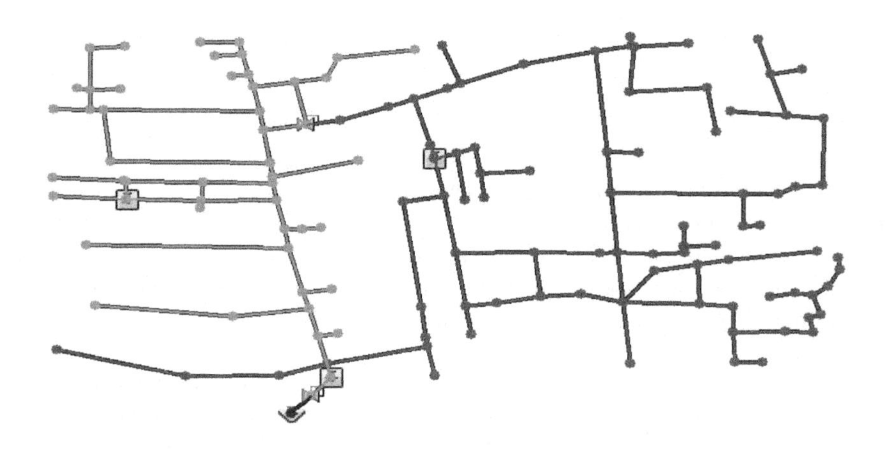

Fig. 6 High pressure zone

5 Conclusion

With the analysis through the surveying it shows that there has been unbalance in the household demand to that of the demand that is being assigned to the study area. Demand is calculated as 85 litre per day that should be achieved anyhow but scenario proves that there has been negative pressure and critical pressure zone has been created with reference to the elevation of the surface. Results provides the visualization of insight into the water distribution network and reason behind the leakage of pipes during the monsoon seasons. Thus efficient model of considering various parameters shall be used for the equivalent distribution

6 Future Study

This study can be further enhanced with the initiative of government towards the smart city venture which covers an important aspect on water distribution. Smart water management with centralized control can be implemented with the integration of GIS and Internet of things with the effective backbone network. This is possible even with on-line meter reading of consumer and troubleshooting of pipe rupture by identifying location using GIS by integrating the pressure sensors and flow meters monitoring them regularly.

References

Gad, A.A., Mohammed, H.I.: Impact of pipes networks simplification on water hammer phenomenon. Sadhana **39**(5), 1227–1244 (2014)

Araujo, L.S., Ramos, H., Coelho, S.T.: Pressure control for leakage minimisation in water distribution systems management. Water Resources Management **20**(1), 133–149 (2006)

Besner, M.-C., Ebacher, G., Jung, B.S., Karney, B., Lavoie, J., Payment, P., Prévost, M.: Negative pressures in full-scale distribution system: field investigation, modelling, estimation of intrusion volumes and risk for public health. Drinking Water Engineering and Science **3**(2), 101–106 (2010)

Chandapillai, J., Sudheer, K.P., Saseendran, S.: Design of water distribution network for equitable supply. Water resources management **26**(2), 391–406 (2012)

Creaco, E., Franchini, M.: A new algorithm for real-time pressure control in water distribution networks. Water Sci. Technol.: Water Supply **13**(4), 875–882 (2013)

Eljamassi, A., Abeaid, R.A.: A GIS-based DSS for management of water distribution networks (rafah city as case study) (2013)

Hopkins, M.: Critical Node Analysis for Water Distribution Systems Using Flow Distribution. Ph.D. thesis, California Polytechnic State University, San Luis Obispo (2012)

Kwon, H.: Computer simulations of transient flow in a real city water distribution system. KSCE Journal of Civil Engineering (2007)

Machell, J., Mounce, S.R., Boxall, J.B.: Online modelling of water distribution systems: a UK case study. Drinking Water Engineering and Science **3**(1), 21–27 (2010)

McKenzie, R., Wegelin, W.: Implementation of pressure management in municipal water supply systems. In: EYDAP Conference Water: The Day After, Greece (2009)

Stoianov, I., Nachman, L., Madden, S., Tokmouline, T., Csail, M.: Pipenet: a wireless sensor network for pipeline monitoring. In: 6th International Symposium on Information Processing in Sensor Networks. IPSN 2007, pp. 264–273. IEEE (2007)

Keyser, G., Tang, T.: Spatial analysis of household water supply and demand in a distributed geographic network in the towns of amherst and clarence, New York. Middle States Geographer **40**, 133–141 (2007)

Wu, Z.Y., Wang, R.H., Walski, T.M., Yang, S.Y., Bowdler, D., Baggett, C.C.: Efficient pressure dependent demand model for large water distribution system analysis. In: Proceedings of the 8th Annual Water Distribution System Analysis Symposium (2006)

Xu, Q., Chen, Q., Ma, J., Blanckaert, K., Wan, Z.: Water saving and energy reduction through pressure management in urban water distribution networks. Water resources management **28**(11), 3715–3726 (2014)

7 Appendix

Table 1 Flow Distribution and Pressure dependent demand

Start Node	Stop Node	Diameter (mm)	Flow (L/s)	Velocity (m/s)	Node	Elevation (m)	Demand (L/s)	Hydraulic Grade (m)	Pressure (psi)
Res1	PJ1	600	475.14	1.68	J13	390	30	455.18	92.7
PJ2	J1	600	(N/A)	(N/A)	J21	385	20	452.13	95.5
J1	J2	600	(N/A)	(N/A)	J27	395	75	448.89	76.6
J2	J3	450	210.14	1.32	J20	410	20	454.55	63.4
J3	J4	450	210.14	1.32	J29	396	10	449.33	75.9
J4	J5	450	210.14	1.32	J32	397	50	449.04	74
J6	Res2	450	210.14	1.32	J34	420	30	449.1	41.4
J5	J6	450	210.14	1.32	J37	355	30	451.35	137
J13	J17	300	53.85	0.76	PJ1	363	0	382.79	28.2
J17	J21	200	37.05	1.18	PJ2	363	0	462.84	142
J21	J26	200	24.49	0.78	J1	412	(N/A)	(N/A)	(N/A)
J27	J26	200	-24.49	0.78	J2	395	0	461.06	94
J22	J21	200	7.44	0.24	J3	395	0	459.96	92.4
J17	J18	300	-13.2	0.19	J4	386	0	459.04	103.9
J18	J22	200	36.45	1.16	J5	380	0	457.56	110.3
J22	J27	200	33.86	1.08	J6	420	0	456.64	52.1
J19	J18	300	(N/A)	(N/A)	J17	380	0	454.56	106
J23	J19	200	(N/A)	(N/A)	J22	395	0	452.28	81.5
J23	J22	200	4.84	0.15	J18	420	0	454.63	49.3
J9	J13	300	83.85	1.19	J14	396	0	455.25	84.3
J9	J14	300	91	1.29	J9	395	0	456.41	87.3
J14	J18	300	66.54	0.94	J26	390	0	450.33	85.8
J23	J24	200	-8.13	0.26	J23	396	0	452.35	80.2
J24	J25	200	-11.04	0.35	J19	435	(N/A)	(N/A)	(N/A)
J25	J33	200	30	0.95	J15	397	0	455.19	82.8
J33	J34	200	30	0.95	J10	395	0	455.84	86.5
J2	J7	450	265	1.67	J11	410	0	455.56	64.8
J8	J9	450	265	1.67	J16	397	0	455.14	82.7
J35	J17	200	-30	0.95	J12	420	0	455.33	50.3
J35	J36	200	30	0.95	J24	397	0	452.42	78.8
J36	J37	200	30	0.95	J25	408	0	452.57	63.4
J23	J30	200	38.79	1.23	J30	396	0	449.7	76.4
J30	J31	200	18.66	0.59	J28	396	0	449.23	75.7
J31	J28	200	6.53	0.21	J31	396	0	449.28	75.8
J28	J29	200	-10.13	0.32	J33	410	0	451.77	59.4
J29	J30	200	-20.13	0.64	J7	395	0	458.16	89.8
J31	J32	200	12.13	0.39	J8	395	0	458.15	89.8
J32	J24	200	-37.87	1.21	J35	372	0	453.49	115.9
J28	J27	200	16.65	0.53	J36	360	0	452.42	131.4
J9	J10	300	90.15	1.28					
J10	J11	300	90.15	1.28					
J11	J15	300	50.98	0.72					
J15	J14	300	-24.46	0.35					
J15	J16	300	17.24	0.24					
J16	J12	300	-39.17	0.55					
J12	J11	300	-39.17	0.55					
J15	J19	300	(N/A)	(N/A)					
J16	J20	300	56.4	0.8					
J20	J19	200	(N/A)	(N/A)					
J19	J24	200	(N/A)	(N/A)					
J25	J20	200	-41.04	1.31					
PJ1	PMP1	600	475.14	1.68					
PMP1	PJ2	600	475.14	1.68					
J7	VLV1	425	265	1.87					
VLV1	J8	425	265	1.87					
PJ2	HT-1	600	475.14	1.68					
J15	ST-1	300	58.21	0.82					
J23	ST-1	200	-35.5	1.13					
ST-1	J24	200	34.96	1.11					
J20	ST-1	200	-4.64	0.15					
HT-1	J2	600	475.14	1.68					
ST-1	J18	300	-16.89	0.24					

Distributed Air Indexing Scheme for Full-Text Search on Multiple Wireless Channel

Vikas Goel, Anil Kumar Ahalawat and M.N. Gupta

Abstract Wireless data broadcast is the most popular method to disseminate frequently requested data efficiently to a large number of mobile devices. Full text search is a popular query type, for retrieving the documents. Many research efforts have been made which focuses how to apply full text search on wireless broadcast. By increasing the numbers of broadcast channels is a logical way to minimize the energy consumption and access latency. In this paper, we further extend the problem of generating a broadcast sequence of data items to facilitate energy efficient full text search on multiple wireless channels. To support our proposed indexing scheme, we propose a data access algorithm and a data broadcast model for full text search indexing scheme on multi channel. Since the energy of portable devices is limited, minimization of energy consumption and access latency for broadcasting are the important issues. The performance of the proposed scheme is further analyzed and compared with existing full text search indexing schemes. The results show the efficiency of our approach with respect to energy consumption and access latency.

Keywords Full-text search · Inverted list · Multiple channel · Distributed indexing scheme · Data broadcast model · Data access algorithm

V. Goel(✉)
Department of CSE, Ajay Kumar Garg Engineering College, Ghaziabad 201009, UP, India
e-mail: rvikasgoel@yahoo.com

A.K. Ahalawat
Department of CA, Krishna Institute of Engineering and Technology, Ghaziabad 201206, UP, India
e-mail: a_anil2000@yahoo.com

M.N. Gupta
Department of CSE, Amity School of Engineering and Technology, Brijwasan, New Delhi, India
e-mail: mngupta@gmail.com

S. Berretti et al. (eds.), *Intelligent Systems Technologies and Applications*,
Advances in Intelligent Systems and Computing 385,
DOI: 10.1007/978-3-319-23258-4_12

1 Introduction

With the rapid development of 4G networks together with smart-phones make mobile computing technology wide spread. These smart-phones prefer to access public information like stock price, real-time traffic, and weather information through a wireless connection [1], [8].

Wireless Data Broadcast is an attractive solution to disseminate data efficiently because of its scalability and flexibility. The main aim of improving broadcast methods is by providing indexes on broadcast channel [7], [11], [12], [15], and [16]. Due to this index information, the clients can swap into doze mode (less energy consumption) most of the time and come into active mode (more energy consumption) only when the desired data reaches on broadcast channels. There are two performance metrics for evaluating the performance of indexing schemes: tuning time and access time. Tuning time is the amount of time spent by a client in actively listening to broadcast channel for finding required data. Access time is the total amount of time from finding index information to download all the data needed by a client.

The existing index techniques for wireless data broadcast [18] [12] [25] [21] [22] indexes the data based on predefined structured data on disk based environment with key attributes. These indexing techniques cannot be directly applied for full-text search that has arbitrary words as search keys. Therefore, there is a requirement of new designs for indexing data that facilitates full-text search in order to ensure both time efficiency and energy efficiency.

The significant work has been done by Chung et al. [3], Yang et al. [10] and Goel et al. [5]. However, Chung et al. [3] does not consider the problem of generating a broadcast sequence of a set of data items on multiple channels. Goel et al. considers multiple channels for generating an index of full text search data [5]. In this paper, we have extended the concept of allocation of index information and data items on multi channel broadcast [5]. In [5], we have proposed a distributed indexing scheme for full text search on multi channel broadcast. A data broadcast model and data access algorithm for full text search scheme on the multi channel broadcast are proposed in this paper. The performance of the proposed indexing scheme is further analyzed and compared with existing full text search schemes. The results show the effectiveness of our proposed work.

The rest of the paper is organized as follows: In section 2, we describe related work about the indexing. In section 3, we discuss proposed full text search indexing scheme for multiple channels. In section 4, we proposed data broadcast model and data access algorithm for the proposed indexing schemes. In section 5, performance analysis & evaluation of the proposed scheme with existing schemes are performed. In section 6, performance comparison and result of the proposed technique are analyzed. At last, section 7 concludes the results drawn.

2 Related Work

For indexing design, Imielinski et al. proposed distributed B+-tree indexing [7]. Later they customized this with the other work in [8]. Xu et al. presented an exponential index scheme in [20]. Chen et al. and Shivakumar et al. discussed how to construct an imbalanced index tree to minimize the average tuning time [2], [9].

Inverted list is one of popular data structures for full text search in document retrieval systems. Tomasic et al. studied the incremental updates of inverted lists by dual-structure index [17]. They developed two approaches for improving the document retrieval efficiency. Zobel et al. explored inverted files for text search engines in [24] by giving a survey. Chung et al. firstly applied inverted list for full-text search in wireless data broadcast [3]. They also combined traditional B+ tree-based indexing technique with inverted lists in the dual structure for full-text query on broadcast documents. However, the construction of a searching B+ tree and the duplication of inverted list will extend the total length of a broadcast cycle, which results more access latency as compared to Yang K. et al [10].

There are some other works focusing multiple channels by generating broadcast program over multiple channels. In [9], a binary search tree Alphabetic Huffman Tree is proposed. Peng and Chen proposed the problem of generating broadcast programs on multiple channels into one of constructing a channel allocation tree with variant fan-out [13]. However, this technique does not consider how to construct index structures. Lo and Chen proposed a solution for optimal index and data allocation over broadcast channel, which minimizes the average access time for any number of broadcast channels [12]. Moreover, in [6][26][9], the issue of allocating dependent data on multiple channels is discussed. Im et. al. proposed a multilevel air indexing scheme in non-flat wireless data broadcast for efficient window query processing [9]. Vishnoi S. and Goel V. proposed a novel table based air indexing technique for full text search over wireless stream [19].

Increasing the number of available broadcast channels is a logical way of increasing throughput. In this approach, we have proposed allocation of index information and data items on multiple broadcast channels by extending the distributed indexing approach for full-text search. We have extended the concept of Chung et al. by introducing the multi channel with two levels of indexing structure, i.e. index tree and an inverted list of the different channels.

3 Full Text Search Indexing Schemes

3.1 "INVERTED LIST + INDEX TREE" Scheme for Multiple Channels

We have proposed the scheme by extending the existing "inverted list + Index tree" scheme over a single channel. In the proposed scheme, server broadcasts index and data over multiple channels consist of index information and data. The broadcasted data are divided over multiple channels: index channel and data channel instead of sending them over a single channel [5] [25] [26]. The first channel: index channel consists of a dual structure index, index tree buckets (T) and inverted list buckets (L). The second channel: data channel consists of only data buckets (D) repeatedly over this channel. [5].

In order to search required data item, mobile client tunes the index channel to access index tree buckets and inverted list buckets. If, accessed bucket is the start bucket, then the client will traverse the B+ tree in depth first search pre-order. Otherwise, it will go to doze mode and wait for the start of index bucket. If a required word is found by traversing B+ tree in pre-order at leaf nodes. Then leaf node of B+ tree has pointer to inverted list. The inverted list has index information about all the words in all documents. If that word is not framed in traversing of the B+ tree, then the client has to wait for the next bcast. The bcast size is reduced significantly because it consists of indexes only. Hence, the average tuning time of the proposed indexing scheme is reduced significantly. After finding required index of data mobile, client is hopped from index channel to data channel. Now, mobile device listens the data channel and accordingly accessed the data. The average access time of the proposed indexing scheme is reduced significantly [5].

4 Proposed Work

4.1 Proposed Data Broadcast Model

This section presents the proposed data broadcast model for describing functions of the proposed indexing scheme over multiple channels. The Fig. 1 depicts all the components of the proposed data broadcast model. The proposed data broadcast model consists of three processes that execute independently. The first process broadcasts each data item, second process generates requests from all mobile users querying for a keyword and third process processes the requests arising from clients within data broadcast range. The components of communication model are described as follows:

Client: After initializing both broadcast server and data source, it starts both processes: broadcasting and request generation.

Broadcast Server: It is a process that continuously broadcasts the data over data channels. It decides the broadcast sequence, size of packet etc.

Data Source: It is the data repository at the server side. It can be understood as a database of documents.

Channel: This is the available bandwidth over which the data is to be broadcasted. The available bandwidth is divided into two separate channels: data channel and index channel. The Broadcast Server processes broadcast information contained in the broadcast Channel.

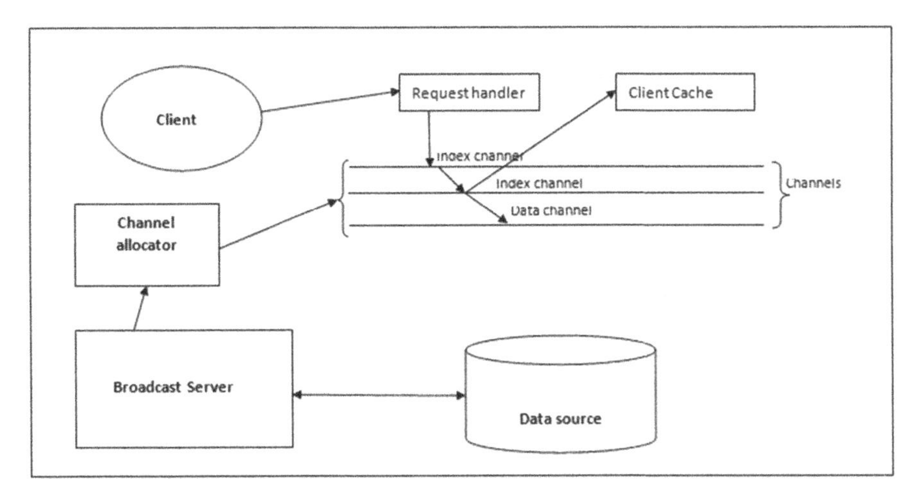

Fig. 1 The data broadcast model

Channel Allocator: It divides the data over an index channel and the data channel from broadcasting server. The channel allocator assigns respective channel to index information or data information. The index tree and the inverted list buckets are broadcasted over index channel. The respective data buckets are broadcasted over third channel: data channel.

Request Handler: The Request Handler extracts, processes access time and tuning time parameters and output them in the proper desired format.

Client Cache: Client cache stores intermediate result as an inverted list so that the client does not require accessing the inverted list channel again and again. The client can access more than one document listed in inverted list. Client side caching significantly improves power saving of the client device.

4.2 Proposed Data Access Algorithm

Fig. 2 describes proposed data access algorithm for distributed full text search indexing scheme over multiple channels. Initially, request array is empty and the requested query word has some value. First, the client sends the request of the query word to request handler. Request handler on behalf of client generates a request to access first index channel. If the first bucket accessed by the client is not the start of index tree area, then it waits for next bcast i.e. start of the next index tree area. Once the client gets start of the index tree, it starts traversing index tree from top to bottom in preoreder by ProcessTree() method. If the last node of index tree bucket is a leaf node and contains the query word, then the client will wait for the index list bucket pointed by leaf node of the index tree. The client hops to next index channel to process inverted list bucket by ProcessList(). After processing the list, the client will come to know the address of query word and all the related documents on data channel. Then, again it hopes to data channel for accessing all documents containing specified query word.

- Client sends request to the request handler.
- Request handler tunes in to index channel.
- Initially

```
          Request= { };
          Queryword=qw;
          Current_position=cp;
```

- Find the queryword by traversing Index tree ProcessTree();
- Loop

```
          If(cp equals to qw)
                    word is forme;
                    return address of corresponding documents;

          else
                    Wait & check new & index tree for word formation;
```

- End loop
- Traverse corresponding inverted list, ProcessList();
- Get the index information of documents on data channel;
- Hop to data channel to access the documents containing the specified word;

Fig. 2 Data access algorithm for multi channel indexing

5 Performance Analysis and Evaluation

5.1 The system model and Parameters

Table 1 Parameters

Parameter	Description
NumberofDocuments	Database on server
DocumentSize	Size of Document
BucketSize	Size of Bucket
NumberofDistinctWords	The no of distinct words in the database
WordSize	The length of the word
AddressTupleSize	The address of the bucket on the stream
H	The height of the tree
N	Fanout
S	Selectivity of a query word i.e. the number of documents matched to a query word
BroadcastSize	The size of the broadcast stream
AvgAT	Average access time
AvgTT	Average tuning time

5.2 "INVERTED LIST + INDEX TREE" Scheme for FULL-TEXT SEARCH on Multiple Wireless Broadcast Channels

We have proposed full text search indexing scheme over multiple channels. In the proposed scheme, the index of index tree and inverted list are broadcasted over the index channel. The data are broadcasted over the separate channel: data channel. Therefore, average tuning time (AvgTT) and average access time (AvgAT) can be defined as follows [5]:

$$\text{AvgAT}= \frac{1}{2} * (\text{SizeOfIndexTree}) + \frac{1}{2} * (\text{SizeofIndexList}) + \left(\frac{s}{s+1}\right) * \text{DBSize} \quad (1)$$

$$\text{AvgTT} = 2+\text{H} \quad (2)$$

6 Results and Comparisons

6.1 Results

In this section, we have made an attempt to analyze the performance of our proposed indexing scheme. We have computed average access time and average tuning time of the proposed scheme on parameters having their values according to table 1. All the related schemes are also analyzed and compared on the basis of performance metrics average access time and average tuning time. For a fair comparison, the values of the parameters are kept same as in [3][10]. For evaluating the schemes, we have taken 10000 no. of documents each of size 1024. The no. of distinct word is 4703 for creating the index tree and inverted list. The total size of the word is 16 and 4 is the address tuple size. The no. of repetition of a word is kept uniform from values (1-5) for a document and the selectivity of the word is 51. The size of the bucket is 1024.

For showing the accuracy and correctness of results, we have performed simulations using NS2, a world widely accepted Network Simulator Tool. The simulations are implemented on NS2.34 using Intel i5 CPU with 2.60 GHZ processing speed, 5 GB RAM and 1 TB HDD with Ubuntu 10.04 as operating system.

6.2 Comparisons

In this section, we have conducted three experiments for comparing performance of the proposed scheme with existing indexing schemes for full text search further.

6.2.1 Comparison with no Replication Schemes

In the first experiment, we have assumed $\alpha=1$ *and* $\beta=1$ which means that there is no replication in indexing scheme.

In Fig. 3, there is an increase in average access time of existing indexing schemes due to the data overhead of the index tree and the inverted list. The data overhead is reduced in the proposed scheme because of the distribution of data and index over multiple broadcast channels.

There is a very sharp fall in the tuning time of the proposed indexing scheme. In multi-channel environment, index and the data are broadcasted on different channels. This reduces the size of the bcast significantly, causes a sharp decrease in tuning time of the proposed scheme.

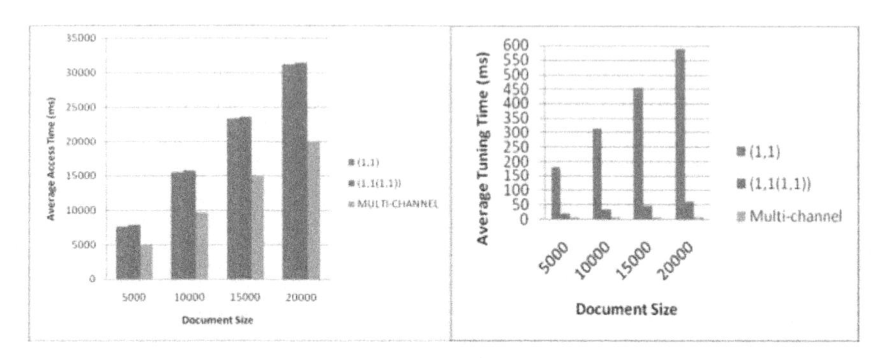

Fig. 3 Average access time and average tuning time of the proposed indexing scheme with existing indexing schemes (*α=1 and β=1*)

6.2.2 Comparison with Optimal Values of Replication

In the second experiment, we have considered replication of index with optimal values. We have considered *α=2* for the inverted list indexing scheme and *α=1* & *β=6* for an inverted list + Index tree indexing scheme.

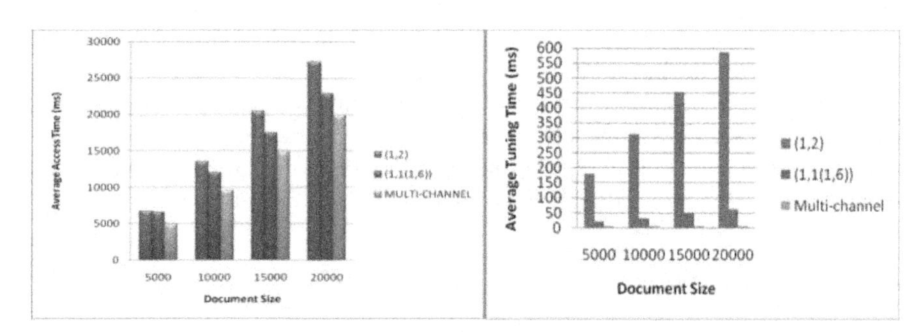

Fig. 4 Average access time and average tuning time of the proposed indexing scheme with existing indexing schemes for optimal values (*α=2 and α=1, β=6*)

In Fig. 4, the average access time and the average tuning time of the proposed indexing scheme is still reduced as compared with existing indexing schemes. Although, there are replications with optimal values in existing indexing schemes. But, there is no replication of data in our proposed indexing scheme. There is a very sharp decrease in the value of average tuning time due to very small bcast size.

6.2.3 Comparison with Full Text Search Indexing Schemes

According to Fig. 5, average access time of the proposed indexing scheme is reduced significantly over existing full text search indexing schemes. Average access time for the proposed indexing scheme is reduced by 36.5% and 30.5% with inverted list and MHash respectively. So, our proposed indexing technique is much faster for accessing data than the available full text search indexing schemes.

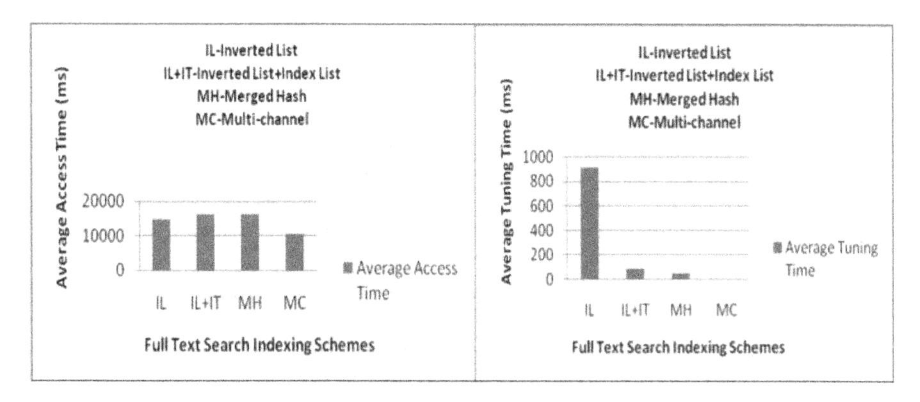

Fig. 5 Average access time and average tuning time of the proposed indexing scheme with existing Full text searching indexing schemes

In Fig. 5, average tuning time of the proposed indexing scheme is reduced by 85% with merged hash having least tuning time among existing indexing schemes for full text search. This factor proves power efficiency of our indexing schemes. The proposed indexing scheme is the most power efficient than full-text search indexing schemes.

7 Conclusion

In this paper, we have explored the problem of generating a broadcast sequence of a set of data items over multiple channels. The scheme allocates index information and data items on multiple broadcast channels by extending the distributed indexing approach for full-text search. To support our proposed scheme, a data access algorithm and a data broadcast model are proposed. The proposed data access algorithm retrieves the set of documents matching to given query word on multichannel environment. The proposed data broadcast model presents a blueprint of a base station that periodically broadcasts data and its index information over multiple channels. Using separate channels for data bucket and index bucket results in faster query resolution, while broadcasting data over the multi channel environment.

Average access time for the proposed indexing scheme is reduced by 30.5% with the most efficient indexing scheme. Average tuning time of the proposed indexing scheme is reduced by 85% with least tuning time indexing scheme. It is found that our proposed indexing scheme is the most power efficient and fast data access indexing schemes for full-text search over wireless channels.

In future, we will consider probabilities of words i.e. non uniformity of words in a document.

References

1. Barbara, D.: Mobile computing and database survey. IEEE Trans. on Knowledge and Data Engineering, 108–117 (1999)
2. Chen, M., Wu, K., Yu, P.: Optimizing index allocation for sequential data broadcast in wireless mobile computing. TKDE **15**(1), 161–173 (2003)
3. Chung, Y.D., Yoo, S., Kim, M.H.: Energy- and latency-efficient processing of fulltext searches on a wireless broadcast stream. IEEE TKDE **22**(2), 207–218 (2010)
4. Goel, V., Panwar, G., Ahlawat, A.K.: Energy efficient air indexing schemes for single and multi-level wireless channels. In: IEEE 3rd International Advance Computing Conference (IACC), pp. 525–530 (2013)
5. Goel, V., Ahlawat, A.K., Gupta, A.: Energy efficient distributed indexing scheme for full-text search on multi channel broadcast. In: International Conference on Communication Systems and Network Technologies (CSNT), IEEE CSNT-2013, pp. 664–669 (2013)
6. Hurson, A.R., Chehadeh, Y.C., Hannan, J.: Object organization on parallel broadcast channels in a global information sharing environment. In: IEEE International Performance, Computing, and Communications Conference (IPCCC), pp. 347–353 (2000)
7. Imielinski, T., Viswanathan, S., Badrinath, B.R.: Energy efficiency indexing on air. In: Proceedings of the International Conference on SIGMOD, pp. 25–36 (1994)
8. Imielinski, T., Viswanathan, S., Badrinath, B.R.: Data on Air: Organization and Access. IEEE Trans. Knowledge and Data Eng. **9**(3), 353–372 (1997)
9. Im, S., Choi, J.T.: MLAIN: Multi-leveled Air Indexing Scheme in Non-flat Wireless Data Broadcast for Efficient Window Query Processing. Computers and Mathematics with Applications, Elsevier Science **64**(5), 1242–1251 (2012)
10. Yang, K., Shi, Y., Wu, W., Gao, X., Zhong, J.: A novel hash-based streaming scheme for energy efficient full-text search in swireless data broadcast. In: Yu, J.X., Kim, M.H., Unland, R. (eds.) DASFAA 2011, Part I. LNCS, vol. 6587, pp. 372–388. Springer, Heidelberg (2011)
11. Lee, W., Lee, D.: Using Signature Techniques for Information Filtering in Wireless and Mobile Environments. J. Distributed and Parallel Databases **4**(3), 205–227 (1996)
12. Lo, S., Chen, A.: Optimal index and data allocation in multiple broadcast channels. In: Proc. 16th International Conference Data Engineering, pp. 293–302 (2000)
13. Peng, W.C., Chen, M.S.: Dynamic generation of data broadcast programs for a broadcast disk array in a mobile computing environment. In: Proc. of the ACM 9th Intern'l Conf. on Information and Knowledge Management, pp. 38–45 (2000)
14. Seifert, A., Hung, J.-J.: FlexInd: a flexible and parameterizable air-indexing scheme for data broadcast systems. In: Ioannidis, Y., Scholl, M.H., Schmidt, J.W., Matthes, F., Hatzopoulos, M., Böhm, K., Kemper, A., Grust, T., Böhm, C. (eds.) EDBT 2006. LNCS, vol. 3896, pp. 902–920. Springer, Heidelberg (2006)

15. Shivakumar, N., Venkatasubramanian, S.: Efficient Indexing for Broadcast Based Wireless Systems. Mobile Networks and Applications, 433–446 (1996)
16. Tan, K., Yu, J.: An Analysis of Selective Tuning Schemes for Nonuniform Broadcast. Data and Knowledge Eng. **22**(3), 319–344 (1997)
17. Tomasic, A., Garcia-Molina, H., Shoens, K.: Incremental updates of inverted lists for text document retrieval. SIGMOD Rec. **23**(2), 289–300 (1994)
18. Vaidya, N.H., Hameed, S.: Scheduling data broadcast in asymmetric communication environments. Journal Mobile Networks and Applications **5**, 171–182 (1999)
19. Vishnoi, S., Goel, V.: Novel table based air indexing technique for full text search. In: IEEE International Conference CICT 2015, pp. 410–415 (2015)
20. Xu, J., Lee, W.-C., Tang, X., Gao, Q., Li, S.: An error-resilient and tunable distributed indexing scheme for wireless data broadcast. TKDE **18**(3), 392–404 (2006)
21. Yao, Y., Tang, X., Lim, E.P., Sun, A.: Energy-efficient and access latency optimized indexing scheme for wireless data broadcast. IEEE TKDE **18**(8), 1111–1124 (2006)
22. Zhang, X., Lee, W.C., Mitra, P., Zheng, B.: Processing transitive nearest-neighbor queries in multi-channel access environments. In: EDBT 2008: Proceedings of the 11th International Conference on Extending Database Technology, pp. 452–463 (2008)
23. Zheng, B., Lee, W.C., Lee, K.C., Lee, D.L., Shao, M.: A distributed spatial index for error-prone wireless data broadcast. The VLDB Journal **18**(4), 959–986 (2009)
24. Zobel, J., Moffat, A.: Inverted files for text search engines. ACM Computer Survey **38**(2) (2006)
25. Zhong, J., Gao, Z., Wu, W., Chen, W., Gao, X., Yue, X.: High Performance Energy Efficient Multi-Channel Wireless Data Broadcasting System, 1–6 (2013)
26. Zhong, J., Wu, W., Gao, X., Shi, Y., Yue, X.: Evaluation and Comparison of Various Indexing Schemes in Single-Channel Broadcast Communication Environment. Knowl. Info. System **40**(2), 375–409 (2014)

Fuzzy Differential Evolution Based Gateway Placements in WMN for Cost Optimization

G. Merlin Sheeba and Alamelu Nachiappan

Abstract Mesh node placement problem is one of the major design issues in Wireless Mesh Network (WMN). Mesh networking is one of the cost effective solution for broadband internet connectivity. Gateway is one of the active devices in the backbone network to supply internet service to the users. Multiple gateways will be needed for high density networks. The budget and the time to setup these networks are important parameters to be considered. Given the number of gateways and routers with the number of clients in the service area, an optimization problem is formulated such that the installation cost is minimized satisfying the QOS constraints. In this paper a traffic weight algorithm is used for the placement of gateways based on the traffic demand. A cost minimization model is proposed and evaluated using three global optimization search algorithms such as Simulated Annealing (SA), Differential Evolution (DE) and Fuzzy DE (FDE). The simulation result shows that FDE method achieves best minimum compared with other two algorithms.

Keywords Differential Evolution · Simulated Annealing · Fuzzy Differential Evolution · Cost minimization · WMN · Mutation · Cross over

1 Introduction

Wireless Mesh Network (WMN)[1] is the recent attractive technology for applications such as metropolitan area networks, campus networks, urban and rural areas, medical systems, surveillance systems etc.[2]. Mesh routers are different

G. Merlin Sheeba(✉)
Department of ETCE, Sathyabama University, Chennai, India
e-mail: merlinsheeba.etc@sathyabamauniversity.ac.in

A. Nachiappan
Department of EEE, Pondicherry Engineering College, Puducherry, India
e-mail: nalam63@pec.edu

S. Berretti et al. (eds.), *Intelligent Systems Technologies and Applications*,
Advances in Intelligent Systems and Computing 385,
DOI: 10.1007/978-3-319-23258-4_13

from other conventional routers in terms of coverage and power constraints. In mesh topology every node is connected in a multihop fashion which enables the information to be transmitted in all available paths redundantly. It is one of the robust features in WMN which make them very reliable in node failures.

The mesh clients on other hand have the capability of acting as routers but not as gateways or bridges. The uplink traffic generated from the Mesh client passes through the mesh router or directly to Internet Gateway (IGW) and also the downlink traffic passes to the client from the IGW.Hence IGW hop is mandatory therefore its placement plays a vital role based on connectivity, fairness, throughput and reliability of the network [3]-[5]. In general many researchers have used the Genetic Algorithm (GA), Simulated Annealing (SA), Tabu Search (TS) Methods for solving many optimization problems for near optimality. The remainder of the paper is organized as follows. In section 2 the related works are elaborated. The problem is described in the section 3.The traffic weight placement of gateways is discussed in section 4.In section 5 the algorithms for optimization such as SA,DE and fuzzy DE(FDE) techniques are discussed. In section 6 the comparative analysis is done for the three methods with implementation in NS2.Finally the paper is concluded.

2 Related Works

Many node placement approaches are formulated by the researchers as it is one of the active research areas .enjia Wu et.al[6] have studied a integer linear program based cost effective node placement problem. The proposed algorithm with gateway deployment constraint is evaluated. It combines the three algorithms (a)MSC-based coverage algorithm(b)weighted clustering algorithm (c)GW-rooted tree pruning algorithm, to determine the positions of mesh nodes. The authors have concluded that the placement algorithm is highly effective.

FatosXhafa[7] have presented the Tabu search method for mesh router placement and shown its efficacy. Mistura L. Sanni et.al.[8]have described the basis of internet gateway deployment optimization problems with multicasting strategies. The gateway placement problem is formulated based on the general assignment problem and travelling sales man problem. Admir Barollia et.al.[9] have presented a WMN-GA system and examined the network performance under four different client distributions. The number of mesh routers were varied and the throughput, packet delivery rate, hop counts are analyzed.

Chun-Cheng Lin[10] has defined the router node placement problem with the mobility of routers and the mesh clients can switch on and off the network access. The author has performed investigations on the placement of network nodes with topology adaptation. Benyamina et.al[11] have evaluated a bi-objective problem formulation model using modified version Multi Objective Particle swarm Optimization(VMOPSO).The two objective functions are installation cost of routers, gateways and standard deviation of traffic flows over the network links.

3 Problem Statement

Metropolitan networks or Campus networks need more number of gateways to improve the coverage for all mesh clients. But the deployment cost also increases. The main objective in this paper is to search for a low cost WMN configuration satisfying the placement constraints and determine the number of used gateways.

3.1 Problem Description

Let G (V, E) denote a network. The nodes represent the number of mesh routers, gateways and clients. Edges represent the communication link between the nodes. Let *U* represent the set of mesh nodes, *T* set of candidate positions. Let V be the set of all nodes (routers, gateways and clients) $V \leftarrow U \cup \mathrm{T}$.

Let C_j be the set of node in the coverage range or communication range of node i.

$$C_j = \{ j \varepsilon V, j \neq i, d(i, j) \leq R \} \forall i \varepsilon U \quad (1)$$

Where $d(i, j)$ denotes the Euclidean distance between the node i and j and R is the maximum communication range of node u.

Let the set of edges *E* be the set of all possible links i.e

$$\mathrm{E} = \{ e(i, j); i \varepsilon \mathrm{U}, \mathrm{j} \varepsilon \mathrm{Ci} \} \quad (2)$$

A cost minimization model is proposed based on the placement cost of mesh routers and mesh gateways

The objective function is given by

$$Minimize \sum_{i,j=1}^{n} (\mathrm{C_{pi}} \, \mathrm{M_{ri}} + \mathrm{C_{pj}} \mathrm{M_{gj}}) \quad (3)$$

Where C_{pi}, C_{pj} are the placement cost of Mesh routers and gateways

M_{ri} is i^{th} Mesh Router, Mgj is j^{th} Mesh Gateway.

$$M_{ri} = \begin{cases} 1 & \text{if mesh router is active} \\ 0 & \text{otherwise} \end{cases}$$

$$M_{rj} = \begin{cases} 1 & \text{if mesh gateway is active} \\ 0 & \text{otherwise} \end{cases}$$

Subjected to

The Euclidean distance (d) between the two gateways i and j

$$d(i, j) \geq G_r \quad (4)$$

Where G_r is the Gateway radius (communication range)

4 Gateway Placement Method

The gateway placement method used in our paper is Multihop Traffic flow Weight (MTW) [4].In this paper the gateway deployment problem is defined as a cost optimization problem. We consider a 6x6 grid network .In each cell or grid a mesh router is placed. The clients are unevenly distributed in each cell as shown in figure 1. The clients are connected to the gateway for internet access through the routers. Based on the traffic weight calculation the mesh router acts as a gateway. The traffic flow weight is calculated by

Gateway radius (G_r)

$$G_r = round \frac{\sqrt{N_r}}{2\sqrt{N_g}} \tag{5}$$

Nr is the number of mesh routers and Ng is number of gateways

Traffic demand D(i)

The traffic demand on each mesh router is calculated from the number of clients connected to the router. The algorithm calculates the weight using D(i) and M_{ri}.

TW(i) calculation:

$$TW(i) = \begin{array}{l}(G_r+1)D(i)+\\ G_r\times(Traffic\ Demand of 1-hop\ neighbours\ of\ M_{ri})+\\ (G_r-1)\times(Trafffic\ Demand \text{ of 2 - hop neighbours of } M_{ri}+\\ (G_r-2)\times(\text{Trafffic demand of 3 - hop neighbours } M_{ri})+\ldots\ldots\ldots\ldots\end{array} \tag{6}$$

After the traffic weight calculation in each cell the router with highest weight will act as gateway i.e the router is replaced with the gateway for internet access. The gateway positions are selected according to the high traffic demand. As gateways are costlier than routers it is a major issue in deployment is to minimize the installation cost of gateways and mesh routers. As the number of gateways increases the traffic is seamlessly distributed among the clients in the network. But the major question is about the cost of deployment. Evolutionary search methods are used to solve this problem such as considering the gateway placement as an optimization problem to minimize the cost. Here SA, DE and FDE methods are used to search for a minimum fitness value in the search space satisfying the placement constraint.

12	6	10	8	5
3	6	6	8	9
9	10	7	8	11
10	9	5	9	9
8	8	8	4	10

159	202	215	210	162
201	261	284	266	218
222	293	312	302	237
212	265	293	275	217
160	206	212	202	165

Fig. 1 Distribution of 200 mesh clients in a 5x5 network with Traffic Weight calculation

5 Optimization Methods for Gateway Placement

5.1 Differential Evolution Algorithm

In DE algorithm [12] there are three operators, the population size P, the scale factor S and the crossover probability CR.The major steps in this algorithm are Initialization, mutation, Crossover and Selection.

Algorithm 1: Pseudo Code for DE based Cost optimization

```
Set generation counter n=0;
Initialize the control parameter, S and CR
Initialize the population size Np
Initialize the population vector of x_n individuals randomly
While stopping condition not reached do
For each individual x_i(n) do
    Evaluate the fitness f(x_i(n))
     Create a mutated vector Y_M using the mutation function
     Create an offspring x_i'(n) by applying cross over
     If f(x_i'(n)) is better than f(x_i(n)) then
     Select f(x_i(n))
End
Else
     Select f(x_i'(n))
     End
     End
End
Return the individual with best fitness solution.
```

For each iteration the above said steps are evaluated one by one and the minimum fitness value is stored in an array. After the specified iteration and generation is completed the global minimum is the best minima.

5.2 Fuzzy Differential Evolution

In this section a fuzzy differential evolution method is used to solve the gateway placement problem. The solution or each chromosome is a matrix which represents the cost value to deploy the mesh nodes. DE strategy involves a fixed initial choices such as Scaling factor $S \in [0.5, 1]$, $CR \in [0.8, 1]$ and population size Np=10 multiplied with dimension of the problem D [13].In a basic fuzzy system there are three steps (1) Fuzzification of Inputs (2) Fuzzy Inference (3) Defuzzification of obtaining the crisp output.

5.2.1 Fuzzification of Inputs and Outputs

The two input variables to be fuzzified are the parameters CR ad S.The linguistic variables selected for fuzzification to frame the fuzzy set based on the knowledge of DE are "low", "average", and "high" .For each variable a degree of membership is calculated for every input. For the output cost the linguistic variables used to frame the fuzzy set are "very low", "low", "medium", "high". Triangular membership functions are shown in fig 2 and 3.

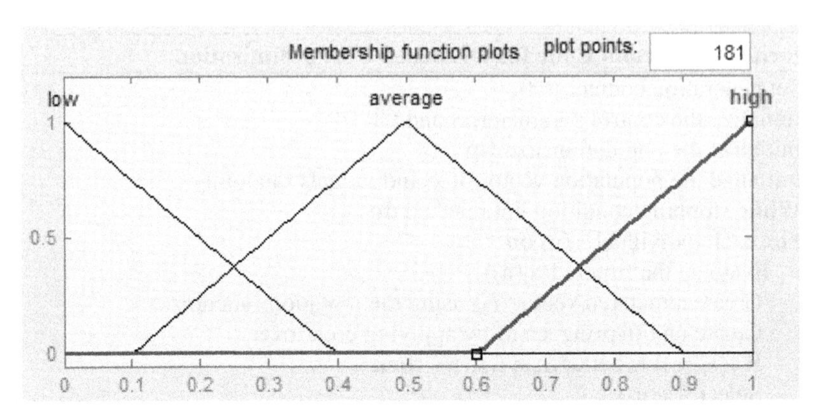

Fig. 2 Fuzzy Membership function for CR and S

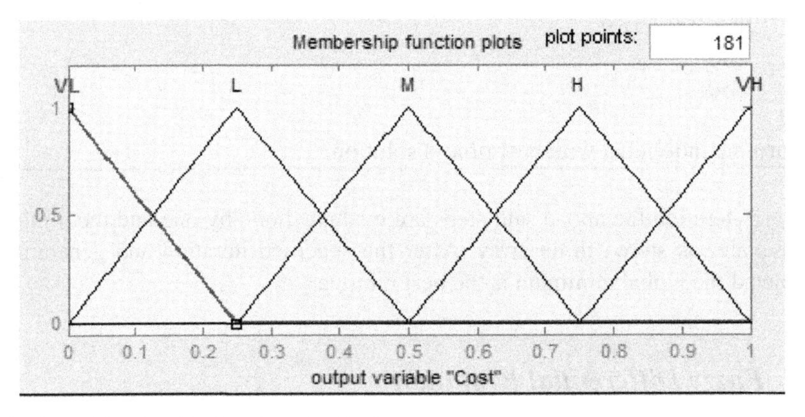

Fig. 3 Fuzzy Membership function for Cost

5.2.2 Knowledge Base Fuzzy Rules

The fuzzy rules follow the IF THEN structure. The fitness value cost is denoted as f(x). A total number of possible fuzzy inference rules will be 9(3*3), hence there are two linguistic states. An example of a rule is as follows. Table 1 shows the fuzzy rule set.

If (CR is low) And (S is low) then 'f(x)is low'

If (CR is average) And (S is high) then 'f(x) is high'

Table 1 Fuzzy rules

Input		Output
CR	S	f(x)
low	low	very low
low	average	low
low	high	medium
average	average	medium
average	high	high
average	low	low
high	high	very high
high	low	medium
high	average	high

5.2.3 Defuzzification

Defuzzification is used to get the crisp values from the fuzzy inference rules. The input fuzzy set μ is defuzzified into crisp value 'c' using centroid technique. For example the linguistic values of CR (low=0.3, average=0.5, high=0.9) and S(low=0.1,average=0.6,high=0.9).the crisp output can be calculated by

$$C=\sum xi\,\mu(\mathrm{xi})/\sum \mu(xi) \tag{7}$$

Where xi is the CR/S selection and $\mu(xi)$ is the linguistic value.The fuzzy ruled parameters and DE method are used.

6 Results and Discussions

To evaluate the proposed FDE method we consider a 6x6 grid network with one mesh routers in each cell, totally 36 mesh routers located in a 1000mx1000m region. The simulation settings are given in table 2. The overall cost includes the positioning of the routers and the installation cost. The Euclidean distance between the gateway positions should always be greater than the coverage radius of the gateway to avoid interference between the gateways; d (i, j)>Gr. A comparative analysis is observed in table 3 using the three methods. The system is validated using NS2 in figure 3.

Table 2 Simulation Settings

Parameters	SA	DE	FDE
Placement of nodes	random	random	random
Population size	100	100	100
CR	Probabilistic	0.5	Fuzzy rule based
S	selection	0.6	selection

Table 3 Comparison between SA, DE and FDE

NO. of routers	No. Of gateways	Installation Cost of Mesh devices(units)		
		SA	DE	FDE
35	1	1490.69	789.6	387.35
34	2	1467.45	743.7	361.46
33	3	1412.63	701.53	338.56
32	4	1389.79	675.94	308.63
31	5	1356.65	638.42	279.57
30	6	1308.65	621.75	262.48
29	7	1229.26	613.83	243.47
28	8	1109.88	601.76	228.54
27	9	967.984	589.95	215.04
26	10	875.789	548.20	195.92

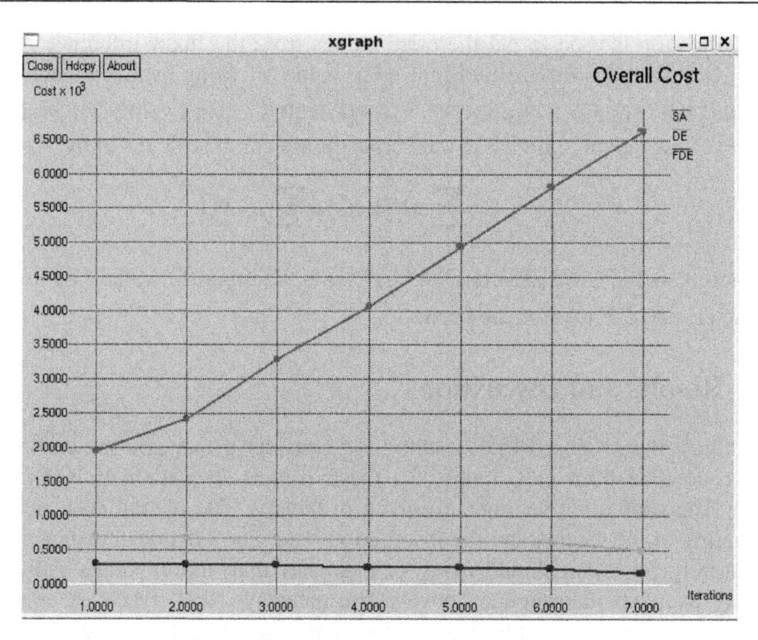

Fig. 4 Comparison of Overall cost Minimization using SA, DE&FDE

7 Conclusion

Efficient design of WMN can be defined as an optimization problem such as computing placement of nodes. The placement of nodes includes the mesh routers, gateways and distribution of clients in the network. Here in our work a single objective cost function is formulated as an optimization problem. The local search method SA is compared with the DE and Fuzzy DE methods. The observed results show that the fuzzy DE based MTW placement achieves best fitness minimum result than the other methods. The work can be further extended with multiobjective optimization problems.

Acknowledgment The authors would like to thank the management of Sathyabama University for providing the necessary facilities for carrying out this research.

References

1. Akyildiz, I.F., Wang, X., Wang, W.: Wireless mesh networks: a survey. Comput. Netw. **47**(4), 445–487 (2005)
2. Chen, C.C., Chekuri, C: Urban, wireless mesh network planning: the case of directional antennas, Tech Report No. UIUCDCS-R-2007-2874, University of Illinois at Urbana-Champaign (2007)
3. Bicket, J., Aguayo, D., Biswas, S., Morris, R.: Architecture and Evaluation of an Unplanned 802.11b Mesh Network. In: Proc. Of MobiCom 2005, August 28–September 2, 2005
4. Zhou, P., Manoj, B.S., Rao, R.: On Optimizing Gateway Placement for Throughput in Wireless Mesh Networks EURASIP Journal on Wireless Communications and Networking **2010**, Article ID 368423, 12 (2010). doi:10.1155/2010/368423
5. Benyamina, D., Hafid, A., Gendreau, M.: Wireless Mesh Networks Design - A Survey. IEEE Communications Survey and Tutorials (2011)
6. Wu, W., Luo, J., Yang, M.: Cost-effective Placement of Mesh Nodes in Wireless Mesh Networks. In: 5th International Conference on Pervasive Computing and Applications (ICPCA). IEEE (2010)
7. Xhafa, F., Sánchez, C., Barolli, A., Takizawa, M.: Solving mesh router nodes placement problem in Wireless Mesh Networks by Tabu Search algorithm. Journal of Computer and System Sciences (2014)
8. Sanni, M.L., Hashim, A.-H.A., Anwar, F., Naji, A.W., Ahmed, G.S.M.: Gateway placement optimisation problem for mobile multicast design in wireless mesh networks. In: International Conference on Computer and Communication Engineering (ICCCE 2012), Kuala Lumpur, Malaysia (July 2012)
9. Barolli, A., Oda, T., Ikeda, M., Barolli, L., Xhafa, F., & Loia, V.: Node placement for wireless mesh networks: Analysis of WMN-GA system simulation results for different parameters and distributions. Journal of Computer and System Sciences (2014)
10. Lin, C.-C.: Dynamic router node placement in wireless mesh networks: A PSO approach with constriction coefficient and its convergence analysis. Information Sciences (2013)
11. Benyamina, D., Hafid, A., Gendreau, M., Maureira, J.C.: On the design of reliable wireless mesh network infrastructure with QoS constraints. Journal of Computer Networks (2010)
12. Merlin Sheeba, G., Nachiappan, A.: A Differential Evolution Based Throughput Optimization for Gateway Placement in Wireless Mesh Networks. International Journal of Applied Engineering Research **9**(21), 5021–5027 (2014)
13. Liu, J., Lampinen, J.: A Fuzzy Adaptive Differential Evolution Algorithm. In: Proceedings of IEEE TENCON 2002, pp. 607–611

Power Efficient Routing by Load Balancing in Mobile Ad Hoc Networks

G. Ravi and Kishana Ram Kashwan

Abstract In Mobile Ad hoc Network (MANET), energy efficient routing in the hybrid domain based on residual energy provides many benefits. The energy efficient routing approach based on load balancing method can spread out data traffic and is still free from being overburdened of nodes. This can be done through creating many alternatives for efficient utilization of available resources. It keeps the network alive and attempts to maximize the network lifetime. The hybrid routing approach is a combination of the good features of reactive and proactive routing approaches in the MANET. An energy efficient routing protocol is proposed to fruitfully accomplish various challenges. Reliable Zone Routing Protocol (RZRP) is one such example. It is designed to minimize energy consumption in proactive routing and mobility prediction in reactive routing approach. RZRP has the ability to discharge the basic requirements for an efficient energy consumption technique. It is reliable and can increase the network lifetime of MANET. The results are analyzed and compared with Zone Routing Protocol (ZRP).

Keywords Energy efficient routing · Hybrid routing protocols · Load balancing method · Mobile ad hoc network · Residual energy · Zone Routing Protocol

1 Introduction

Mobile Ad hoc Network (MANET) is normally defined as an infrastructureless network. The main drawback of it is the nonexistence of base stations that makes communication among various nodes in the network extremely complicated. In order to meet diverse challenges in MANET, each node must be designed as intelligent one to handle various operations of communication such as

G. Ravi(✉) · K.R. Kashwan
Department of Electronics and Communication Engineering,
Sona College of Technology, Salem, TN, India
e-mail: {raviraj.govind,kashwan.kr}@gmail.com

S. Berretti et al. (eds.), *Intelligent Systems Technologies and Applications*,
Advances in Intelligent Systems and Computing 385,
DOI: 10.1007/978-3-319-23258-4_14

transmission, reception of messages and other signalling operations between nodes. Besides, MANET has basically many difficulties to establish required communication between different nodes in the network in contrast with cellular networks. There are many possibilities for MANET applications due to the growing demand of more efficient communications which possess the tendency to be used as future trends. The movement of nodes makes the routing task difficult [2, 18]. Considering future aspects, it is important to meet the challenges of MANET to have the best communication in an integrated way which is generally found in cellular networks that are capable to sustain high data traffic of future expectations [1, 19]. A hybrid model of communication requires a great deal of research work and a balanced approach [10] must be taken into consideration for easier routing. One of the principal solutions is energy efficient routing approach for considering residual energy [5, 16] which has the capability to handle the challenges of MANET. These include reliability, connectivity, mobility, adapting dynamic changing of topology and so on. The residual energy of nodes and routing technique can help to manage available energy. It can be achieved by balancing load and by not allowing being overused. So, it leads to maximization of the network lifetime. Moreover, this method can offer an efficient communication in an integrated manner. For various devices used and increasing large number of nodes in MANET, it is much more appropriate to go for energy efficient routing in an integrated approach in the hybrid domain [13]. It is better compared to proactive and reactive domains [6] of MANET. The main focus is on hybrid domain. The several performance related factors of MANET like network size, coverage area, transmission range, speed **No table of contents entries found.** and ability of the nodes, bandwidth utilization, and battery power are considered. Wireless based hybrid mobile network has the special provision or a feature that combines both mobile ad hoc network and well established infrastructure network such as cellular network. It has better throughput, load balancing and coverage area [3, 7].

1.1 Hybrid Routing Protocols in MANET

Hybrid Routing protocols are effectively utilized for the two main routing classification of MANET, topological and geographical routing for recent times. It comes under routing information update mechanism to manipulate topological, geographical and hierarchical information. Hybrid routing approach in essence depends on zone routing concept which has three design selection criteria, namely the size of the zone, position of nodes in the network and construction of path from source to destination [4]. In fact Hybrid approach is a good attempt for providing effective communication in mobile ad-hoc networks in an integrated manner. It has improved reliability, mobility, connectivity, scalability, multi-hop, network lifetime, security, quality of service and low energy consumption in MANET [9, 12].

As far regarding to reduction in energy consumption, energy efficient routing is the key solution in which route to the destination is found out by energy consumption at an end-to-end packet transmission. Energy efficient routing can help to find the routes which fulfill the basic requirements of MANET such as less energy consumption, reliability and maximizing network lifetime [16, 12]. Energy efficient routing offers valuable results of being reliable and conserving remaining energy of nodes. The Load balancing method takes care of energy expenditure both at node level as well as at the transmission level by using power aware routing and following traffic regulation [8, 15].

2 Zone Routing Protocol (ZRP)

Zone Routing Protocol (ZRP) is one of the hybrid routing protocols that is conceived to tackle the demerits of reactive and proactive routing approaches. In reactive routing, the route is found out only whenever it is demanded whereas in proactive routing, the route isn't demanded and it is immediately available because of periodically tabulated relevant information of each node in the network. The routing structure of ZRP is flat in which all nodes have equal potential to play and very convenient to the low-density network. In a high-density network, scalability factor is removed, but offers congestion affordability, link state routing and facilitates quality of services. Maximum energy consumption is due to frequent update of information for all nodes in the network and being short of mobility prediction which causes delay of data transmission to the required destination. These are the two main setbacks of proactive and reactive routing approaches respectively. A ZRP utilizes zone concept in which proactive routing is exploited within the zone and reactive routing is employed out of the zone. Because of confined limit of the zone, the job of route detection, maintenance, updates and requests is made very simple. The construction elements of ZRP are as follows:

(i) Neighbour discovery protocol at MAC layer
(ii) Inter-Zone Routing Protocol (IERP)
(iii) Intra-Zone Routing Protocol (IARP)
(iv) Border- cast Routing protocol (BRP)

NDP transmits "HELLO" messages at periodic intervals. Upon receiving a hello message, the neighbour's table is updated. The Neighbours, for which no message has been received within a particular time interval, are removed from the table. IERP and IARP try to find the best routes to the required destination outside and inside of the zone respectively. BRP acts as an intermediate role which helps to forward IERP route queries to the peripheral nodes [11, 14, 15, 17]. The route discovery practice must be related to multiple paths and the criterion for choosing the excellent path may be least delay path, the shortest path and so on.

3 Reliable Zone Routing Protocol (RZRP)

Reliable Zone Routing Protocol (RZRP) has been preferred for achieving better performance than Zone Routing Protocol (ZRP). In RZRP, the route from one node to another node within and beyond a zone is made with the maximum possibility of connectivity and alternate path is immediately assigned if the foremost assigned path is disturbed. RZRP mainly focuses on the principal needs of MANET as energy efficiency, reliability and increasing network lifetime by energy efficient routing with the load balancing approach. It makes an attempt to reduce energy consumption through sleep state mechanism and traffic density measures in proactive approach. It locates the position of a node under mobility prediction approach in a reactive approach.

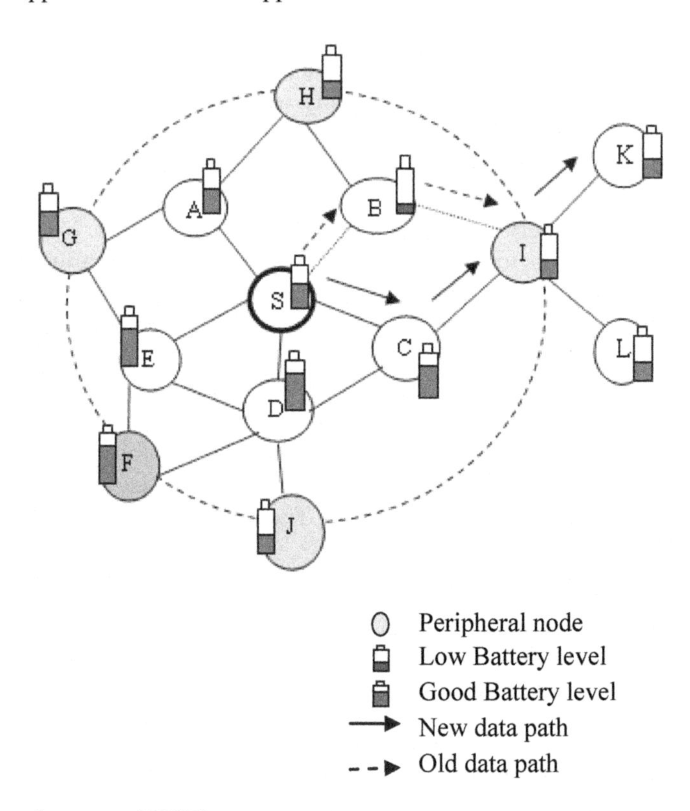

Fig. 1 Route discovery of RZRP

Fig. 1 shows a possible network topology resulting from node B which has low battery level. It means that it enters into the sleep state. The new path S–C–I–K from source to destination includes nodes with good battery levels. Selection of this path is assigned to the routing protocol without node B. It illustrates that how load is distributed by using nodes with good battery levels.

3.1 Algorithm of RZRP

The challenge is that which nodes are going to act as a source node and as a destination node. If it is identified, the next is to identify the destination node location to find whether it is within the zone or out of the zone. If the destination node is within zone, apply IARP (Proactive routing), else apply IERP (Reactive routing).

3.1.1 Proactive Routing

The objective of proactive routing of RZRP is to extend the lifetime of the network through minimizing energy consumption by the load balancing method. While periodically updating topological information at every node, the severity of data traffic is addressed to help to choose the route to destination with less traffic load. It leads to have least delay path in the network. Sleep state procedure is applied to appropriate overloaded nodes as a positive measure before a path link is broken. It is mostly applicable to the nodes in inactive state of communication within the zone. RZRP concentrates on how to accomplish data transmission at efficient manner with respect to node's battery power. No definite route provision to destination is followed. The most preferable route, in a given situation, is followed. If it is unable to adapt the most preferred path, it is directed to adapt the next level of preference. As such kind of ideas, energy efficient routing becomes to be a dynamic one.

3.1.2 Sleep State Procedure

Certain conditions are defined for applying sleep state procedure and under that situation, nodes are assumed to obey the following.

1. If node tends to reach severe data traffic which result more usage of energy that is known from topological information, then the node should convey to all other relevant nodes as it is going to sleep state.
2. If nodes are at inactive during active communication stage, then apply sleep state procedure.
3. The topological information should have the following metrics:
   - (i) Compute nodes' current energy
   - (ii) Compute energy at each forwarding instance
   - (iii) Update remaining energy of nodes periodically
   - (iv) Each node must maintain minimum level of battery power
   - (v) Traffic density is measured at each node

3.1.3 Set Sleeps State Condition

$$\frac{\text{Node current battery power}}{\text{Node initial battery power}} < \Delta \qquad (1)$$

Where Δ is 30 % of battery power at the node for participating in communication, indicated in equation (1), above.

If node is in data forwarding state and reached severe data traffic, it sends sleep state message to other relevant nodes. The other nodes stop sending packets and send sleep approval message to that requested node. The request node completes all packets transmission from its queue to channel and then goes to sleep state. After accomplishment of data transmission, request node disconnects all links from route cache table and informs neighbours accordingly. This stops data transmissions for that node in a route. Here, it must be very clear on this point that sleep state procedure is applied to prevent for excess amount of energy of node. Whenever it happens to a node, the current performing task should be shifted to its immediate neighbour node. This attitude substantially helps to minimize the excess amount of energy of a node while at data transmission.

3.1.4 Reactive Approach

Reactive approach is followed when the destination node is identified at the outside zone. Having identified the destination node, the reactive approach keeps the following steps.

(i). Source updates destination node speed periodically to contact next zone neighbours.
(ii). Periodically validating next zone neighbours' information and their speed.
(iii). If node moves within coverage range based on receiving signal strength, it can easily compute the distance of it.
(iv). If shortly a node is likely to move to near border of zone point, apply prediction method to find location or position.

The following computation methodology is performed as indicated by the positions of the respective nodes.

(i). Predicted position of $X' = X + (\text{current time} - \text{last time}) \times \text{speed}$
(ii). The predicted position of $Y' = Y + (\text{current time} - \text{last time}) \times \text{speed}$

Where, X' is new position and X is the current position, similarly, Y' is new position and Y is the current position,

The distance of next zone node is computed as per following expression.

$$D = \sqrt{(X_1 - X_2)^2 + (Y_1 - Y_2)^2} \tag{2}$$

If a node is moving out of transmission range, it stops transmitting data packet and applies path switching to choose next node from routing table. Each node maintains a routing table for updating routing information. In case a node is reaching border area, there comes a need to determine errors. If node speed is greater than error determination speed, choose an alternative node to choose next hop towards destination. In ZRP, it has the tendency to be lack on reactive method due to mobility of nodes. RZRP has tried to overcome this issue as possible by predicting the position of nodes.

4 Simulation Setup

To ensure the basic requirements of energy efficiency, reliability, and maximizing life time in MANET, some standard metrics like Control Overhead, Packet Delivery Ratio (PDR) and Throughput are taken for analyses. RZRP is without deficiencies of ZRP. It provides the opportunities for valuable communication in an integrated way. The two Hybrid Routing Protocols namely ZRP and RZRP are simulated to test the results by NS - 2 Software.

The parameters are chosen for the experimental setup.

(1) Simulation range: 1000 m X1000 m
(2) Transmission range: 250 m, thus the maximum number of hops to reach anywhere is 10.
(3) Protocol: ZRP, RZRP
(4) Number of nodes: 50 to 100 nodes in steps of 10.
(5) Zone Radius: 2
(6) Movement: Nodes follows random way point model with the speed set at 5 m/s.
(7) Traffic 20 % connection is simulated with CBR packets. The packet size is 2000 kb
(8) Packet interval: 0.1.
(9) Simulation time: 200 seconds and the results provided here are the averages of 20 simulations.
(10) Energy: Every node is provided with an initial energy of 100 joules

5 Results and Discussion

It is observed, as illustrated in Figures 2, 3, 4 and 5, that if the numbers of nodes increase, traffic of data is well controlled and regulated in RZRP compared to ZRP due to energy efficient routing. It intends to reduce energy consumption considering residual energy and immediate identification of node in the zone as a primary step followed in RZRP. Due to this factor, data packets are easily sent to the destination that ensures basic needs of MANET which is illustrated in the above said Figures.

Reliable Zone Routing Protocol (RZRP) always keeps consistency for ensuring basic requirements of MANET through load balancing method than Zone Routing Protocol (ZRP). It is, however, very difficult to maintain and produce consistence performance if the number of nodes tend to increase in the network. RZRP fulfils the most of the requirements of MANET by using a balanced approach.

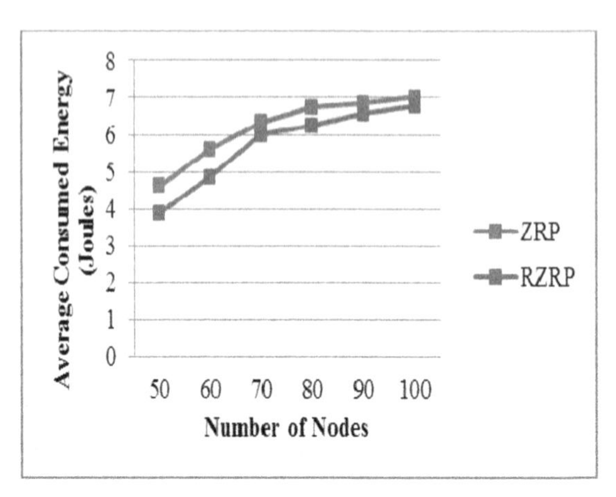

Fig. 2 Average Consumed Energy

The Fig. 2 shows that the average consumed energy is lower for RZRP compared with ZRP alone. If the number of nodes increases, the energy consumption also increases. It is evident that ZRP consumes 4.6 J of energy at low density environment whereas RZRP consumes only 3.88 J of energy. It extends the network lifetime by distributing the load amongst different nodes based on energy level.

The Fig. 3 shows the delivery ratio of ZRP and RZRP for the range of nodes between 50 and 100. The number of nodes increases with PDR value. The higher value of packet delivery ratio indicates the improvement of reliability. RZRP has produced better PDR values compared to ZRP.

Fig. 4 shows the control overhead values for the different number of nodes. The RZRP has lower overhead compared to that of ZRP.

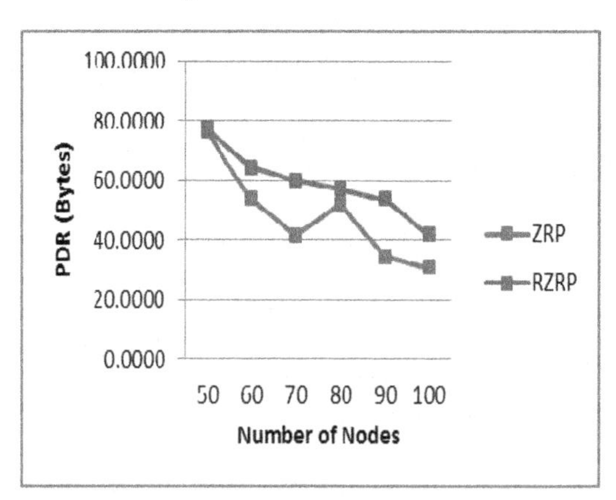

Fig. 3 Packet Delivery Ratio

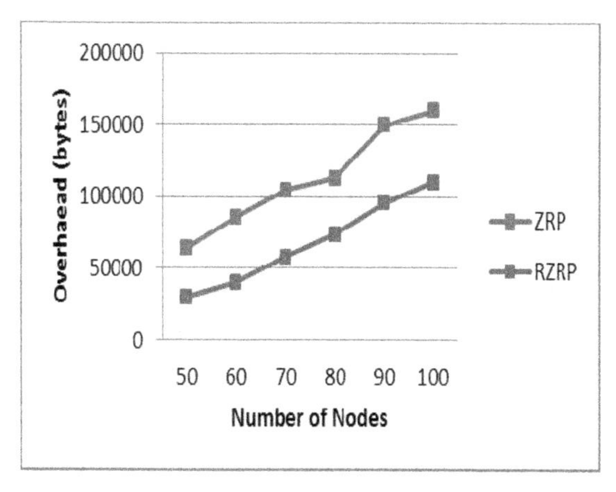

Fig. 4 Control Overhead

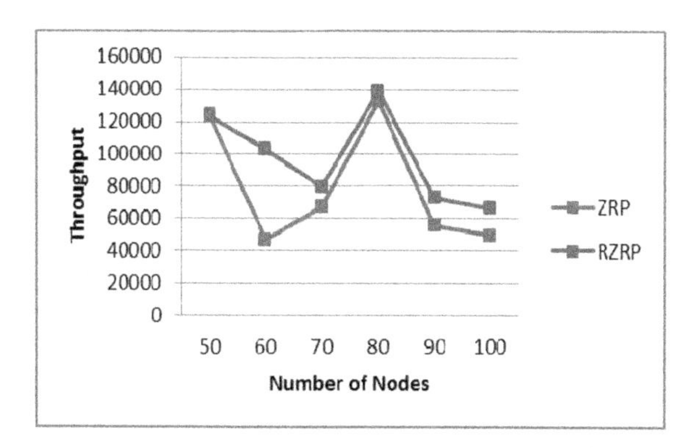

Fig. 5 Throughput

Fig. 5 shows the comparison of throughput for various numbers of nodes. The throughput of the RZRP is larger than the ZRP. This result is attributed by balance the load. Fig .6 shows the packet loss of both protocols.

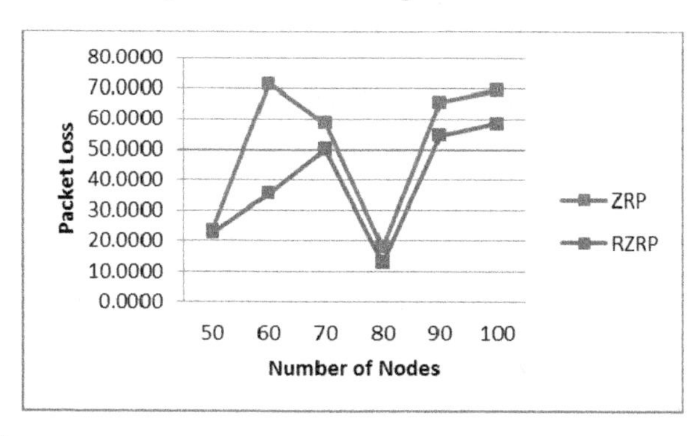

Fig. 6 Packet Loss

6 Conclusion

Energy efficient routing offers various research opportunities to overcome on challenges in mobile ad hoc networks, particularly in limited energy resource environment. The future trends show tendency to use the concepts of embedded systems. However, the improvements in energy efficient routing should not affect basic communication as and when required. An inclusion of new methodology must be supportive and a logical extension to proceeding methodologies. The limitations of ZRP are overcome by a new method of RZRP. It works through insertion of energy preservation techniques during inactive state of communication and mobility prediction at very frequent intervals. The notable observations include that RZRP doesn't affect the basic operation of MANET. The hybrid routing is an approach for using a variety of devices for the diversified applications. The isolated use of reactive and proactive routing may have random benefits but hybrid approach has wide benefits. The modern smart devices are on increasing demands of communication at multiple interface environments for intelligent usage. This needs a hybrid approach to meet the challenges of complex communication protocols.

References

1. Ravi, G., Kashwan, K.R.: Energy Aware Zone Routing Protocol using Power Save Technique AFECA. International Review on Computers and Software **8**(10), 2373–2378 (2013)
2. Ravi, G., Kashwan, K.R.: A New Routing Protocol For Energy Efficient Mobile Applications For Ad Hoc Networks. Computers & Electrical Engineering (2015). doi:10.1016/j.compeleceng.2015.03.023
3. Toh, C.: Maximum Battery Life Routing to Support Ubiquitous Mobile Computing in Wireless Ad Hoc Networks. IEEE Comm. Mag. **39**, 138–147 (2001). doi:10.1109/35.925682
4. Cheng, H., Cao, J.: A Design Framework and Taxonomy for Hybrid Routing Protocols in Mobile Ad Hoc Networks. IEEE Communications Surveys & Tutorials, 62–73 (2008). doi:10.1109/COMST.2008.4625805
5. Zhu, J., Qiao, C., Wang, X.: On Accurate Energy Consumption Models for Wireless Ad Hoc Networks. IEEE Trans. Wireless Commun. **5**, 3077–3086 (2006). doi:10.1109/TWC.2006.04566
6. Fotino, M., Gozzi, A., Cano, J.C., et al.: Evaluating Energy Consumption of Proactive and Reactive Routing Protocols in a MANET. IFIP International Federation for Information Processing on Wireless Sensor and Actor Networks **248**, 119–130 (2007)
7. Vassileva, N., Arroyo, F.B.: A Survey of Routing Protocols for Maximizing the Lifetime of Ad Hoc Wireless Networks. International Journal of Software Engineering and Its Applications **2**(3), 77–90 (2008)
8. Pi-Cheng, H., Tei-Wei, K.: A Maximum-Residual Multicast Protocol for Large-Scale Mobile Ad Hoc Networks. IEEE Transactions on Mobile Computing, 1441–1453 (2009). doi:10.1109/TMC.2009.54

9. Joshi, R.D., Rege, P.P.: Distributed energy efficient routing in ad hoc networks. In: IEEE Fourth International Conference on WCSN 2008, pp. 16–21 (2008). doi:10.1109/WCSN.2008.4772674
10. So, J., Vaidya, N.F.: Load-Balancing Routing in Multichannel Hybrid Wireless Networks With Single Network Interface. IEEE Trans. Veh. Technol. **56**, 342–348 (2007). doi:10.1109/TVT.2006.889569
11. Sree Ranga Raju, M.N., Mungara, J.: Enhanced ZRP Protocol for Mobile Ad-hoc Networks. International Journal of Wireless & Mobile Networks, 160–173 (2011). doi: 10.5121/ijwmn.2011.3411.16
12. Banerjee, S., Misra, A.: Minimum energy paths for reliable communication in multi-hop wireless networks. In: Proc. ACM Mobi Hoc, pp. 146–156 (2002)
13. Ramasubramanian, V., Haas, Z.J., Sirer, E.G.: SHARP: a hybrid adaptive routing protocol for mobile ad hoc networks. In: Mobi Hoc Proceedings of the 4th ACM International Symposium on Mobile Ad Hoc Networking & Computing, pp. 303–314 (2003)
14. Saigal, V., Nayak, A.K., Pradhan, S.K., Mall, R.: Load balanced routing in mobile ad hoc networks. Computer Communications **27**(3), 295–305 (2004)
15. Hoang, V.D., Shao, Z., Fujise, M.: Efficient load balancing in MANETs to improve network performance. In: IEEE 6th International Conference on ITS Telecommunications Proceedings, pp. 753–756 (2006). doi:10.1109/ITST.2006.289010
16. Vazifehdan, J., Prasad, R.V., Niemegeers, I.: Energy-Efficient Reliable Routing Considering Residual Energy in Wireless Ad Hoc Networks. IEEE Trans. Mobile Comput. **13**, 434–447 (2014). doi:10.1109/TMC.2013.7
17. Wang, L., Olariu, S.: A Two-Zone Hybrid Routing Protocol for Mobile Ad Hoc Networks. IEEE Trans. Parallel Distrib. Syst. PP **PP**, 1105–1116 (2004). doi:10.1109/TPDS.2004.73
18. Huang, X., Du, X., Li, X., Bian, K.: Efficient Communications in Mobile Hybrid Wireless Networks. IEEE GLOBECOM (2012) doi:10.1109/GLOCOM.2012.6503765
19. Qin, Y., Wen, Y.Y., Ang, H.Y., Gwee, C.L.: A routing protocol with energy and traffic balance awareness in wireless ad hoc networks. In: IEEE 6th International Conference on Information, Communications & Signal Processing (2007). doi:10.1109/ICICS.2007.4449643

Network Optimization Using Femtocell Deployment at Macrocell Edge in Cognitive Environment

Joydev Ghosh, Subham Bachhar, Uttam Kumar Nandi, Ajit Rai and Sanjay Dhar Roy

Abstract This research focuses on the problem of cell edge user's coverage in the context of femtocell networks operating within the locality of macrocell border where pathloss, shadowing, Rayleigh fading have been included into the environment. As macro cell edge users are located far-away from the macro base station (MBS), so that, the underprivileged users (cell edge users) get assisted by the cognitive-femto base station (FBS) to provide consistent quality of service (QoS). Considering various environment factors such as wall structure, number of walls, distance between MBS and users, interference effect (i.e., co-tier and cross-tier), we compute downlink (DL) throughput of femto user (FU) for single input single output (SISO) system over a particular sub-channel, but also based on spectrum allocation and power adaptation, performance of two tier network is analyzed considering network coverage as the performance metric. Finally, the effectiveness of the scheme is verified by extensive matlab simulation.

Keywords Cognitive-Femtocell networks · Okumura-Hata propagation model · DL throughput · Network coverage

J. Ghosh(✉) · U.K. Nandi · A. Rai
ETCE Department, The New Horizons Institute of Technology, Durgapur-8, WB, India
e-mail: {joydev.ghosh.ece,nandi783,ajit393.rai}@gmail.com

S. Bachhar
Dr. B.C. Roy Engineering College, Durgapur, WB, India
e-mail: subham4792@gmail.com

S.D. Roy
ECE Department, National Institute of Technology, Durgapur-9, WB, India
e-mail: s_dharroy@yahoo.com

S. Berretti et al. (eds.), *Intelligent Systems Technologies and Applications*,
Advances in Intelligent Systems and Computing 385,
DOI: 10.1007/978-3-319-23258-4_15

1 Introduction

In a double-layer heterogeneous network (HetNet), macrocell characterized by its high transmit power and bigger coverage area co-exit femtocells characterized by their small transmit power and limited coverage area. Femtocells are low power invention that comes up with better coverage to portable users through femtocell access point (FAP) at the indoor scenario. The access node named as FAP, perform as the base station (BS) for the femtocell and take help of internet as backhaul network to get connected with MBS. Co-layer interference is a performance limiting parameter for the HetNet which can be reduced by introducing cognition into the femto access points (FAPs) [1]. The salient features and the potential of the Orthogonal Frequency Division Multiple Access (OFDMA) scheme makes it perfectly suitable for the CR based transmission system. The prime idea of OFDMA technology is that the available spectrum split into several orthogonal sub-channels and permits provision to multiple users to transmit data simultaneously on the different sub-channels. Combining cognitive radio and OFDMA is one of the best feature in the future mobile networks due to its flexibility in allocating resource among cognitive radio (CR) users. Here, we address quite a few shortcomings still holding in the current stage of networking and emphasis on some design considerations of 5G embracing femtocells and cognitive radio technology. "Femto Cell" is a low power radio access point which operates in both licensed and unlicensed spectrum and it offers network coverage from 10 meters to several hundred meters. Fractional frequency reuse (FFR) has been developed as fruitful inter-cell interference coordination (ICIC) technique in OFDMA based wireless networks [5],[6]. The usage of FFR in cellular networks results in natural compromise between data rate and coverage in cell edge users and overall cell throughput and spectral efficiency. If the deployment of FBSs increases, the tendency of allocating the same frequency spectrum to two adjacent FAPs femtocells increases, and the inter-FAPs' interference becomes more significant [9].

In this paper, we introduce Azimuth angle of sectored sites which takes an important role in case of antennas with rather small horizontal beam width to optimize the network coverage. Note that antenna direction can be achieved by the azimuth angle. In practical downlink transmission most of the handsets can only accommodate one or at most two antennas. Here, Single input single output (SISO) system. In single input single output (SISO) system, the transmitter has a single antenna and the receiver also has single antenna. In contrast to Omni-directional transmissions, beam forming reduces interference, allowing more concurrent transmissions in the network. The intention of considering beamforming is to improve signal to interference plus noise ratio (SINR). Additionally, by focusing the transmission energy in a particular direction, beamforming induces a signal that is in order of the magnitude stronger than that of the signals in undesired direction. This technique is utilized here to enhance the coverage of a specific zone or data rate or channel capacity of the system [8].

In particular, the major contributions of this paper are highlighted below-

- A simplified network model of resource allocation in CR based femtocell network has been introduced.
- The transmitting signal ($T_{x(MBS)}$) from MBS weakened and worsen quicker once the signal reaches indoors. Femtocells furnish way out to the difficulties present in macrocell edge area by means of reducing the number of outage. So that, FBS network coverage is one of the prime concerns in indoor environment to get good quality of service (QoS).
- We consider Azimuth angle as a degree of freedom for optimization of network coverage. Here, difference between antenna gains towards the main lobe and the half angle between adjacent sectors is relatively large and the cells associated to neighboring sites are so adjusted to obtain maximum network coverage.
- We present comprehensive numerical outcome to vindicate the developed simplified network model and to exhibit the effectiveness of our proposed scheme.

The remainder of this paper is organized as follows. In Section 2, we describe system model to define propose network. In Section 3, the assumptions considered for execution of exact scenario in the simulation model. In Section 4, we present and analyse the numerical results. Ultimately, we finish this work in Section 5 with conclusion.

2 System Model

We consider a scenario where femtocells are deployed over the existing macrocell network and share the same frequency spectrum with macrocell. Here we focus only on downlink scenario. The downlink communications in a network with one macrocell and N_F=4 number of femtocells is as shown in Fig. 1 in which 4 femtocells are located at (2, 2), (-2, 2), (-2,-2), (2,-2) within the macrocell coverage. In a macrocell, total N_{MUE} user equipments (UEs) are randomly distributed within its coverage area.

In theoretical analysis, we assume that the OFDMA based dual tier network consists of N_M number of hexagonal gride macrocell and N_F number of femto cells in each macrocell. Total bandwidth associated with the macrocell edge regions is split up into 3 sub-bands by applying FFR scheme. A sub-band containing N_{sc} number of sub-channels that are available to provide service to the users located at the cell-centre area and the corresponding cell edge area. Besides, we also consider that the channel is slowly time varying and follows the Rayleigh multipath fading distribution. Three kinds of possible links in dual-layer networks are as follows: MBS to outdoor user's link, FBS to indoor user's link, MBS to indoor user's link.

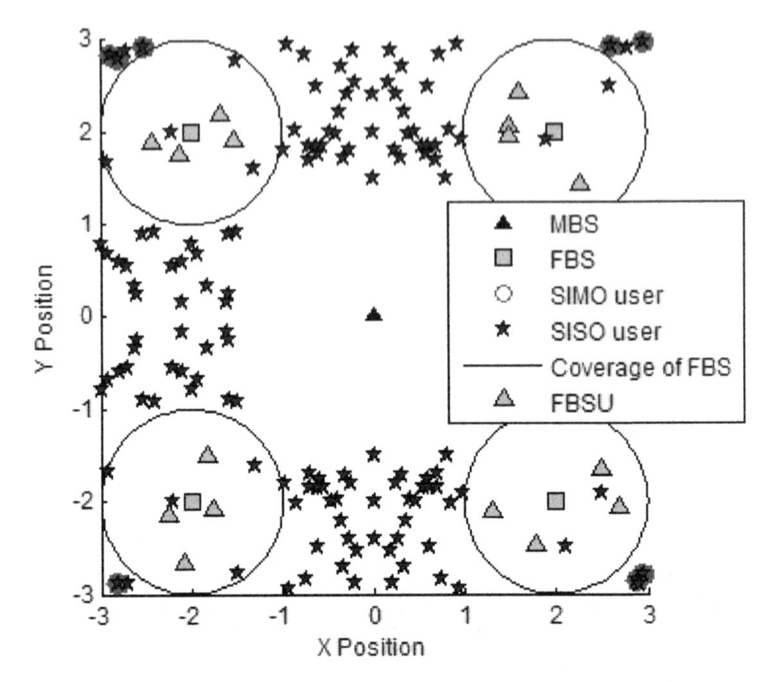

Fig. 1 Co-existance scenario of FAP and Macrocell in our proposed network.

2.1 Propagation Path Loss Model (Empirical Model)

As per our system model Okumura-Hata propagation path loss model for urban area is chosen to do the evaluation of network performance.

- *Macrocell Path Loss Model-*

  Empirical model is nothing but an excessive computational effort. In Okumura-Hata propagation model the important parameter for us is to know how much overall area is covered. Okumura takes urban areas as a reference and applies correction factors [2]:

$$L_{dB} = A + Blog_{10}(d) - E \tag{1}$$

where $A = 69.55 + 26.16log_{10}f_c - 13.82log_{10}h_b$

$B = 44.9 - 6.55log_{10}h_b$

$E = 3.2\left(log_{10}(11.7554h_m)\right)^2 - 4.97\ for\ large\ cities, f_c \geq 300Mhz$

Table 1 Some Values Related to Okumura-Hata model

Parameters	Symbol	Range
Carrier Frequency	f_c	$150 \le f_c \le 2000$(MHz)
BS Antenna Height	h_b	$30 \le h_b \le 200$(m)
MS Antenna Height	h_m	$1 \le h_m \le 10$(m)
Distance Between BS and MS	d	$1 \le d \le 20$(m)

- **Femtocell Path Loss Model**

Two different path loss exponents and small breakpoint distance of 100 meters are used to characterize the femtocell path loss propagation model [4].
Path Loss:

$$\begin{aligned} L &= 10 * \propto_{p1} log_{10} r + L_1 \qquad for\ r < r_b \\ &= 10 * \propto_{p2} log_{10}(r/r_b) + 10 * \propto_{p1} log_{10} r_b + L_1\ for\ r > r_b \end{aligned} \tag{2}$$

where L_1 = Reference Path Loss at r=1m.
r_b =Breakpoint distance
$\alpha_{p1=}$ Pathloss exponent for $r \le r_b$
$\alpha_{p2=}$ Pathloss exponent for $r \le r_b$

To avoid sharp transition between the two regions:

$$L = L_1 + 10 * \propto_{p1} log_{10} r + 10 * (\propto_{p2} - \propto_{p1}) log_{10}(1 + {}^{r}/_{r_b}) \tag{3}$$

2.2 *Radio Channel Model*

Only MBS can cause interference for femto base station user (FBSU) receiver or FBS transmitter. The position of MBS, macro base station user (MBSU), FBS, and FBSU are shown in Fig. 1. The utilization of spectrum resources can be improved by cognitive radio (CR) technology. CR facilitates SU to access the spectrum bands licensed to PU as long as PUs QoS is not affected. Here, MBSU is treated as PU and FBSU is treated as SU. PU is power controlled by the BS. The link gain between MBS and u-th UE can be expressed as [2]:

$$G_{p2p,u}(r_u, \theta_u) = d_u^{-\alpha_p} 10^{\xi_s/10} |h_{p2p,u}|^2 \tag{4}$$

where d_u (r_u, θ_u) is denoted as the distance of a MBSU /PU at a location (r_u, θu) with respect to the MBS and let $|h_{p2p,u}|^2$ denote the channel gain between the MBS and its associated MUE. We have neglected the co-tier interference as FBSs are deployed in such a manner that it eliminates all the difficulties of overlapping of FBSs. The link gain between j-th FBS and i-th FBSU can be expressed as:

$$G_{s2s,ji} = d_{s2s,ji}^{-\alpha_p} 10^{\xi_s/10} \left| h_{s2s,ji} \right|^2 \quad (5)$$

where $d_{s2s,\ ji}$ is the distance between j-th FBS and i-th FBSU. ξ_s (in dB) is a Gaussian random variable with zero mean and variance, σ^2, due to shadowing in the channel and let $\left| h_{s2s,ji} \right|^2$ denote the channel gain between j-th FBS and its asso ciated i-th FBSU. A FBSU is assumed to be within a circle of radius, r_f r_f around a FBS. If the position of a FBS is (r_j, θj) and the position of a FBSU is (r_i, θi). Then,

$$d_{s2s,ji} = \sqrt{(r_j cos\theta_j - r_i cos\theta_i)^2 + (r_j sin\theta_j - r_i sin\theta_i)^2} \quad (6)$$

If the position of a MBS is (r_k, θ_k) then the distance of it from FBSU may be expressed as-

$$d_{p2s,ki} = \sqrt{(r_k cos\theta_k - r_i cos\theta_i)^2 + (r_k sin\theta_k - r_i sin\theta_i)^2} \quad (7)$$

The link gain from k-th MBS to i-th FBSU may be expressed as:

$$G_{p2s,ki} = d_{p2s,ki}^{-\alpha_p} 10^{\xi_s/10} \left| h_{p2s} \right|^2 \quad (8)$$

where $d_{p2s,\ ki}$ is the distance between k-th MBS to i-th FBSU. All the channel gains are assumed to be independent and the channels have a coherence time greater than or equal to a time slot. Here $\left| h_{p2s,ki} \right|^2$ denote the channel gain between k-th MBS and its associated i-th FBSU. For analysis, Rayleigh fading is included with pathloss and shadowing. Moreover, Rayleigh fading gives tractable results which assists understanding of the system response to a particular situation.

We use the notation x to denote the serving network entity for a generic user. That is, x=a if the user is associated to a FAP and x=b if the user is associated to a MBS. Without any loss of generic laws, the analysis is conducted on a typical user located at the origin. Therefore, the signal to interference plus noise ratio (SINR), $\gamma_{n,k,i}^{x}$ at the typical user located at the origin (which also holds for any genericuser) served by an MBS or FAP (MBS/FAP) is given by [7]

$$\gamma_{n,k,i}^{x} = \frac{P_{n,k,i}^{x} G_{n,k,i}^{x}}{I_{n,k,i}^{x} + I_{n,k,i}^{\prime x} + \sigma_{n,k,i}{}^2} \quad (9)$$

where $G^x_{n,k,i}$ is the wireless link gain between the user to the serving network entity (i.e. an MBS or a FAP) over the n-th sub-channel. Here $P^x_{n,k,i}$ is designate -ed as the proportion of total transmit power by an associated serving network entity over the particular sub-channel. Likewise, the channel gains from a generic location $x \in \mathbb{D}^2$ to the FAP, a_i and the MBS, b_i are denoted by

$h_{a_i} \sim \sqrt{X^2_{a_i} + Y^2_{a_i}}$ and $h_{b_i} \sim \sqrt{X^2_{b_i} + Y^2_{b_i}}$, respectively, where X_x, Y_x are

independent gaussian random variables with zero mean and desired variance ${\sigma_{n,k,i}}^2$ is the noise power of zero-mean complex valued additive white Gaussian noise (AWGN).

2.3 *Outage Probability for Downlink Transmission of Macro User*

A FUE experiences two kinds of outages. The first is due to the channel unavailability because of the opportunistic channel access, and the second is SINR outage. According to practical scenario, there is existence of SISO outage users and SIMO outage users. A user is said to be an outage if SINR of the user is less than a SINR threshold.

The outage probability of a SISO / SIMO may be expressed as:

$$P_{out,SISO} = Prob\left\{\gamma^x_{n,k,i} \leq SINR_{thd(SISO)}\right\} \tag{10}$$

$$P_{out,SIMO} = Prob\left\{\gamma^x_{n,k,i} \leq SINR_{thd(SIMO)}\right\} \tag{11}$$

2.4 *Throughput of Macrocell Network*

The reachable throughput, T_p of an user can be calculated from Shannon's theorem. Considering a bandwidth W is assigned to a sub-channel. We have

$$T_p = W log_2(1 + \gamma^x_{n,k,i}) \tag{12}$$

where γ_x -Signal to Interference plus Noise Ratio.

3 Simulation Model

The simulation is developed in MATLAB. In our simulation, parameters mentioned in Table 1 have used. For simplicity of analysis we consider the pathloss only.

3.1 Implantation of Macrocell (Outdoor), and Femtocell (Indoor)

The following steps are followed to implant macrocell and femtocell successfully under indoor and outdoor scenario.

1. Each powerful MBS is located at the center of macrocell which is having the coverage of 3Km radius. The outdoor network incorporates of circular gride of 19 MBS. The transmitting signal ($T_{x(MBS)}$) from MBS attenuated while the signal reaches indoors. The network performance is under serious measure especially due to poor coverage and underutilize spectrum in border area.

2. The small coverage femto base stations (FBSs) are densely deployed in that area where macro base station (MBS) unable to provide QoS and spectrum also not been utilized efficiently. The location of implantation of low power FBS is strictly inside the indoor area. The indoor network incorporates of circular gride of 300 FBS.

3.2 Generation of Users' Location, and Interference

The generation of the users' locations and interference power is carried out considering the following steps.

1. A fixed number of outdoor users' ($N_{MUE,ku}$) and indoor users' ($N_{FUE,ji}$) is generated and they are randomly distributed within their own coverage area . N_{UE} includes all MBSUs /PUs ($N_{MUE,ku}$) and FBSUs/SUs ($N_{FUE,ji}$) which means $N_{UE} = N_{MUE,ku} + N_{FUE,ji}$. Here, $j \in \mathcal{N}_F = \{1,2,.......,N_F\}$; $k \in \mathcal{N}_\mathcal{M} = \{1,2,.......,N_M\}$; $N_{MUE,ku} = \{1,2,3,...... ,u\mathcal{N}_\mathcal{M}\}$, u $\in$ any large integer value; $N_{FUE,ji} = \{1,2,3,...., i\mathcal{N}_F\}$, $i \in$ any large integer value .

2. The interference on k-th user over the n-th sub-channel are executed as below [3]-

$$I_{k,i}^{f} = \sum_{l=1}^{N_M} P_{i,l,n}^{m}\, G_{i,l,n}^{m} \quad \forall l \in \{1,2,3 \ldots \ldots N_M\} \tag{13}$$

$$I_{k,i}^{\prime f} = \sum_{j=1, j\neq i}^{N_M x N_F} \beta_j^n\, P_{i,j,n}^{f} G_{i,j,n}^{f} \quad \forall j \in \{1,2,3 \ldots \ldots N_F\} \tag{14}$$

$$I_{k,i}^{m} = \sum_{l=1, l\neq i}^{N_M} P_{i,l,n}^{m}\, G_{i,l,n}^{m} \quad \forall l \in \{1,2,3 \ldots \ldots N_M\} \tag{15}$$

$$I_{k,i}^{\prime m} = \sum_{j=1}^{N_M x N_F} \beta_j^n\, P_{i,j,n}^{f} G_{i,j,n}^{f} \quad \forall j \in \{1,2,3 \ldots \ldots N_F\} \tag{16}$$

where $P_{i,l,n}^{m}$ and $P_{i,j,n}^{f}$ indicate the transmit signal powers over the n-th sub-channel of MBS l and FBS j, respectively; $G_{i,l,n}^{m}$ and $G_{i,j,n}^{f}$ indicate the corresponding path gains for MBS l and FBS j, respectively; if indicator function for femtocell resource allocation, If $\beta_j^n = 1$ indicates n-th channel is assigned to femtocell j; otherwise $\beta_j^n = 0$

3. The received signal strength (RSS) is evaluated from PU/MBSU or SU/FBSU at the reference MBS or FBS.

4. Next, the SINR for a PU/Macro user and/or a SU/Femto user are computed.

4 Results and Discussions

The main parameters of the simulation framework are set as shown in Table 1 and separately simulated response of Rayleigh fading component is tested by comparing it with analytical results from Rayleigh fading equation [14] before including it into the networks

In Fig. 2 (a), the simulated PDF from X_x , Y_x samples and analytical PDF from the Rayleigh fading equation are compared and from there the Rayleigh fading expression used in the environment is validated.

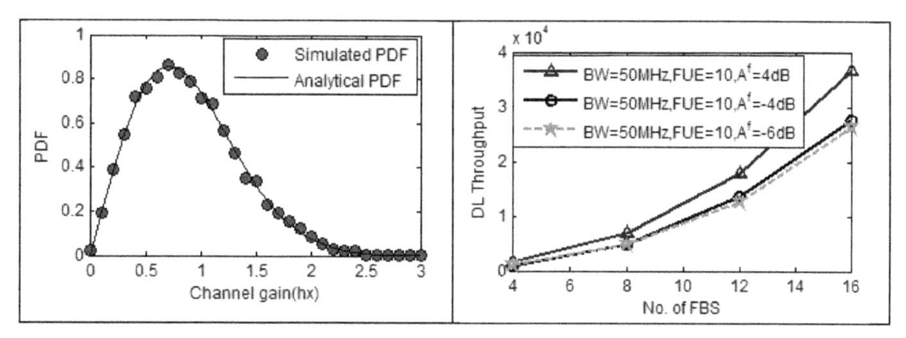

Fig. 2 (a) Variation in PDF of Rayleigh fading with channel gain (h_x) (b) DL throughput of a FUE as a function of number of FBS with $\boldsymbol{A^f}$

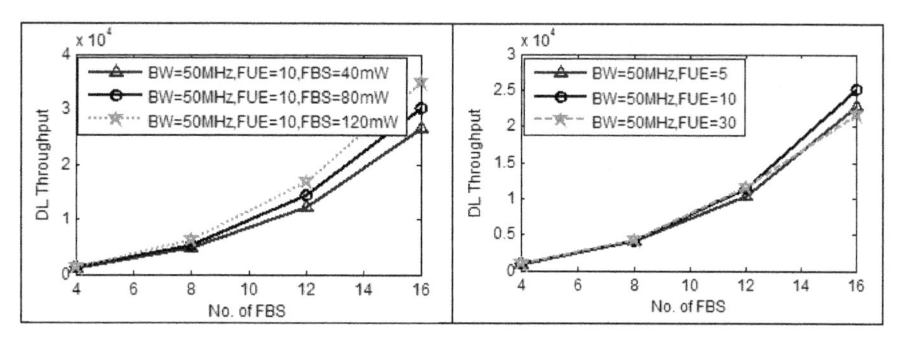

Fig. 3 (a) DL throughput of a FUE as a function of number of FBS with number of FBSs simultaneously transmitting (b) DL throughput of FUE as a function of number of FBS with FUE

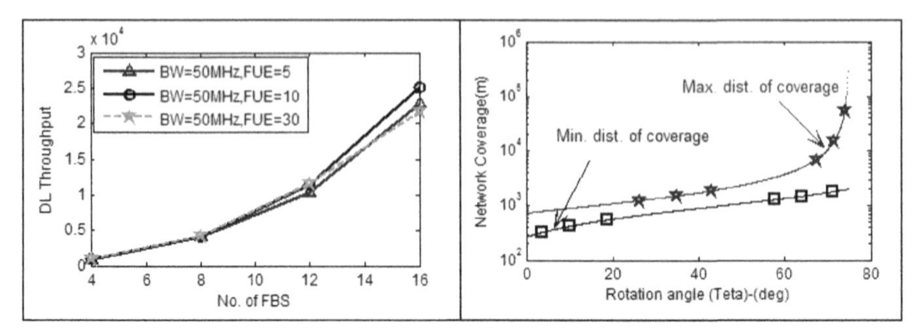

Fig. 4 (a) DL throughput of a FUE as a function of number of FBS with BW of *n*-th sub-channel (b) Network coverage as a function of rotation angle with max. and min. dist. of coverage.

Fig. 2 (b) indicates that as more and more FBSs enter the system, DL throughput increases monotonically. The maximum is 38.2 kb/s in a system with A^f=4dB. Fig. 3(a) & (b) depicts the same information as in Fig. 2 (b). It shows DL throughput as a function of number of FBS, when FBS (T_x) =40 mW and FUE=10 for max. data rate, respectively. The four curves in Fig. 4(a) are segregated by different BW allotment to the *n*-th sub-channel. Fig. 2(b), 3(a) & (b), 4(a), tell us that beamforming parameter (antenna gain) is more effective variable next to BW of *n*-th sub-channel compare to other parameters. Fig. 4(b) shows the relationship of network coverage to rotation of angle for systems with max. dist. and min. dist. of coverage, when node distance=1km and azimuth angle= 30^0 A FUE gets the advantage of optimum network coverage by means of quality of service for the rotation angle more than 60^0.

5 Conclusion

In this paper, we develop a novel simulation testbed model to demonstrate various aspects in terms of substantial parameters in connection with network coverage and DL throughput of a FUE for dual-tier cognitive femtocell networks. Beamforming mechanism with azimuth angle integrated at serving network entity (FBS) aims to reduce user effective interference. There is appreciable impact on network coverage in terms of max. and min. dist of coverage for a FUE. DL throughput is focused in this particular research work to improve network performance by improving instantaneous data rate of a FUE that are generally located at cell edges.

References

1. http://www.smallcellforum.org
2. Saunders, S.R., Carlaw, S., Giustina, A., Bhat, R.R., Srinivasa Rao, V., Siegberg, R.: Femtocells: Opportunities and Challenges for Business and Technology. John Wiley & Sons Ltd. (2009), ISBN: 978-0-470-74816-9
3. Roy, S.D., Kundu, S.: Performance Analysis of Cellular CDMA in presence of Beamforming and Soft Handoff. Progress in Electromagnetics Research, PIER **88**, 73–89 (2008)
4. Ranvier, S.: Path loss models: S-72.333 Physical layer methods in wireless communication systems. Helsinki University of Technology, Smarad Centre of Excellence, November 23, 2004
5. Dong-Chanm, O., Yong-Hwan, L.: Cognitive Radio Based Resource Allocation in Femto-cell. Journal of Communications and Networks **14**(3), June 2012
6. Novlan, T.D., Ganti, R.K., Ghosh, A., Andrews, J.G.: Analytical Evaluation of Fractional Frequency Reuse for Heterogeneous Cellular Networks. IEEE Transactions on Communications **60**(7), July 2012
7. Ahmed, A.U., Islam, M.T., Ismail, M.: A Review on Femtocell and its Diverse Interference Mitigation Techniques in Heterogeneous Network. Wireless Personal Communications 78(1), March 2014
8. Namgeol, O., Sang-wook, H., Kim, H.: System capacity and coverage analysis of femtocell networks. In: Proc. of WCNC 2010 (2010)
9. Lopez-Perez, D., Valcarce, A., de la Roche, G., Zhang, J.: OFDMA Femtocells: a Roadmap on Interference Avoidance. IEEE Communications Magazine, June 2009

Power Budgeting and Cost Estimation for the Investment Decisions in Wireless Sensor Network Using the Energy Management Framework Aatral with the Case Study of Smart City Planning

Muthuraman Thangaraj and Subramanian Anuradha

Abstract Energy engineering study in the field of Wireless Sensor Network (WSN) attracted many researchers in the last decade. The growing interest of researchers has contributed a variety of energy optimization solutions in the field for WSN. There is a need for consolidating all these energy efficiency initiatives at hardware, software, protocol level and algorithmic and architectural corrections and publish them as services for the energy management. The challenge is how this independent energy management framework helps in monitoring, optimizing and coordinating with the energy harvesting units of a typical WSN application bed set up and facilitate the entire energy management. One step further, can the energy management framework, keep track of benchmarks of energy usage for a typical WSN profile and help in recording the operating cost of the WSN application bed in terms of energy is the quest behind this framework Aatral. The independent energy framework Aatral helps not only managing the energy auditing, optimization, harvesting associated with the Wireless Sensor Network but also keeps track of the operating cost, cost estimations and helps in deciding the investments by its special module called Energy Economics Calculator. This paper explains the architecture and design principles of the energy management framework and its functionality of power budgeting, cost estimation, investment decision with a use case of smart city planning with building depreciation sensors, traffic sensors, temperature sensors, intruder detection sensors, monitoring sensors, current leakage sensors.

M. Thangaraj · S. Anuradha(✉)
Department of Computer Science, MKU, Madurai, India
e-mail: thangarajmku@yahoo.com, sa_radha@hotmail.com

S. Berretti et al. (eds.), *Intelligent Systems Technologies and Applications*,
Advances in Intelligent Systems and Computing 385,
DOI: 10.1007/978-3-319-23258-4_16

1 Introduction

Internet of Things (IOT) will be the future network where the small wireless devices with the limited power supply are getting integrated with the social media and the big data platform to take meaningful decisions and control the external environment with the automated operations. Smart homes, industry automation, precision agriculture, wild fire detection are some of the application fields of WSN where wireless node senses the environment factor and propagate the aggregated data to the system, helps in taking decisions and in turn performs the decided operation. For Eg. The temperature sensor senses the temperature and informs the control board as well closes the furnace on a particular degree of temperature. Embedded coding and the artificial intelligent robotic systems with sensor make the world completely automated.

The challenge here is the limited battery in the sensor nodes. The network life time is very critical factor in this kind of sensitive automated operations. Conscious decision on energy optimization using the energy aware techniques in aggregation schemes, processor scheduling for state changes and the routing mechanism can greatly extend the lifetime of the network [1]. But, still we need applications for monitoring and auditing the energy in the wireless sensor network.

It is very obvious that we need the framework for the energy management. Aatral is the energy management framework for the WSN developed in the department of computer science, Madurai Kamaraj University, Madurai for the monitoring, opti-mizing and harvesting the energy in the wireless sensor networks. The scope has been extended for finding the energy variances, energy use indices and the energy peak index. Energy consumption of the node units, network operations, communica-tion protocol layers are recorded with the different time stamps for later references. The energy benchmarks for different WSN profile has been derived as the base for the comparison.

Fig. 1 WSN Application Bed Setup for Aatral

The best aspect of the framework is the energy economics calculator which tracks the operational energy cost.

2 Energy Engineering Systems

The battery voltage, the current consumed at any point of time with any of the nodes is visualized and can be stored with the timestamp. An energy measuring circuit helps in tracking the energy levels of the node. The different energy readings can be compared with the benchmarks and the variances are presented in the form of a graph. Energy Optimization system helps to optimize the energy corrections suggested by the auditing system on the user confirms. The decision making system feature associated with the energy economics calculator helps in evaluating the cost spent on energy and energy improvements. From the historical data, for the different capacity levels, the profitability can be derived and the investment decisions can be made. The energy harvesting system attached can be helpful for getting the extended lifetime from the energy harvested from the external sources. It is the MATLAB/ LABVIEW/Mobile web applications are helping to constitute the presentation layer. LPC expresso IDE is used with an LPC link with the LED display for writing the embedded coding.

3 Programmable City or Smart City

The growing size of the population and conversion of urban and semi urban into smart cities is creating the new issue of planning and maintaining the densely populated cities with all the infrastructures and safety [2]. The technology for democracy principle takes the satellite, mobile and sensors for the day to day affairs or smart living. Social media data, surveillance swipe cards, radar scans, satellite zooms, GPRS systems are known avenues in the smart city projects.

But, the quest to localize the details lead to the installing compound sensors (all in one different types of sensors in one simple sensor unit with more transmission range and highly secured data propagations) in the smart city management projects [3].

How grand it will be to have a single dashboard or control board to have the control over the city happenings and the direct report being sent to the concerned department. The leakages in the drainage and water supply, vigilant surveillance, theft and the traffic in the street are getting automatically updated or informed, to the concerned departments and public with restriction [4]. Multi sensors deployment with various purposes, configuring them and controling the city using the deployed sensors is the second stage.

The first stage is to plan for it and find the setup and variable cost out of it to decide on the benefits. There is no doubt that the information out of city deployed

sensors are going to make the day to day affairs of each department smooth and fast, the city safer . Further, the citizens' waiting time for services is going to be reduced and the city corporations will have the complete control over the city at any point of time [5]. But, the question is whether the operating and maintenance cost can be afforded by the government budget for the service so that it can be covered in the taxation or the concerned department is going to bear the cost of their own information.

In both the cases, what is the threshold limit of usability versus cost. Based on the decision the 50%, 80% or 100% capacity can be planned. Here, one of the major operating cost is the cost of power.

The energy spent on the sensors is to be tracked first for the accurate power cost and then the other possibility of scavenging, optimizing or harvesting is to be con-sidered for the extended life of the smart city sensors. This will help us in greatly for the power budgeting regarding the sensors for future affairs.

The critical factor is initially the smart cities are considered with legacy systems, whereas the data produced out of smart sensors are 2.2 Zetta bytes for smaller city from thousands of servers [5]. So, the platform could only be the biggest data platform with foolproof on cyber security.

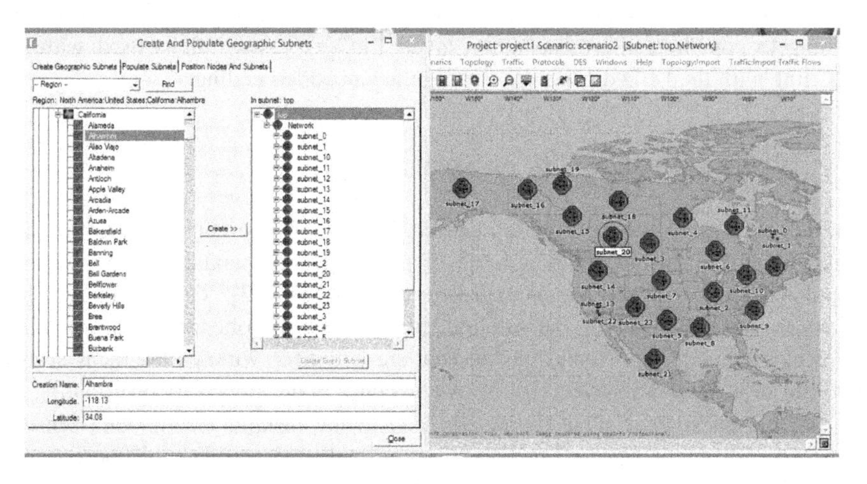

Fig. 2 Smart City Planning

3.1 Challenges in Smart City Projects

(i) Geometrical modelling of the space and regions. Classification of regions into urban, semi urban, city roads,public places, private premises and based on the need deploying the wireless nodes with the adequate sensors [6].

Table 1 Spectrum consideration as per standard

	Urban Area LOS	Urban area NLOS
Band (MHz)	0.2	0.2
Distance D (m)	75	75
Center frequency (MHz)	873	873
Path loss attenuation factor	2	3.5
EIRP (dBm)	14	14
Path loss (dB)	72.17	96.89
Shadowing (dB)	0	10
RX antenna gain	0	0
Average RX power (dBm)	-58.17	-92.89
Noise Power (dBm)	-120.98	-120.98
Noise Factor (dB)	8	8
Implementation loss (dB)	2	2
Required Eb/No (dB)	9	9
Sensivity (dBm)	-101.98	-101.98
Link Margin at 14 dBm EIRP	43.81	9.09
Link Margin at 20 dBm EIRP	49.81	15.09

Here we consider the Line Of Sight (LOS) and Non Line Of Sight(NLOS) both in urban area. When it comes to the NLOS situation, that can occur when a taller building sits between the devices, or when the router is installed below roof level (as installation at the top of the roof is often impossible), ITU (International Telecommunication Union) classically consider an increased but reasonable attenuation factor (3.5) and add a 10 dB of shadowing. All other hypotheses remain similar to the LOS case. Under these conditions and despite the short range used (75 m), the situation is far less comfortable. The link margin is reduced to 9 Db, which is below the commonly accepted value of 15 dB for a reliable link. A margin of 15 dB can only be achieved with 100 MW.

(ii) The telecommunication backbone and the social, technical infrastructure to hold the system in place. Standard bodies to clear the development and instalments in time and in budget.

(iii) Data driven cities lead to the pool proof of big data handling platforms. Power supply to the remotely deployed sensors. The sensor replacement on sensor life time.

(iv) Funding and accountability, as most of the smart city projects are managed by the public private partnership funds which need to be repaid by the service cost. The paying capacity of the users determines what can be digitized

3.2 Factors Considered in Smart City Projects

(i) The combo sensors, the combination of multiple type of sensors is required to monitor different types of environment factors like temperature, pressure, objects movement, humidity, traffic etc.

(ii) The area, data can be classified as per the priority and depends on the priority on the hour the baud rate, propagation delay can be decided.

(iii) The more localized and relevant information is to be provided for the citizens.

(iv) The forecast and projections are always under vigilance, need accuracy and refinement based on the recent factors.

(v) Self charging from the renewable energy, energy harvesting units, energy optimization is considered to improve the communication and routing methods for the extended life of the sensors in the smart city deployment of sensors.

(vi) The threshold and benchmarks are always set as per the latest and relevant statistics and any across limit benchmarks are programmed for immediate escalation by automated report channels and alerts.

(vii) The Green city mission is in the picture to reduce the less carbon-di-oxide emission and installing the zero carbon emission city infrastructure installation is also considered in smart city planning [7].

(viii) Long term sustainability, improved economics, easy operation and maintenance and less man power involvement are taken into account in the design.

(ix) As the sensitive information about the city and the way the city is getting operated is in the data that is propagated by the sensors, security is the main consideration. Device authentication, user authentication algorithms are considered for the security mechanism.

(ix)To reduce the cost as well get benefited by the expertise spread across, usually an open source platform is set by an agreement and bandwidth with multiple application development vendors in the smart city projects. Existing smart city projects are explored as the test bed for the further improvement and planning in the smart city projects [8].

(xi) SMART METERING
Wireless technologies that are used for smart metering include cellular, ZigBee, Wireless M-Bus, WiMax and other mesh radio technologies. Some meters include more than one transmitter. For example, a 900 MHz band transmitter for connection to the monitoring network and a 2.4 GHz transmitter module for connection to wirelessly enable equipment in the home. Figure 2 shows how data from smart meters can be transferred via a long-term evolution (LTE) wireless network to a server that is accessible by the end consumer over the Internet.

(x) Organizing data with multiple variations, velocity and volume in big data platform with decision making engines are the key factors in the design.

(xi) For virtual server management and have an integrated security, the cloud enablement with servers and mirror servers on different geographical locations are considered for the enhanced back up and the improved security.

(xii) The whole system is modeled and designed first as the conceptual framework and implemented in stages with a better plan to decide on the timely installation and usage [9]. Based on the funding, density of population, information required in that area, the bandwidth and the protocols are decided.

(xiii) Existing smart city setup is considered as a test bed for the study and implementation in other cities.

(xv) Sensors are grouped by the function or the geographical location. The sensors propagate data to the nearby or logical cluster head with the sensed data on the regular interval set by the scheduling mechanism. The event happens in the environment also triggers the data propagation. Some of the recent sensors have the data logger – an internal cache of recent recordings. If the sensed data is different from the prior reading average or the change point decided, then only the propagation is considered as the communication of data is the costly affair than the sensing. The cluster head aggregates the data to get the accuracy on the local data. The gateway nodes get the aggregated data from different cluster head siblings under them. The data classification method or the semantic tag association or tabulating data for inference are some of the mechanisms introduced at the cluster head and gateway nodes for getting the meaningful inferences. Semantic descriptions for data, streams, devices (resources) and entities that are represented by the devices, and description of the services. Then comes the step of annotating data, indexing them to get published. The published data can be queried and linked to other sources to give more meaning. The application part helps us to visualize the data.

3.3 Economics and Funding with Smart Cities

The most common applications of the smart city projects are traffic control, environmental monitoring, irrigation, education, accident detection and prevention, crime detection and investigation, security and reporting. The funding is from municipal corporation or from the private sector and now a days government allots the fund in the budget itself for smart cities [10]. Models for early demonstration and deployment of innovative solutions using a grant, guarantee and loan blending mechanism.

1. Project financing
2. Spread share holding
3. Smart bonds
4. Crowd finance
5. Energy performance contracting for energy efficiency

In January 2014, Milton Keynes received a 16 million grant from the Higher Education Funding Council for England (HEFCE) to take forward a Smart City, Big Data (also called MK:Smart) project [11]. Working with IBM, MPD has enhanced its crime fighting techniques with IBM predictive analytics software and reduced serious crime by more than 30 percent, including a 15 percent reduction in violent crimes since 2006. MPD recorded an 863 percent ROI in just 2.7 months, an average annual benefit of $7,205,501. In the Smarter, Sustainable Dubuque Water Pilot Study, IBM technology helped reduce water utilization by 6.6 percent and increased leak detection eight fold [12]. The China Smart City Industry Alliance, this year announced a 50 billion ($8 billion) fund to invest in smart city research and projects [13].

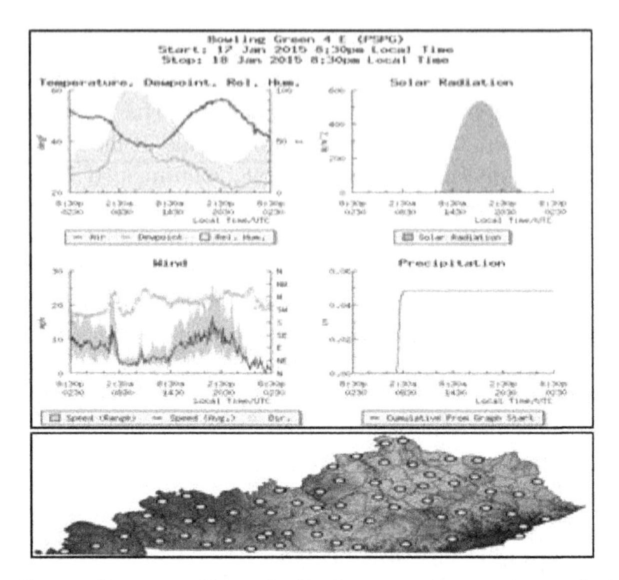

Fig. 3 Warren Country Report on Kentucky Environmental smart city test bed

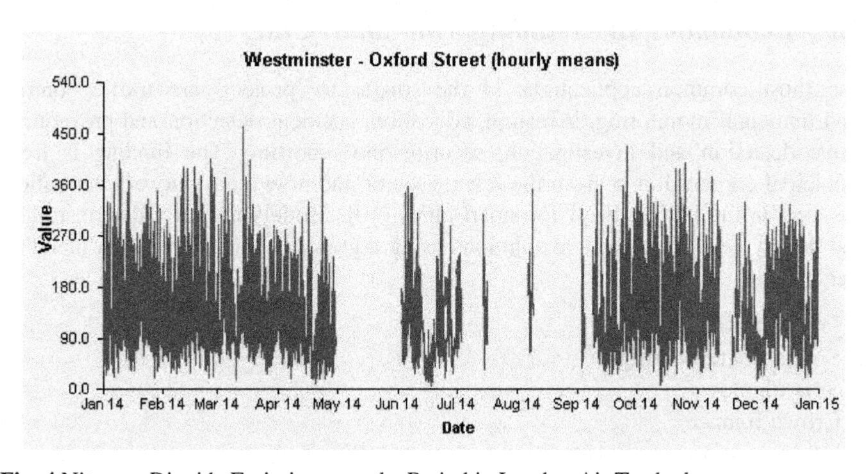

Fig. 4 Nitrogen Dioxide Emission over the Period in London Air Testbed

4 Case Study of New York City

The 13th biggest city in the world is Newyork with 1214 Kilometers square which means, 13067386740 Sq.ft. In the case study of analysis, the factors of study are taken into account are energy, water, assisted living, traffic control, smart parking, goods tracking, in vigilance, environment monitoring. There is need for six types of sensors, namely - temperature, humidity, motion detection, RFID, vigilance camera and traffic sensor.

Based on the baud rate and frequency of the processor, bandwidth of the network links, the node distribution is decided. With the recent trends, transmission range of the nodes is taken into account for the decision of the distance between the sensor unit.

The combo sensors are deployed between 400 square feet, as the transmission range of sensors in recent technological advancement is on an average 440 square feet. Even if the cost of the sensors is reduced to the greater extent, as we are considering the combo sensors, the cost per sensor unit is decided as 100$. So, the total cost of the sensors is around 19 billion dollars.

On an average we get 30 data messages from a node per second. It means 2,592,000 messages per day. With the sensors count decided, the count will be Fifty quadrillion, eight hundred and five trillion, eight hundred and twelve billion, nine hundred and ninety-five million messages per day. If we consider that the messages are of 128 byte size, then the average size of data generated by the sensor setups will be Six quintillion, five hundred and three quadrillion, one hundred and forty-four trillion, sixty-three billion, three hundred and eighty-five million, six hundred thousand.

That means 0.006 Zetta bytes of data per day is 0.18 Zetta bytes per day; taking into account the control packets and query retrievals, the data rate is 2.2 Zettabytes of data per month is estimated for the simple operations in the smart city projects in Newyork [14]. The average storage cost per gigabyte is 0.05$ (if the overall storage exceeds 100 TB). The monthly storage cost is around 12 million dollars.

The archiving and backing up with summarization is to be considered with the desummarization options. The application hosting, services and DB hosting all together create the storage cost of 1.2 billion dollars in the smart grids. The most advanced WiMax, Zigbee, blue tooth technologies and 4G spectrum of wireless communications are available for the M2M, M2H communications [15]. The cost of internet link is by the dedicated line for the IOT of Things in the city.

There are fixed costs of installation and deployment. Every year there is an enhancement considered for the improvement of the smart city projects. There are huge manpower involved in the development, hosting and maintenance of the application. Multiple department and user groups access the system from different devices of smart phones, web or mobile microsites or desktop systems. Based on the traffic of a particular application, variation on load sharing can be introduced in later stages. It is declared in the USA government budget that 4.8 trillion dollars will be considered for a smart city project [16]. Here are the tentative budget and cost sheet work-outs for that.

5 Summary of Findings

On an average, even in the optimized mode the power consumption of the node is around 16KW. One KW costs around 2$, the yearly cost of the power is considered here.The administrative and office cost will be lesser as most of the applications are meant for the concerned department and citizens. The department coordination and labor cost on deployment could vary depends on the country, region [17].The greatest challenge could be the node replacement cost, usually it is considered as part of yearly enhancement and maintenance. The average life of the nodes varies from 4-20 years. The battery charging happens at harvest energy supply or remote charging features. The battery replacement usually happens with 2-3 years. The population of New York city is around 8 million as per 2014 statistics [18]. So, the average per head capita cost will be 14 thousand dollars. If the government grants, taxation, private sectors and some of the department taking the 50 percentages , the **cost per head captia** will be around 30$.

5.1 *Results and Discussion*

The typical power consumption tracking with the indices and the monthly average of consumption is depicted here. It is the average consumption of power of the nodes per day by considering the 100 nodes in the network. The cost not only involves just the power consumption reported, but also the harvested units which supported the system to survive. The power cost per kilowatt is configured as 2$. The power cost along with the other standard operational cost gives the clear picture of the variable cost.

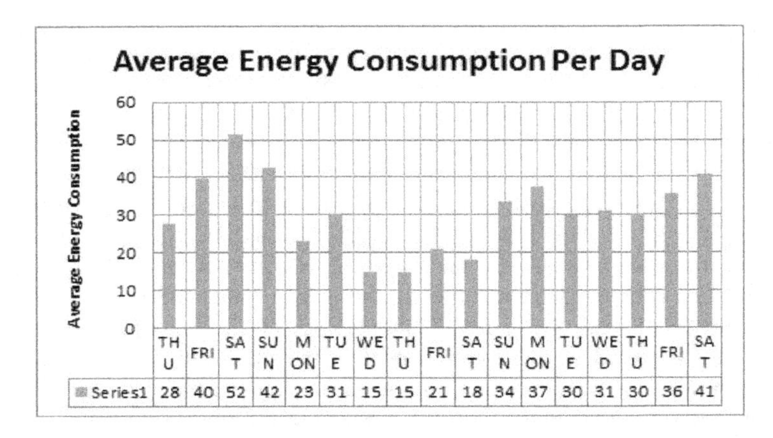

Fig. 5 Average Energy Consumption Per Day

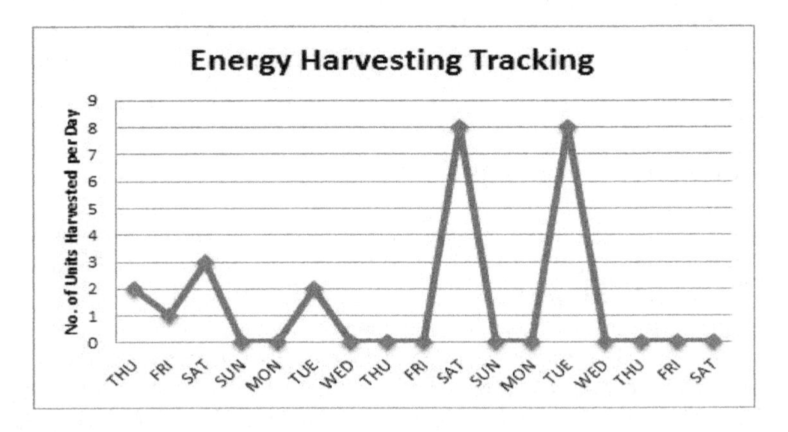

Fig. 6 Typical Energy Harvested Tracking

Smart City Planning for Newyork City					
Sensor Requirements and Cost Derivation					
			400 Sq.ft = 1 Sensor Unit		
			1 Unit = 6 Types of Sensors		
		1 Km Squre = 10763910 Sq.ft			
		1214 KM Square = 13067386740 Sq.ft			
		400 Sq.ft = 6 Sensors			
		13067386740 Sq.fts =	? Sensors		
		13067386740 X 6 X 400	196,010,801 Sensors		
		Total No. of Sensors = 196,010,801 Sensors			
		Cost of 1 Sensor	100$		
		196010081 Sensor Cost =	**19,601,008,100$**		
Operating Cost Sheet					
(A) Fixed Cost					
	Investment Cost(Billion $)				
1.	Sensor Node Cost			19	
2.	Server Cost (IOT subnet servers, sublinks)			10	
3.	Soft ware Development Cost				
	- Application - Web		8		
	- Application - Mobile		1.5		
	- Data base		2		
	- Testing		4		
	- Cloud enables		0.5		
	- internet connectivity estabishment		0.45		
	- Security & Software Licenses		0.12	16.57	
4.	Installation Cost			0.35	
5.	Office Premises Cost (Own Building)			4	
			(A)		49.92
(B) Semi Variable / Semi Fixed Cost (in Billion $)					
1	Hosting Server Maintenance Cost - Own Server			15	
2	General Testing Charges at frequent intervals			0.75	
3	Reporting Cost			0.5	
4	Enhancement Cost			0.25	
5	Office furnishing & Others			2	
			(B)		18.5
(C) Variable Cost (in Billion $)					
1	Staff Salary (Managerial)			10	
2	Internet Usage Cost			2	
3	Hosting Server maintenance - Hire Charges (Hire Server) –Smart Grids			1.2	
4	Sensor maintenance			1	
5	Power & Lighting Expenses			**22.50**	
6	Rent for Office Premises - (Rented Building)			0.25	
			(C)		36.95
	Total Cost		(A) + (B) + (C)		105.37
	10% of Total Cost provided for contingency				10.53
	TOTAL COST				**115.9**

Fig. 7 Power Budget For New York

SMART CITY POWER COST - OPERATION COST TRACKING											
Average Energy Consumption in Kilo Watt			Energy Expenditure in Kw		Battery Replacement	Energy Measures				Average Operation Cost
Max	Min	Avg	Cosumption	Harvested by Solar Panel	Max	Average Residual Energy(%)	Des. Rate in %	EUI	EPI	
38.7	16.7	27.7	27.7	2	1	45	12	0.65	0.172	59.4
41.4	37.5	39.5	39.5	1	0	80	14	0.72	0.92	81
62.4	40.8	51.6	51.6	3	0	78	18	0.55	0.27	109.2
60.1	24.7	42.4	42.4	0	0	68	4	0.92	0.28	84.8
29.4	17	23.2	23.2	0	0	64	32	0.84	0.76	46.4
35.8	25.1	30.5	30.5	2	0	65	23	0.65	0.17	65
25.3	4.2	14.8	14.8	0	1	63	8	0.73	0.28	29.6
28.5	1.6	15	15	0	0	82	4	0.85	0.25	30
28.1	14.2	21.2	21.2	0	0	80	6	0.72	0.76	42.4
31.5	5.2	18.3	18.3	8	0	78	13	0.28	0.92	52.6
45.5	21.5	33.5	33.5	0	0	82	3	0.34	0.54	67
42.9	32	37.4	37.4	0	1	78	62	0.65	0.64	74.8
31.9	27.8	29.9	29.9	8	0	87	44	0.78	0.76	75.8
34.8	26.9	30.9	30.9	0	0	78	23	0.87	0.82	61.8
33.3	27.3	30.3	30.3	0	0	76	67	0.76	0.82	60.6
48.3	23.1	35.7	35.7	0	0	74	45	0.65	0.91	71.4
57.5	23.9	40.7	40.7	0	0	72	33	0.74	0.92	81.4
39.7	21.7	30.7			89	73.53	24.18	0.6882	0.6	
			522.6	0						1093.2

Fig. 8 Typical Power Cost Tracking Sheet

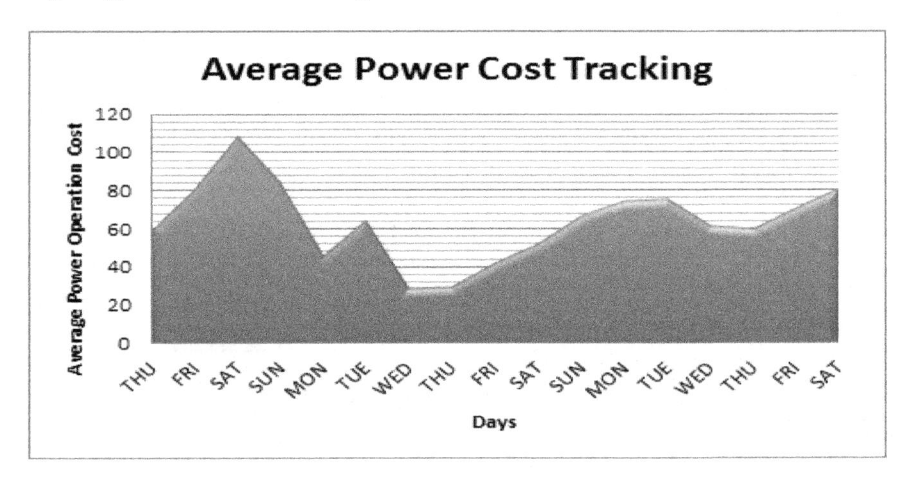

Fig. 9 Power Cost Tracking

6 Conclusion and Future Direction

Planning for the smart city, fund allocation coordination between departments, the working model, application platform and an energy management scheme could be based on the learning curves and statistics collected from the data on the already implemented city models. One of the main variable cost is the cost of the energy or battery and the big data handling cost.

The budget for power with energy harvesting units, energy optimization based on the benchmarks are to be considered in the cost reduction of the power cost. Here the smart city projects create the smart assisted living and reduction, decreases the crime rate, increases the traffic regulation. But, not directly giving any revenue. Even it is unavoidable growth with the current population and city life styles, the cost and funding need clarity because the returns on investment and internal rate of return is not very clear. The government investment strategy needs the techno-financial expert's advice and greater exploration.

With the energy economics calculator feature of Aatral, the following features can be utilized for financial budgeting, cost sheet, cash inflow, out-flow and expert judgement, audit reports on smart city projects.

There are BIGDATA platforms for IOT (Internet Of Things) are available for establishing the better end to end integration and voluminous big data processing. These platforms simplify the connectivity to the devices of IOT, protocol stacks, data processing, data reporting into one platform. Arm mbed and thing speak platforms solves the purpose. Sitewhere, Glass beam, Zetta are some of the other IOT BigData platforms. Efficient data handling of data in WSN with the big data platform invariably improves the energy and gives us the insight on business intelligence of power budgeting. In the future, the BigData platform for IOT will be considered for the smarter WSN setup.

References

1. Chong, S.K., Gaber, M.M., Krishnaswamy, S., Loke, S.W.: Energy-aware data processing techniques for wireless sensor networks: a review. In: Hameurlain, A., Küng, J., Wagner, R. (eds.) Transactions on Large-Scale Data- and Knowledge-Centered Systems III. LNCS, vol. 6790, pp. 117–137. Springer, Heidelberg (2011)
2. Santiago, G., Martnez-Mart, F., Martnez-Garca, M.S.: Embedded sensor insole for wireless measurement of gait parameters. Australasian Physical Engineering Sciences in Medicine (2013). Springer
3. Gabrys, J.: Programming environments: environmentality and citizen sensing in the smart city environment and planning. D: Society and Space (2014)
4. Saiki, S., Nakamura, M, Yamamoto, S., Matsumoto, S.: Using materialized view as a service of scallop4sc for smart city application services. Soft computing in big data processing. Advances in Intelligent Systems and Computing (2014). Springer
5. Cagnetti, M., Leccese, F., Trinca, D.: A smart city application: A fully controlled street lighting isle based on raspberry-pi card. a ZigBee Sensor Network and WiMAX. Sensors (2014)
6. Devigili, F., Andreolli, M., De Amicis, R., Prandi, F., Soave, M.: Services oriented smart city platform based on 3D city model visualization.isprs. In: ISPRS Technical Commission IV Symposium on Annals of the Photogrammetry, Remote Sensing and Spatial Information Sciences (2014)
7. Walker, S., Chourabi, H., Nam, T.: Understanding smart cities: an integrative framework. In: 45th Hawaii International Conference on System Sciences: 2012. IEEE Computer Society (2012)

8. Fu, Z., Lin, X.: Building the co-design and making platform to support participatory research and development for smart city. In: Rau, P. (ed.) CCD 2014. LNCS, vol. 8528, pp. 609–620. Springer, Heidelberg (2014)
9. Liang, X., Li, M., Zhou, J.: Modeling and description of organization-oriented architecture. Journal of Software (2014)
10. Al-Hezmi, A., Elmangoush, A., Steinke, R.: On the usage of standardised m2m platforms for smart energy management. In: The International Conference on Information Networking (ICOIN 2014). IEEE, Thailand (2014)
11. Saint, A.: The rise and rise of the smart city [urban britain]. Engineering Technology (2014)
12. Klauser, F., Paasche, T.: Smart cities as corporate storytelling. City: Analysis of Urban Trends. Culture, Theory, Policy, Action (2014)
13. Huang, D., Zhang, X., Wang, W., He, Z.: Research on service platform of internet of things for smart city. In: ISPRS Technical Commission IV Symposium, Suzhou, China: 2014. The International Archives of the Photogrammetry, Remote Sensing and Spatial Information Sciences: 2014 (2014)
14. Gnay, F., Arsan, T., Kaya, E.: Implementation of application for huge data file transfer. International Journal of Wireless Mobile Networks (2014)
15. Cao, Y., Fang, M., Jin, Y., Rui, G., Wang, M., Tian, W.: Renewable energy and environmental technology. Applied Mechanics and Materials (2013)
16. Reddick, C.G.: Open government opportunities and challenges for public governance. Public Administration and Information Technology. Springer, New York (2014)
17. Christen, P., Perera, C., Zaslavsky, A., Georgakopoulos, D.: Sensing as a service model for smart cities supported by internet of things. Transactions on Emerging Telecommunications Technologies (2014)
18. Remington, P.L., Alghnam, S., Palta, M.: The association be-tween motor vehicle injuries and health-related quality of life: a longitudi-nal study of a population-based sample in the United States. Quality of Life Research (2013)

Part II
Intelligent Distributed Computing

Path Planning of a Mobile Robot in Outdoor Terrain

S.M. Haider Jafri and Rahul Kala

Abstract In this study, we discuss the path planning of mobile a robot using an aerial image. Many times mobile robots are to be deputed to go to far off lands on a mission over uneven outdoor terrains. The aerial image available either as satellite images or produced by aerial drones can be used to construct a rough path for the navigation of the mobile robot. First Gaussian Process Bayesian classifier is used to classify the different classes of terrain. Next each class is associated with a cost denoting the cost of traversal of a unit distance in that particular domain. The costs account for the energy costs, risk of accidents, etc. These numerical values corresponding to each location are called as a costmap, and that array of costmap is passed to the A* algorithm which is a graph search algorithm. The A* algorithm gives the optimal path. The final result is shown in the form of the path over the aerial image with different resolutions and samples of the image.

Keywords Robot motion planning · Outdoor robotics · A* algorithm · Graph search · Gaussian process bayesian classifier

1 Introduction

Robotics is a highly multidisciplinary area of research that attracts people from different domains with a common aim to make the robots operate autonomously. In the area of mobile robotics the most challenging problem is how to navigate from one place to another, using less memory and time with no collision from the obstacles. To fulfill this requirement hardware engineers are working very hard to get processors with very high processing speeds and large amount of capabilities, but after all, hardware areas have lots of limitations. This necessitates the design of algorithms which work well with limited resources.

S.M. Haider Jafri(✉) · R. Kala
Indian Institute of Information Technology, Deoghat, Jhalwa, Allahabad, Uttar Pradesh, India
e-mail: {sayedhaiderjafri6,rkala001}@gmail.com

S. Berretti et al. (eds.), *Intelligent Systems Technologies and Applications*, Advances in Intelligent Systems and Computing 385,
DOI: 10.1007/978-3-319-23258-4_17

Path planning [1-2] is a big challenge in the area of mobile robots as it requires complete information about the operational area in the forms of a map. Through this map the robot can navigate from the initial position to the goal position by avoiding collision while going through the shortest path. In path planning it must be ensured that the algorithm always returns the path if it is available in the map, takes less execution time and the returned path is optimal.

Many times mobile robots are asked to travel on long missions across geographical locations in outdoor, hostile and diverse terrains. It may be suited to send robots on such missions, not requiring humans to travel for prolonged hours in uncertain terrains, further risking their life. In either case it is important to plan the route of the robot. Unlike planning for autonomous vehicles [3] where a structured road-network graph is readily available, here all one may have is a satellite image or an aerial image taken by some drone deputed for surveillance. The images are usually of low resolution and or noisy. The absence of a structured road structure makes navigation a challenging problem. The purpose of this paper is to design a route planning algorithm in such a context.

In this paper it is assumed that an aerial image of the navigation area is available. The source and the goal is assumed to be known. Terrains are marked on the aerial image using a Gaussian Process Bayesian classifier [4]. Each class is associated with a cost denoting the relative expense of navigating through a region of that class. The costmap is reduced in dimensionality by the application of bilinear interpolation. The costmap hence generated is used by A* algorithm [5] to get a path from the source to the goal.

The idea of path planning for mobile robots is not new, a lot of work has been done on using grid cells, but less on real world imagery with a different algorithms. Lee's algorithm [6], was based on breadth–first search algorithm, it was a complete algorithm and guaranteed to find a goal, if it exist but also highly inefficient in terms of memory and time. In the 'Quadtree' cell decomposition approach given by Kambhampati and Davis [7], the map was decomposed into quarter cell. If the decomposed cells are mixed with the obstacle and free cells, it will again be decomposed till a completely free space cell is obtained. After the decomposition, a graph based search algorithm was applied to find the optimal path.. Alexopoulos and Griffin [8] used visibilty graph assuming that all obstacle are polygonal in shape.

Izraelevitz and Carlotto [9] used planning on aerial image. They used a Laplacian operator to detect the road from a map by assigning some threshold value. If the aerial image is noisy then it becomes difficult to detect the road class. To overcome this problem they operated on the whole map by dividing in small regions with an assumption that the small regions are linear. Sofman et al. [10] showed the notion of guiding the ground vehicle through path planning based on an aerial image. They classified the image using neural network, created a costmap, and applied D* algorithm for path planning.

Gaussian process is also widely used in robotics. O' Callaghan et al. [11] classified the operational area of robots into occupied and not occupied. To analyze the capability of Gaussian process classification Bazi and Melgani [12] also classified using the remote sensing image.

Section 2 describes the problem statement and the solution design. Section 3 describes the technique for creating a deterministic costmap. In section 4 we show how to use the deterministic costmap by A* algorithm for path planning . Section 5 presents the simulation results and analysis. Section 6 gives the conclusion remarks.

2 Problem Definition and Solution Design

The problem is to navigate a robot from a specific start point (*S*) to a specific goal point (*G*). We are given an aerial image *I* of the navigation area. It is assumed that the different types of terrains *C* are known in advance i.e. shrubs, obstalces, roads, fields, forests, etc. Each terrain C_i is associated with a cost denoting the expense of navigating through that region, say $cost(C_i)$. Each pixel of the image (x,y) may be representing any of the known terrain classes *C*. The data are assumed to be noisy. The result is a trajectory of the robot τ: $[0,1] \rightarrow R^2$, $\tau(0)=S$, $\tau(1)=G$, which represents the approximate path to be followed by the robot.

In this paper, we use a graph based search algorithm A* [5] to find the best path in a static environment (τ). The novelty of this work that we apply the A* algorithm in the real world by using aerial images of the operational area. To apply the A* we have to pass a costmap, *map*: $R^2 \rightarrow R^+$ to the A*. The costmap denotes the relative cost (denoted $cost(x,y)$) of navigating a unit distance in the area (x,y). Since A* algorithm is a single objective algorithm, all factors contributing to the overall path cost need to be integrated to a single cost value per cell. The factors may include robot safety, fuel economy, navigation speeds, etc. The A* algorithm uses the costs indicated by the costmap as step costs which contribute to the overall path cost of the different competing paths.

To construct costmap first we have to divide the operational area available as an aerial image (*I*) into grid cells $I(x,y)$ and then classify each grid cell into either of the different terrain classes (*C*). Each terrain (C_i) is associated with a deterministic cost value, $cost(C_i)$. Here we have used the Gaussian Process Bayesian classifier [4] to classify the operational area with different number of samples and compared the final result in each sample.

A* algorithm is highly sensitive to resolution. A large resolution results in better results, however the computational expense increases exponentially. Hence we apply bilinear interpolation to reduce the dimensions of the original image and increase execution speed of the software. The overall process is shown in Figure 1.

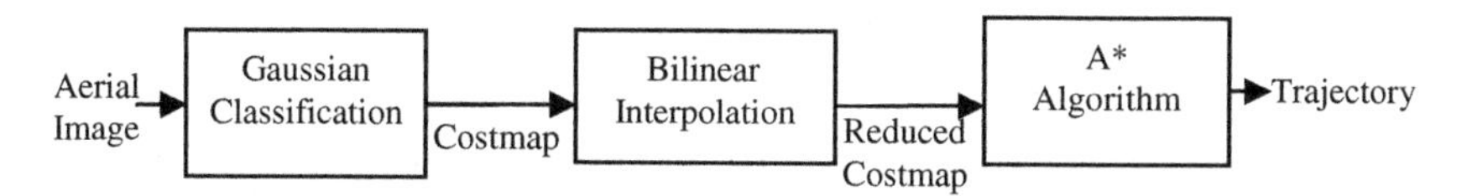

Fig. 1 General Working Methodology

3 Deterministic Costmap Using Gaussian Processes

To create a costmap of aerial image we consider each pixel of the image $I(x,y)$ as one of the grid cells of the operational area. This operational area is classified in either of the different terrain classes C. We assign a scalar cost to each terrain class ($cost(C_i)$) and ultimately the grid ($cost(x,y)$). This cost states the traversing cost of a robot in a particular terrain. As an example if a cell belongs to obstacle class, which the robot cannot traverse, the cost associated will be infinite. Gaussian Processes Bayesian classifier is used to map every pixel to a terrain class and ultimately the cost.

Bayesian classifier is a statistical approach to classify the different classes using the Bayes theorem. To classify the terrain we need to select some samples from each terrain and on the basis this feature we classify into pre-defined classes. This serves as the learning data for the Bayes classifier, based on which the classifier model is made and used for testing of unknown data. Let RGB denote the intensity value of the terrain classes and variables C_i denote the different terrain classes.

Where the symbols have their usual meanings. The model for our specific problem is given by equation (1).

$$P(C_i \mid RGB) = \frac{P(RGB \mid C_i) \times P(C_i)}{\sum_i P(RGB \mid C_i) \times P(C_i)} \tag{1}$$

Here $P(C_i|RGB)$ is the posterior probability of occurrence of class C_i given the intensity level of the pixel as RGB. $P(RGB|C_i)$ is the likelihood of getting RGB intensity level for the class C_i. $P(C_i)$ is the prior probability of getting the class C_i. $\sum P(RGB|C_i) * P(C_i)$ is the total probability of evidence RGB.

Any given input associated with intensity value of RGB is labeled by the class having the largest probability of occurrence. From equation (1), we calculate the posterior probability of pixel with respect to each class and find maximum between them.

Intuitively, Gaussian process [4] extracts the spatial structure from operational area to predict the map. The Gaussian process requires prior knowledge of operational area to calculate the covariance function. The covariance and mean of sample data are required to calculate the likelihood (probability distribution) of the grid cell. The likelihood probability is given by equation (2).

$$P(RGB \mid C_i) = \frac{\exp\left(-\frac{\|RGB - \mu_i\|^2}{2\sigma_i^2}\right)}{\sqrt{2\pi\sigma_i^2}} \tag{2}$$

Here RGB is the input vector, μ_i is the mean RGB intensity value of the class C_i, σ_i is the covariance of sample points of class C_i and $\|.\|$ is the Euclidian norm.

Correspondingly, the value of prior is given by equation (3).

$$P(C_i)=\frac{|samples(C_i)|}{\sum_i |samples(C_i)|} \tag{3}$$

Here $|samples(C_i)|$ is the total number of samples corresponding to class C_i. Using these equations, we calculate the probabilities of each grid cell, compare these probabilities, find the class with the highest probability, label the grid and assign the cost to produce the costmap used by the A* algorithm.

4 Graph Search

The A* algorithm [5] is a graph search algorithm to find an optimal path between a pre-specified source and goal. It is a resolution optimal and a resolution complete algorithm, meaning that both optimality and completeness can only be guaranteed for the current resolution of operation. The higher resolutions result in better paths, however with an alarmingly high computational costs. The algorithm maintains a frontier of nodes which are to be expanded, and a closet set of nodes which have been expanded. Initially the frontier consists of only the source. At any iteration the node with the lowest cost metric is taken from the frontier, expanded and added to the closed set of nodes. All neighboring nodes generated in expansion are added to the frontier if they are not present in closed set of nodes and not already present in the frontier with a better cost value.

The costmap generated from section 3 is first compressed to reduce its resolution using a bilinear interpolation technique. Each grid in the smaller resolution costmap is a state of the A* algorihtm. Each state is assumed to be connected to the 8 neighbouring states by an edge. This makes the complete search graph.

The A* algorithm is associated with three costs. The historic cost $g(n)$ of a node n is the cost incurred in reaching the state from the source. The heuristic cost $h(n)$ is an estimated cost to reach the goal from the node n. This heuristic value is not the actual value of the cost to the goal which can only be computed by computationally expensive searches, it is rather the estimated value. The correctness of the estimate are not guaranteed. The A* algorithm is optimal only when the heuristic is admissible or the heuristic function always returns an estimate lesser than the actual cost to goal. The total cost $f(n)$ is the sum of historic $g(n)$ and heuristic $h(n)$ costs and estimates the cost incurred to reach the goal from the source via node n. The costs for a node n generated from a parent n' are given by equations (4-6).

$$g(n)=g(n')+\|n-n'\|.\text{cost}(n) \tag{4}$$

$$h(n)=\|n-G\|.\min_i(\text{cost}(i)) \tag{5}$$

$$f(n)=g(n)+h(n) \tag{6}$$

Here ||.|| is the Euclidian norm. *cost*(n) is obtained from the cost and denotes the cost of traversing a unit step in the region.

As we consider our operational area as the an array of grid cells, so we can say that each node of the above graph a grid cell. The path computed by the A* algorihtm is in the form of the low resolution costmap representation. The path is scaled up to the original costmap. The path is printed over the aerial image.

5 Results and Analysis

In order to test the algorithm, we took a large number of images from [13-15]. To save space, the results to only one such scenario is discussed. The original image is shown in Figure 2(a). The figure shows an aerial image of an airport area. As it can be seen the scenario consists of obstacles which cannot be traversed by the robot whose cost is taken as infinity; well-built roads which are the easiest and safest to travel around associated with cost 1; and grasslands which can be traversed but with some difficulty and have a cost of 10.

5.1 *Results of Classification*

We used Gaussian Process Bayesian classifier to classify the terrain in different classes. We classified the input aerial image in three classes, i.e. road, grass and obstacle. We randomly sample out 30 samples per class to train the classifier. The variance and mean of each class were noted and used to calculate the likelihood and the prior probability, which was used to classify all points in the image by computing the posteriors. The classified image is shown in Figure 2(b). It can be clearly seen that most of the pixels of the image were correctly classified by the algorithm.

From Figure 2 it can be seen that even though the classification was largely corect, there were some errors with one class being confused with the others due to the noise in the image. These are uncertainties in each class, but if we increase the number of samples from each class we can avoid this uncertainty by compromising with its execution time. The results for higher number of samples are shown in Figure 2(c-d). We can see that the uncertainty decreases with an increase in the number of samples. It can be seen that the uncertainty decreased, as in grass class there are very few marks of black and white and the same condition for road class and obstacle class.

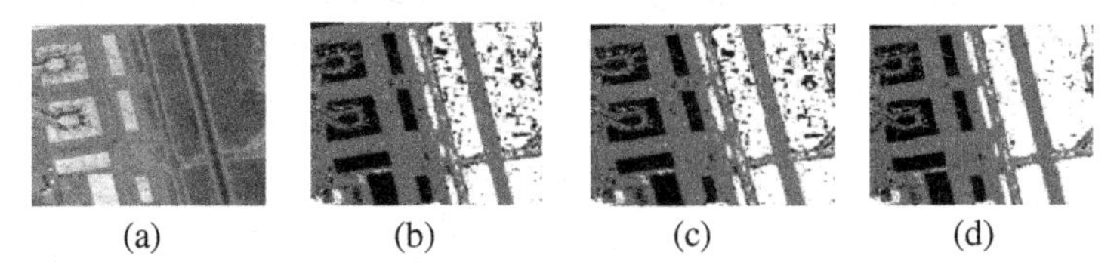

(a) (b) (c) (d)

Fig. 2 Classification Results (a) Input Aerial image, (b) Classified image in three classes, with 30 samples per class black represent obstacles, gray represent roads and white represent grass, (c) 50 samples per class (d) 100 samples per scale.

After classification of the operational area, we make a deterministic costmap denoting the cost of each classified class in the original image. The costmap is of the same dimensionality as the operational area. Each pixel in the image is mapped to a class and correspondingly to a cost which is pre-assumed for every class. The cost map is used by the A* algorithm.

5.2 Results of A* Algorithm

The costmap is used by the A* algorithm for the generation of the path. First the resolution of the costmap needs to be reduced in order to get results in small computational times. The original image is of resolution 512×512, which is converted into a lower resolution costmap using bilinear interpolation. The lower resolution costmap is then used for planning the path of the robot using A* algorithm.

Experiments are performed by using different resolution settings. In general, a higher resolution is associated with a larger computational expense and better paths. Correspondingly the factor of sample size as studied above also plays a similar role, larger number of samples resulting in better classification and better paths, at the same time resulting in higher computational times. To best study the parameters, 3 resolutions were considered which are 25×25, 50×50 and 100×100. For each of these, 1 sampling sizes per class were considered which are 30, 50 and 100. The paths are plotted for each combination of parameters and are shown in Figure 3. Figures 3(a-c) show the results for a resolution of 25×25 with 30 samples, 50×50 with 50 samples and 100×100 with 100 samples.

From Figure 3 it can be seen that low resolutions result in coarser cells of the costmap, visually appearing as unit grids with larger cell sizes. This forces the robot to take large unit steps and the path of the robot cannot be finely tuned by the A* algorithm, hence implying sub-optimality. Similarly larger number of samples result in slightly better classification and hence the paths change a little.

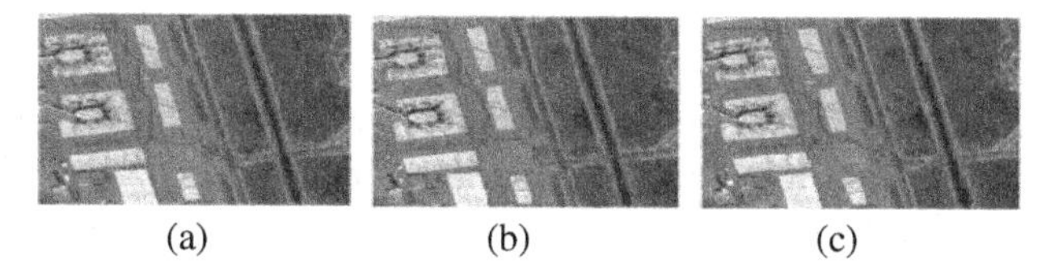

(a) (b) (c)

Fig. 3 Experimental results (a) Resolution: 25×25 with 30 samples, (b) Resolution: 50×50 with 50 samples, (c) Resolution: 100×100 with 100 samples.

Based on Figure 3 it can also be seen that the lower resolutions do not sight the narrow road region connecting the airbase area to the runway area, which should be easier for the robot to navigate also resulting in a smaller path cost. This is the narrow corridor problem in the case of the A* algorithm, wherein the A* algorihtm cannot sight the narrow corridor because the corridor width is larger than the size of the unit grids. But if we increase the resolution of our costmap, this narrow corridor problem can be avoidable as shown in Figure 3(a-c).

The larger resolution, however compromises with the speed of execution. This execution speed is also an important factor, therefore we cannot make the resolution very high. The resolution settings require a tradeoff between the opposing factors of computational time and path optimality. Figure 4 shows the execution time with respect to an increase in resolution.

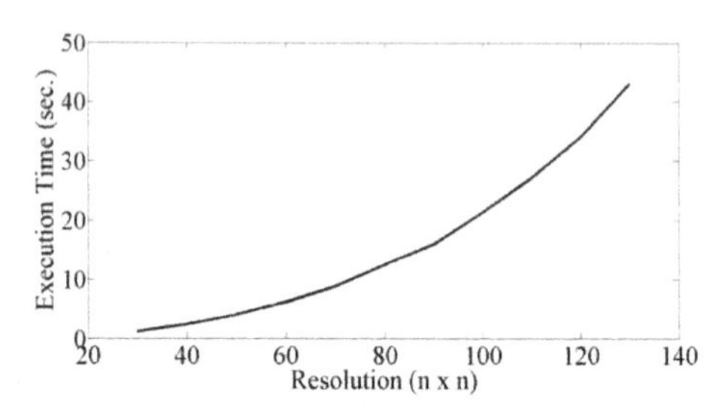

Fig. 4 Plot of execution time of A* algorithm v/s resolution ($n \times n$).

6 Conclusions

In this paper, we presented an approach of constructing a deterministic costmap using an aerial image and path planning using that costmap. The Gaussian process classifier was used to classify the aerial image into different classes. On the basis of classification we produced a costmap for each of the terrain classes. To make a better costmap, it is required that classification be good. It was shown that the costmap improves on increasing the number of samples drawn. This costmap was used by a graph search algorithm (A*) as the unit cost of each node and the graph search algorithm created the best suitable path for mobile robots.

In the future, we will generalize the graph search to accommodate for cost uncertainities, making the search on a probabilistic costmap. We will create a probabalistic costmap stating the probability density function of cost of each node, on which the A* algorithm will operate. Coventional A* algorithm cannot handle probabilities, which is a challenge to address. Further we need to create a dataset of different terrain types to eliminate the need of a human to tag initial few terrain types used for classification. Better modelling of the terrain costs needs to be done. The approach needs to be extended to physical testing on robots.

References

1. Kavraki, L.E., Svestka, P., Latombe, J.C., Overmars, M.H.: Probabilistic roadmaps for path planning in high-dimensional configuration spaces. IEEE Transactions on Robotics and Automation **12**(4), 566–580 (1996)
2. Tiwari, R., Shukla, A., Kala, R.: Intelligent Planning for Mobile Robotics: Algorithmic Approaches. IGI Global Publishers, Hershey (2013)

3. Kala, R., Warwick, K.: Planning Autonomous Vehicles in the Absence of Speed Lanes using an Elastic Strip. IEEE Transactions on Intelligent Transportation Systems **14**(4), 1743–1752 (2013)
4. Rasmussen, C.E., Williams, C.: Gaussian Processes for Machine Learning. MIT Press, Cambridge (2006)
5. Nilsson, N.J.: Principles of Artificial Intelligence. Tioga Publishing Company, Palo Alto (1980)
6. Lee, C.Y.: An Algorithm for Path Connections and Its Applications. IRE Transactions on Electronic Computers **EC-10**(3), 346–365 (1961)
7. Kambhampati, S., Davis, L.: Multiresolution path planning for mobile robots. IEEE Journal of Robotics and Automation **2**(3), 135–145 (1986)
8. Alexopoulos, C., Griffin, P.M.: Path Planning for a Mobile Robot. IEEE Transactions on Systems, Man, and Cybernetics **22**(2), 318–322 (1992)
9. Izraelevitz, D., Carlotto, M.: Extracting road networks from low resolution aerial imagery. In: Proceedings of the SPIE Visual Communication and Image Processing Conference, Cambridge, MA (1988)
10. Sofman, B., Bagnell, J.A.D., Stentz, A.T., Vandapel, N.: Terrainclassification from aerial data to support ground vehicle navigation. Technical Report No. CMU RI-TR-05-39, Robotics Institute, Carnegie Mellon University, Pittsburgh, PA (2006)
11. O'Callaghan, S., Ramos, F.T., Durrant-Whyte, H.: Contextual occupancy maps using gaussian processes. In: Proceedings of the 2009 IEEE International Conference on Robotics and Automation, Kobe, Japan, pp. 1054–1060 (2009)
12. Bazi, Y., Melgani, F.: Gaussian Process Approach to Remote Sensing Image Classification. IEEE Transactions on Geoscience and Remote Sensing **48**(1), 186–197 (2010)
13. Dai, D., Yang, W.: Satellite Image Classification via Two-layer Sparse Coding with Biased Image Representation. IEEE Geoscience and Remote Sensing Letters **8**(1), 173–176 (2011)
14. Xia, G.S., Yang, W., Delon, J., Gousseau, Y., Sun, H., Maitre, H.: Structrual high-resolution satellite image indexing. In: Proceedings of 2010 ISPRS, TC VII Symposium, Part A: 100 Years ISPRS - Advancing Remote Sensing Science, Vienna, Austria (2010)
15. Yang, W.: High-resolution Satellite Scene Dataset. Signal Processing and Modern Communication Lab, Wuhan University. http://dsp.whu.edu.cn/cn/staff/yw/HRSscene.html (accessed September 18, 2014)

Forensic Framework for Skype Communication

M. Mohemmed Sha, T. Manesh and Saied M. Abd El-atty

Abstract Skype is a secure internet telephonic application which establishes connection between its clients through a peer-to-peer architecture. The connection between Skype client to its server and other clients uses an encrypted channel that uses Transport layer Security (TLS) protocol. At the same time, connection between Skype client and Public Switch telephone Network (PSTN) gateway is accomplished through unencrypted digital channel using Voice over Internet Protocol (VoIP). The encrypted channels in the Skype communication make forensic analysis frameworks to work badly in decrypting the traffic and procuring critical forensic details of the network stream against intruders and cyber criminals. Furthermore, policy violations and unbound usage of Skype VoIP communication over PSTN users waste the network bandwidth. Here we propose a sophisticated Skype forensic framework that collects forensic information by decrypting the Skype client-server communication along with recreating voice content in the Skype to PSTN VoIP communication. We also propose an efficient packet reconstruction algorithm powered by time stamping technique for regenerating malicious content from payloads of the Skype network stream followed by supporting prosecution of policy violators and cyber criminals in the court of law.

Keywords Skype Forensic Analysis · TLS · Skype VoIP · Packet reordering · Pcap file

M. Mohemmed Sha(✉) · T. Manesh(✉) · S.M. Abd El-atty
Department of Computer Science and Information,
Prince Sattam Bin Abdul Aziz University, P O Box. 54, Al-Kharj 1199, Saudi Arabia
e-mail: {sahalshas,maneshpadmayil,sabdelatty}@gmail.com

T. Manesh
Shankara Research Centre in Information Science,
Adi Shankara Institute of Engineering and Technology, Kalady 683574, Kerala, India

S.M. Abd El-atty
Faculty of Electronic Engineering, Menoufia University, Menouf 32952, Egypt

S. Berretti et al. (eds.), *Intelligent Systems Technologies and Applications*,
Advances in Intelligent Systems and Computing 385,
DOI: 10.1007/978-3-319-23258-4_18

1 Introduction

Skype VoIP technology now play an important role in today's communication arena with much convenience to our daily activities. It is completely a decentralized and distributed ad-hoc communication on internet. It works similar way as many file sharing networks. Skype is very common among VoIP user because of its three main services [2]. At first, Skype allows VoIP communication and media streams between Skype clients after proper authentication with the Skype server. Second, its instant messaging service allows two or more Skype clients to exchange real time text messages. Third, Skype clients allows to initiate and receive calls via regular telephone numbers through VoIP-PSTN gateways.

1.1 Challenges of Skype Forensic Frameworks

Now Skype client uses strong encrypted channels for establishing connection between its server through Transport Layer Security (TLS) and with other Skype clients through Secure RTP (SRTP) for media streams. As a result, most of the forensic analysis tools and framework fails in getting forensically important credentials and regenerating the actual content to trace out the cyber criminals and policy violators (individuals or users in LAN) that misuse Skype platform to exchange their malicious contents. The rise of Skype VoIP has also led to challenges for traditional law enforcement interception, since there is no single physical point of interception (like a telephone exchange) where call traffic can be centrally identified and monitored.

1.2 Our Contribution

Our major goal towards forensic analysis of Skype communication involves the extraction and decryption of Skype handshake connection between suspected Skype client and Skype server to trace out the credentials such as IP and Port numbers used for communication along with malicious user details. We are successful in decrypting the forensically rich Skype handshake connection followed by regenerating the voice data between Skype client and PSTN user which is over an unencrypted channel. In this paper, we are not focusing on regenerating the content between two Skype clients as it uses dual encryption using TLS and SRTP for encrypting both SIP messages and RTP media streams as part of forensic analysis. Rather, we present a novel forensic framework which decrypt the Skype's initial SSL handshake connection to trace forensic details against malicious and policy violators that use Skype VoIP over PSTN and also successful in recreating the original Skype VoIP sessions. We also propose an efficient packet reconstruction algorithm powered by time stamping technique for regenerating malicious content from network stream. Proposed framework truly depends on our own packet capturing mechanism. Previous works depends on packet capturing third party softwares like Wireshark. Our mechanism is able to capture and save more than 100000 packets via wireless and LAN adapters.

Rest of the paper is organized as follows; section 2 deals with related work, section 3 describes architectural framework of Skype communication, Section 4 unveils experimental setup for proposed Skype Forensics, Section 5 provides results and discussions of the proposed framework, Section 6 outlines our future works, Section 7 gives conclusion followed by references.

2 Related Work

Many of the previous work deal with the analysis and behavior patterns of Skype communication and collects forensic details from the available network packet streams. Here we outline the important previous works in Skype protocol analysis which inspired us with necessary basic information to introduce our new methodology to decrypt Skype SSL handshake traffic between Skype client and server followed by session regeneration between Skype VoIP client and PSTN user.

Baset S. A. et al. (2006) presented analysis of Skype login, NAT and firewall traversal, call establishment, media transfer and codecs etc. This work helped us to study the entities involved and Skype VoIP sessions. [1]

Chun-Ming Leung, et al. (2007). Presented an approach for detection, blocking and prioritization of Skype traffic in enterprises and ISP networks. This work helped us to various stages involved in Skype communication [3].

Sandor Molnar, et al. (2007) us with various identification methods to detect several types of communications initiated by Skype clients. This work helped us in understanding the Skype communication between clients and its UDP relations [4].

Ronald C.D. et al. (2008) proposed a fingerprint analysis of Skype installation on client system by accessing registry and file system locations. This work aided us with tracing login process and file systems used by Skype Client [5].

Rossi D. et al (2009) presented the working of Skype signaling, voice streams and settings of codec etc. This work helped us to formulate strategies to analyze Skype VoIP streams and identification of necessary codecs. [6].

Gao Hongtao (2011) proposed forensic methods involving VoIP crime followed by analysis on host computer. This work aided us to design proposed forensic method for procuring digital evidence against and malicious use of VoIP infrastructure [11].

Ahmad Azab et al. (2012) presented the latest Skype version and how its components are analyzed, in terms of network traffic behavior for logins, calls establishment, call answering and the change status phases. This helped us to understand call processing in Skype VoIP infrastructure [15].

3 Architectural Framework of Skype Communication

Here we outline the major entities and components which take part in proposed forensic investigation of Skype communication. Key components and functionalities involved in Skype communication are well described in previous works [1, 2, and 3]. Here we skeleton the how a Skype client (supposed to be malicious) initiates connection between its server and PSTN user.

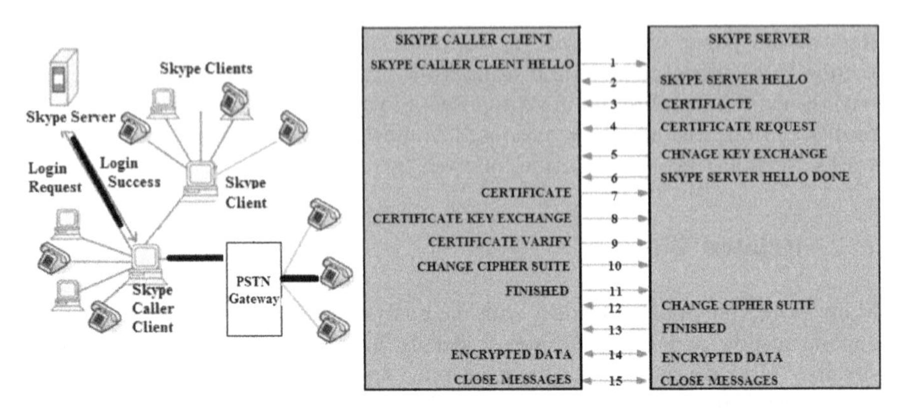

Fig. 1 Skype communication network **Fig. 2** Skype Handshake Connection

Essentially before starting communication through Skype, the Skype caller client login to Skype server for authentication. Once login process success, Skype caller client selects a PSTM user from its contact list and initiates contact request through available VoIP infrastructure. This contact request is propagated by Session Initiation Protocol (SIP) all the way over peer-to-peer overlay network till it identifies PSTN user. This session layer SIP protocol make use of transport layer User Datagram Protocol (UDP) for switching and propagation. Once the PSTN user is located, Skype client establish a Real Time Transfer Protocol (RTP) which is a basic TCP connection [14]. The proposed work analyzes the network traffic between Skype caller client to Skype server and Skype caller client and PSTN user as shown in figure 1 with bolded black lines.

3.1 Skype Caller Client

A Skype caller client is an end user who uses computer system installed with Skype. He possess an authorized account in the skype.com.

3.2 PSTN User

A PSTN user is an end user who uses telephone or mobile over PSTN. A Skype client listens on particular ports (which may vary from 1024 and 65535) for incoming calls by maintaining a table of other online high bandwidth Skype clients (Skype Super Node) called a host cache.

3.3 Working Ports Used in Skype

A Skype client opens a UDP port at the login with Skype server and a TCP port for connection registration and authentication process. Thus, Skype client opens TCP through port 443 which is used to for SSL encrypted traffic between Skype caller client and Skype server.

3.4 *Working of Skype's Transport Layer Security*

Secure socket Layer (SSL) technology provides security to the transport layer. SSL is an enterprise security standard adopted by millions of websites to safeguard their online transaction. It ensures the privacy and integrity of transmitted data during the transaction process. Skype also make use of TLS-SSL for encrypting both SIP messages and RTP media streams between Skype caller client and server [14]. As described in the figure 1, the connection request for startup will be initiated by Skype caller client using an unencrypted SIP message which propagated normally through UDP packet stream to accomplish Skype TLS handshake process. Once the TLS handshake is completed, connection between Skype caller client and Skype Server will be encrypted. So it is necessary and crucial to extract and dissect Skype TLS handshake process. From the handshake process, we get important forensic information like session key (used for establishing SSL session and is calculated from private key of the certificate used for secure connection) used for a particular Skype traffic, cipher suite and protocol version used for session, Skype client and server random numbers, premaster secret, Skype client's key exchange message and its public key etc. The Skype SSL handshake parser in the experimental setup of Skype forensic analysis shown in figure 3 well describes how these critical details are dissected and extracted from network packets. Public and private key is used to establish a secure channel using SSL. The SSL handshake processes between Skype caller client and server are shown in figure 2. These handshake messages till step 10 are exchanged using unencrypted as plaintext using Session Initiation Protocol (SIP). The same has been successfully observed in our experiments. These unencrypted SIP messages make use of port no 5060 or 5061 for the transfer. Its filtration and extraction are well described as part of our experimental setup as shown in figure 3. Skype uses RC4 encryption for Handshake exchange and AES for exchange data. Our experiments are on the Skype version 6.14.0.104.

4 Experimental Setup for Proposed Skype Forensics

The proposed framework shown in figure 3 is supposed to perform offline network forensic analysis related to Skype client to PSTN user VoIP traffic. It recreates the actual voice data between Skype client and normal user over PSTN. The specific Skype traffic will be captured and monitored when some suspicious network usage or communication is reported in proxy controlled LAN environment. The precaptured network session is then fed through various forensic analysis stages in our proposed framework for the finding evidences against any illegal network activity or malicious communication [12, 17]. The ultimate goal is to provide sufficient evidences to allow the perpetrator to be prosecuted by the court of law. Major modules in the proposed work are well described in following sections.

4.1 Forensic Data Mining Module

This module is supposed to collect the available forensic information of the network interface card (NIC), wireless card or system log file whichever is applicable in forensic environment along with network packets stored as Pcap files.

Pcap File: Pcap file is collection of network packets saved with file extension "<filename>.pcap". Packets can be captured through the proxy servers or through NIC of suspected individual's computer system or wireless cards. The "Pcap" file is main source of input to our forensic framework. This Pcap file undergoes series of changes and stages in our framework.

4.2 Network Session Reconstruction Module

This is the heart of the investigation as it recreate original network session and content analysis from an encrypted network traffic between Skype client server and unencrypted VoIP traffic between Skype Client and PSTN user. This module traces source and destination IPs and port numbers involved in a specific traffic and decrypting the Skype traffic between client and server followed reconstructing actual content (credentials and voice data) from packets. Major activities performed at each module are well described in following sections.

Packet Analysis Strategy (Filter IP, TCP and UDP Packets): This module filters all the internet Protocol (IP) packets. From these IP packets, analyzer further filters all TCP and UDP packets. Since Skype make use of UDP stream both for handshake between Skype Client and Server and VoIP transfer to users over PSTN. We developed our own packet filters with help of Java and Jpcap library functions.

Protocol Tracer (Source and Destination IPs and Port Numbers): Protocol tracer traces packet parameters like IPs, port numbers involved in filtered session consist of UDP streams. It fetches all combinations of IPs and port numbers in the specific Skype traffic. Protocol analyzer traces SSL encrypted SIP between Skype client and server and RTP between Skype client and PSTN gateway.

Separate Encrypted UDP Stream (Skype Client to Server): Skype uses port number 443 for incoming and outgoing encrypted SSL for SIP based UDP stream. Once it filters all the SSL packets through source port 443, It filters all encrypted SIP packets with source port 5060 and destination port 5061.These SIP packets are then saved as Pcap file for further processing by certificate handler and TLS Parser. The Skype SSL traffic between client and server is decrypted by novel Skype SSL analyzer that we have developed. This SSL analyzer is new version towards our previous work [17] which was successful in decrypting HTTPS network traffic.

Separate Unencrypted UDP Stream (Skype Client to PSTN): This section filters all unencrypted UDP packets with application layer RTP Protocol. These unencrypted packets follow VoIP conversation between Skype Client PSTN users. These RTP packets are filtered against the ports generally between 10000 and 20000. Thus to identify the useful ports, this section conducts a recursive filtering to identify most frequently used port number for RTP VoIP traffic. Once RTP packets are filtered, they will be saved in another separate Pcap file for further processing by RTP analyzer.

Skype TLS Parser: Primary objective of TLS parser is to extract the protocol version, cipher suite, random numbers of server and client, encrypted PreMasterSecret etc. from the Initial Skype handshake communication [10]. As shown in figure 2, the first two steps agree on TLS version and list of supported cipher suites exchanged between Skype client and Skype server. In the step 3, Skype server provides Skype client with its x.509 certificate for verification which allows Skype client to transfer a secret to the Skype server. Through step 4, Skype client send a certificate to Skype server for mutual authentication. In the step 5, the Skype server sends client key exchange method consists of public key information of the server. In step 8, the Skype client send a certificate key exchange which consists of Premaster Secret with its public key as requested by Skype server in Step 7. The step 8 is the most important among all other steps as the traffic will be protected depending on the security of PreMasterSecret [17].

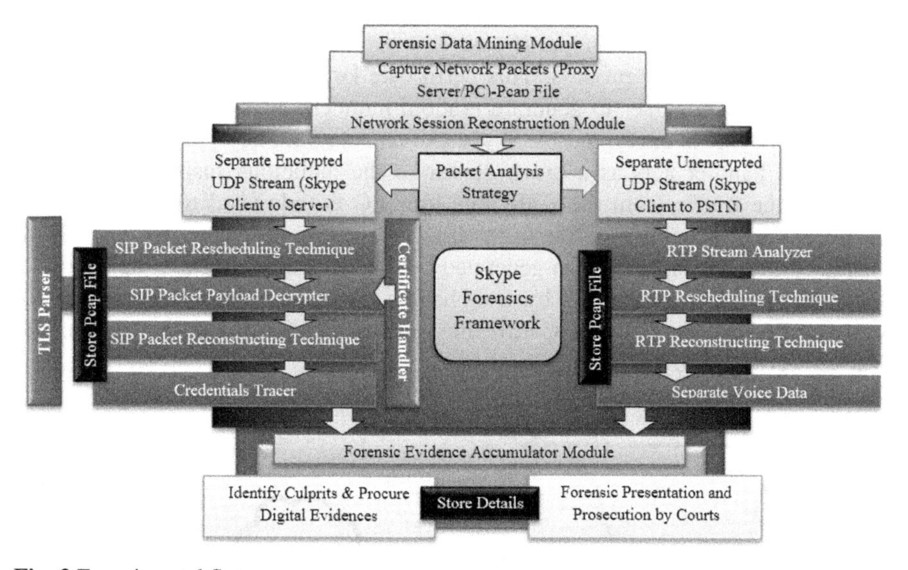

Fig. 3 Experimental Setup

Once both the Skype client and Skype server agree on PreMasterSecret, further communications will be in encrypted mode using public and private key parameters presented using certificate and PreMasterSecret. Extraction of key exchange messages containing public and private key pairs of Skype Server and Skype Client along with certificate management are implemented using our novel TLS Skype traffic parser that we have developed after through study of various procedures involved in decrypting SSL Skype traffic. TLS parser accomplish above mentioned procedures by reading the precaptured Pcap file containing the separated encrypted network session from Skype client to server. From this encrypted UDP streams, TLS parser extracts unencrypted network packets with SIP protocol through port number 5060 or 5061. Then it traces "Skype Client Hello" message by reading the each packet's header part using the specially designed function created using Jpcap library. The same way it also identifies "Skype Server Hello". Once it locates these hello messages, it extracts the client random and server random numbers used to create MAC keys as mentioned in the Certificate handler.

Certificate Handler: The certificate handler is designed for processing Skype SSL certificate exchanged as part of initial TLS handshake process. Authentication and encryption of Skype handshake process is accomplished with help of Transport Layer Security (TLS). Implementation of Skype TLS uses X.509 secure key exchange [10]. Public key certificates and validation is defined with aid of x.509 public key infrastructure. The PreMasterSecret identified and extracted in step 8 of Skype SSL handshake using TLS parser. SSL make use of MAC keys which are used for encrypting the SSL channel. These MAC keys are created from PreMasterSecret, Skype server's and client's random numbers present in Certificate Key exchange message in the step 8 of Skype TLS handshake process. This certificate key exchange message is being extracted by TLS parser to generate random numbers. It is well mentioned that PreMasterSecret is encrypted by Skype Server's private key [10, 17]. Proposed certificate handler acts as a proxy server to the malicious suspected user PC. Thus all the traffic from the malicious user PC is redirected to the certificate handler. Now actual certificate from SSL server is collected by the Certificate handler while returning its own generated certificate. Thus certificate handler will catch Skype TLS handshake by establishing two SSL connections separately between Skype client to the Certificate handler and Certificate handler to the Skype Server as shown in figure 4.

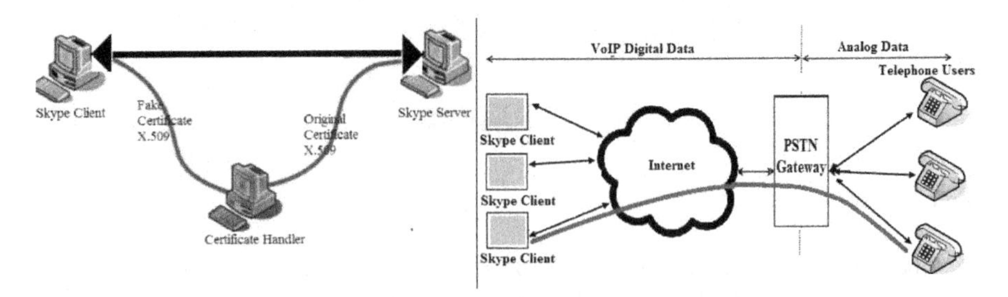

Fig. 4 Skype Certificate Handler

Fig. 5 Skype client to PSTN gateway

Decoding PreMasterSecret of SSL Handshake: The PreMasterSecret is decrypted with help of Certificate that we have extracted from the Skype handshake communication. The private key and certificate X.509 are bounded together using PKCS12 file format [10]. We identify this file and extract the private key by an exhaustive search of packet payload using the hex code equivalent of the texts "BEGIN PKCS12" and "END PKCS12" throughout the handshake stream. We then dissect this file to extract private key along with determining the algorithm used for encoding the PreMasterSecret from the X.509 cipher suite. Thus we decrypt the PreMasterSecret.

Expansion of Key Material Used in TLS: Expansion of Key material is critical step in identifying the keys used to encrypt the traffic [10,17]. Key materials are expanded using the PreMasterSecret along with Skype client and server numbers using client key exchange message extracted from the handshake packet stream. Master key is derived using message digest algorithm (MD5) using PreMasterSecret, Skype client and server random numbers. Where secure hash represents secure hash algorithm (SHA-1). Once Master key is derived, key blocks are created using PreMasterSecret, Skype client and server random numbers. This is followed by generating Skype server and client MAC keys, write keys etc. This process is shown in the following figure 6.

SIP Packet Rescheduling Technique: Now our framework reorders network packet stream saved as Pcap file based using the packet rescheduling algorithm that is well described in the section. This is a prior step before performing the decryption of Skype handshake stream.

SIP Packet Payload Decryption: Once SIP stream is reordered, payload of each packets are decrypted using necessary keys derived in above section. All the decrypted packets are again saved as Pcap file for the reconstruction of all the packets to generate initial stages of Skype handshake communication to identify the IP addresses, parameters and credentials of suspected user PC.

RTP Stream Analyzer: This process is done in order to analyze the unencrypted network packet stream (Skype Client to PSTN) so as to extract the details about the streaming media and its parameters. The figure 5 shows the connection between a Skype client and a telephone user in PSTN user. The connection between Skype client and PSTN gateway will be accomplished through VoIP protocol which is obviously an unencrypted digital connection [9, 13]. The communication path which we consider is marked with red line in the figure. At this stage we are not worried about how call is managed in PSTN network. Rather we collect all VoIP packets at the Skype client with to reproduce the actual communication as voice data.

Tracing the Ports Used in RTP: Initial Skype handshake connection is analyzed to identify the ports used by the Skype client for RTP transfer [13]. Once ports are identified, all such VoIP packets are analyzed and extracted. The details about the ports will be described in the SDP messages attached to the INVITE and STATUS 200 OK messages [9]. RTP Stream Analyzer extract these messages from the packets and a portion of the parsed "INVITE" and "STATUS 200" information is shown figure 7. From the above parsed messages, the forensic analyzer traces the IP addresses and ports involved in the VoIP conversation. The message part "m=audio" followed by the number gives the ports assigned by that client for that particular communication. The "From" and "To" fields contains a 4 digit number followed by '@' symbol this gives the IP address of the Skype client who is performing the communication. Since the forensic analyzer starts filtering the RTP packets used for the communication using IPs and ports using UDP stream followed by successive reordering of RTP stream.

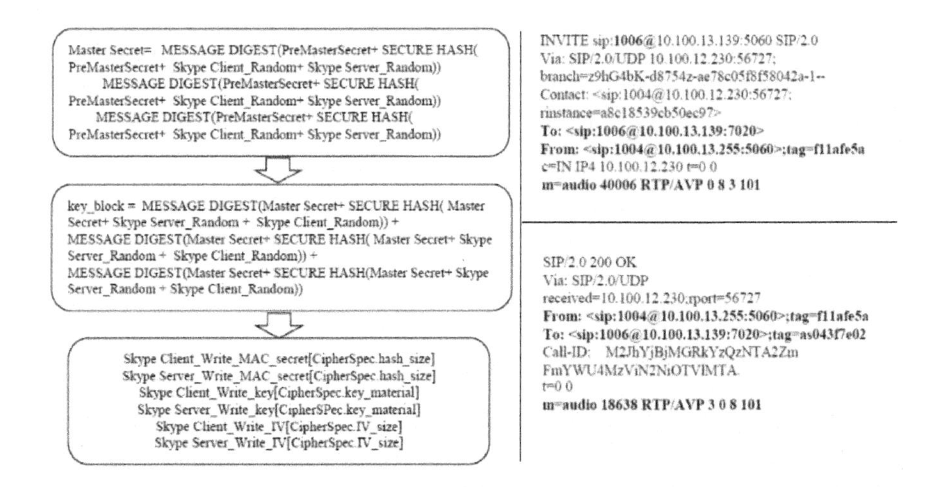

Fig. 6 Key Expansion Process

Fig. 7 INVITE, STATUS OK Messages

RTP Rescheduling and Reconstruction Technique: This technique reschedule the unencrypted RTP stream using all combination of IP addresses and RTP ports. Once it finishes the RTP rescheduling, the payload and header parts of RTP stream will be separated followed by combing RTP payloads [7, 12]. The payload section contains the voice data, our RTP reconstruction technique will extracts data from the packet and write it in to the temporary Pcap file for future analysis works. Thus the actual voice data is reconstructed. The reconstructed stream is then forensically stored for further activities by forensic presentation module of the proposed framework. This activity is accomplished with help of UDP packet reordering algorithm given below.

Algorithm 1. Separate duplicate and retransmitted packets
Input: Pcap file, Source and Destination IPs, Source and Destination ports
Output: Pcap file of retransmitted packets

```
Initialize next=1, time_seq=0;
While packet!= null do
    Read the packet from the Pcap file
    if packet= EOF,  Filter the packets based on IP and Ports end if
    if next.flag then seq=packet. time_seq, seq2=packet. time_seq-seq end if
      While  next==1 do // this is the first packet
          current=seq2, next =current + packet.datalength end while
       if seq2>=next, next=seq2+Packet.datalength
          else Write that packet to the temporary Pcap file
end while
```

Algorithm 2. Packet reordering
Input: Pcap file, Source and Destination IPs, Source and Destination ports
Output: Reordered Pcap file,

```
Initialize next=0, time_seq=0,
While packet!= null do
  Read the packet from the Pcap file
  if packet= EOF or null then exit from the loop
    if packet.source ip==source_ip and packet.source port=source port
     if next.flag is set seq=Packet. time_seq, seq_relative=packet. time_seq-seq
end if
    end if
    if next=0, Write reordered file, next=time_seq relative+packet.datalength end
if
      if seq_relative= next do following two steps, Write packet to the reordered
file,
        next=time_seq relative+packet.datalength  end if
      if  seq relative >next, Read each packet from the temporary Pcap file
        if packet.time_seq-seq=next, write the packet to the reordered file,
          next=next+packet.datalength end if
    end if
end while
```

Algorithm 1 performs the extraction of duplicate and retransmitted packets and store it as a temporary Pcap file. Algorithm 2 performs the reordering of packets. In the proposed algorithm, "next" indicates next packet to be processed, "time_seq" timestamp generated while saving the packets. "packet" represents current packet. "EOF" indicates end of the packet. "seq_relative" indicates next expected time stamp. "datalength" gives length of packet. "next.flag" indicates offset value in UDP packets. Here we provide overall structure of procedures. While implementation, we assume necessary packet parameters required. These two of algorithms reorder Skype VoIP stream between all combinations of IPs and port numbers even though Skype uses port randomization. Once we reorder the packets, its payload and header parts are separated followed by combining the payloads of packets to regenerate the content.

4.3 Forensic Evidence Accumulator Module

During this final stage, all the forensic information about the Skype client and VoIP network stream will be processed to ensure and locate the culprits involved in any malicious activities. This module not only collects and organizes the forensic details, but also process the actual digital evidences such as voice data from the VoIP communication etc. It also will generate the forensic report which involves forensic details like, IP address of Skype client, port numbers, username of the user, time of network activity with procured digital evidences. At this stage, the suspected user or users undergo investigation by the cybercrime police to evaluate the forensic report. Once cybercrime police confirms malicious activity and culprits, they will be prosecuted by local judiciary system as per the IT laws in a specific country.

5 Results and Discussions

This section delivers account of significant achievements with its GUI developed for forensic analysis of VoIP communication through Skype clients. We also incorporates performance analysis of proposed UDP packet rescheduling and reconstruction technique with existing frameworks of similar kind. The developed framework is entitled as Skype Forensic Analysis Framework (SFAF).

5.1 Skype SIP Decrypter

This main process of proposed work which display the decrypted SIP traffic between Skype client and server. It consists of hex and ASCII codes as displayed in the figure 8. For each file the hex code details starts with the file name and the number starts with 0000 and each line contains 16 bytes of the data. This hex code is providing the pattern of the data in the corresponding file. A part of decrypted Skype initial handshake which revels the name of the computer entity (marked red line) used for Skype activity is also shown in the following figure 8.

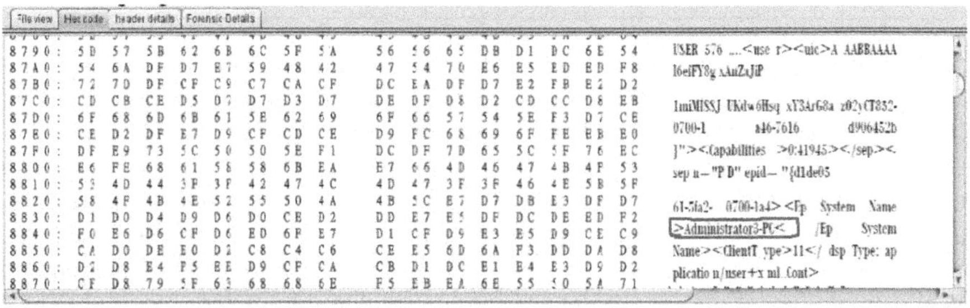

```
File view | Hex code | header details | Forensic Details

8790:  5D 57 5B 62 6B 6C 5F 5A   56 56 65 DB D1 DC 6E 54
87A0:  54 6A DF D7 E7 59 48 42   47 54 70 E6 E5 ED ED F8
87B0:  72 7D DF CF C9 C7 CA CF   DC EA DF D7 E2 FB E2 D2
87C0:  CD CB CE D5 D7 D7 D3 D7   DE DF D8 D2 CD CC D8 EB
87D0:  6F 68 6D 6B 61 5E 62 69   6F 66 57 54 5E F3 D7 CE
87E0:  CE D2 DF E7 D9 CF CD CE   D9 FC 68 69 6F FE EB E0
87F0:  DF E9 73 5C 50 50 5E F1   DC DF 7D 65 5C 5F 76 EC
8800:  E6 FE 68 61 58 58 6B EA   E7 66 4D 46 47 4B 4F 53
8810:  53 4D 44 3F 3F 42 47 4C   4D 47 3F 3F 46 4E 5B 5F
8820:  58 4F 4B 4E 52 55 50 4A   4B 5C E7 D7 DB E3 DF D7
8830:  D1 D0 D4 D9 D6 D0 CE D2   DD E7 E5 DF DC DE ED F2
8840:  F0 E6 D6 CF D6 ED 6F E7   D1 CF D9 E3 E5 D9 CE C9
8850:  CA D0 DE E0 D2 C8 C4 C6   CE E5 6D 6A F3 DD DA D8
8860:  D2 D8 E4 F5 EE D9 CF CA   CB D1 DC E1 E4 E3 D9 D2
8870:  CF D8 79 5F 63 68 68 6E   F5 EB EA 6E 55 50 5A 71

USER 576 ....<use r><uic>A AABBAAAA
I6eIFY8g xAnZxJiP

1miMISSJ UKdw6Hsq xY3ArG8a z02yCT832-
0700-1 a46-7616 d906452b
]"><Capabilities >0:41945></sep><
sep n-"P D" epid- "{d1de05

61-5fa2- 0700-1a4><Ep System Name
>Administrator3-PC< /Ep System
Name><ClientT ype>11</ dsp Type: ap
plicatio n/user+x ml Cont>
```

Fig. 8 GUI of Skype SIP Decrypter

5.2 VoIP Header Extractor and Forensically Regenerated Files

The header extractor as shown in figure 9 provides forensic details about the initial handshake connection established between the Skype client and the VoIP user. This header details contains the phone number of the client who initiates the call and accepting the call and many other important information about the call including the details of the "codec" used and all. Forensically regenerated VoIP communication through all available ports is shown in figure 10.

5.3 Packet Reordering and Performance Comparison

Efficiency of the proposed packet reassembly and reconstruction technique in our work is compared with other widely used tools. The other tools are not purely used for forensic analysis, but used for packet analysis. We used Wireshark, Ethercap and Packetizer tools. Our framework, SFAF is flexibly providing more forensic details with less time (shown in figure 11) than the other tools which is mainly due to the proposed fast UDP packet reordering process powered by time stamping procedure The tools like Wireshark and Ethercap etc. could collect and dissect Skype VoIP packets only and does not support any Skype VoIP stream reconstruction. The result produced better yield in the management of retransmitted and duplicate packets in Skype VoIP sessions which are crucial in reconstructing VoIP sessions as shown in figure 12. Wireshark and Ethercap do not support Skype VoIP session reconstruction facility but manages duplicate and retransmitted files.

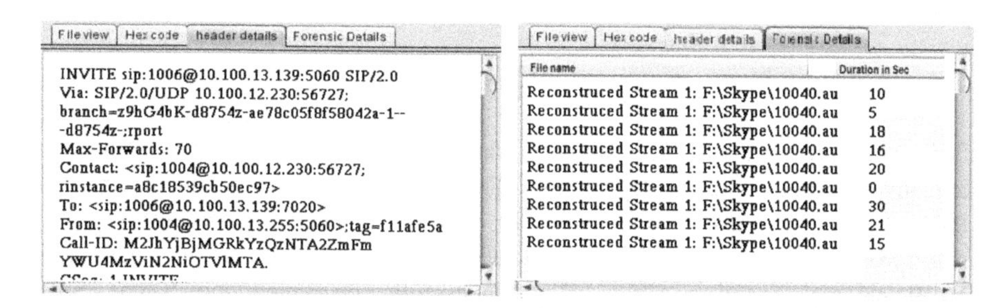

Fig. 9 GUI-VoIP Header Extractor

Fig. 10 GUI- Regenerated Details

Our work thus not only produce better Skype VoIP session reordering but also the Skype VoIP session reconstruction with less time and huge number of retransmitted (shown inside brackets) along with actual no. of packets as shown in figure 12 shows its credibility in finding VoIP digital crimes.

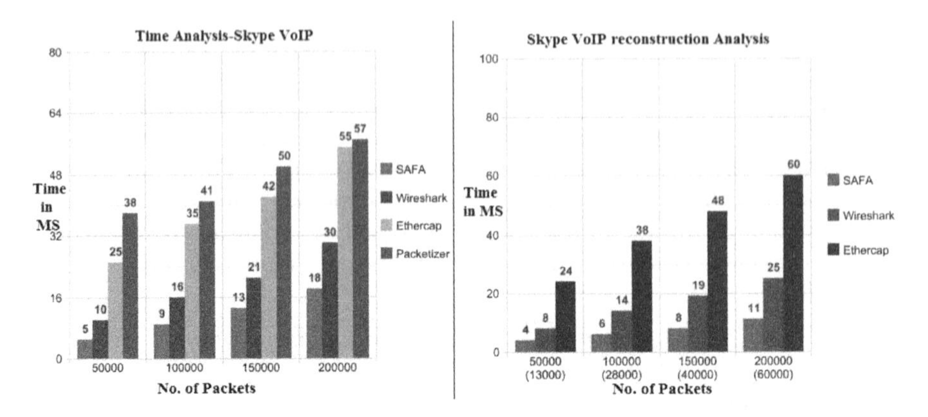

Fig. 11 Time Analysis for Skype VoIP

Fig. 12 Reconstruction time for Skype VoIP

6 Future Work

As described earlier, The Skype and its latest versions now use SSL v3.0 (TLS) for encrypting Skype client to server SIP communication. After authentication Skype client establish connection with other clients though SRTP for encrypting media such as audio and video etc. We are currently focusing on working of this dual security using TLS and SRTP for subsequent decryption of Skype network stream. This would enable forensic investigator to completely regenerate the content present in Skype communication which provide more credible forensic evidences against cyber criminals and network policy violators.

7 Conclusions

This work implements a user-friendly framework for network forensic analysis and investigation Skype VoIP Streams. The concept of model is conceived from the process of retrospective network analysis or network forensics. This framework collects forensic information about sources of any unauthorized Skype VoIP network activities. We have successfully decrypted the Skype SSL handshake to locate and confirm cyber criminals forensically. Once the source of such activity is traced along with the substantial regenerated content, it is possible to prosecute the malicious user involved in that session by the cyber police and court. Results shows that new packet reconstruction mechanism which is developed to address duplicate and retransmitted packets is functioning commendably. Our future work involves complete Skype communication analysis by decrypting SRTP communication between clients.

References

1. Baset, S.A., Schulzrinne, H.: An analysis of the skype peer-to-peer internet telephony protocol. In: IEEE INFOCOM 2006, pp. 1–11, April 2006
2. Guha, S., Daswani, N.: An an experimental study of skype peer to peer VoIP system. In: IPTPS 2006, pp. 10–16, February 2006
3. Leung, C.-M., Chan, Y.-Y.: Network forensic on encrypted peer-to-peer VoIP traffics and the detection, blocking, and prioritization of skype traffics. In: 16th IEEE International Workshops on Enabling Technologies, pp. 401–408, June 2007
4. Molnar, S., Perenyi, M., Gefferth, A., Trang, D.D.: Skype traffic identification. In: IEEE Global Telecommunications Conference, pp. 399–404, November 2007
5. Ronald, C., Dodge, J.R.: Skype fingerprint. In: IEEE Proceedings of the 41st Annual Hawaii International Conference on System Sciences, pp. 484–452, January 2008
6. Rossi, D., Mellia, M., Meo, M.: Evidences behind skype outage. In: IEEE International Conference on Communications, pp. 1–6, June 2009
7. Tinta, S.P., Wong, J.L.: Characterizing end-to-end packet reordering with UDP traffic. In: IEEE Symposium on Computers and Communications, pp. 321–324, July 2009
8. Molnar, S., Perenyi, M.: On the identification and analysis of skype traffic. In: International Journal of communication systems, pp. 97–117, April 2010
9. Alshammari, R., Halifax, N.S.: An investigation on the identification of VoIP traffic: case study on Gtalk and Skype. In: IEEE ICNS-10, Management, pp. 310–313, October 2010
10. Wu, L., Duan, H.-x.: SSL-DP: a rootkit of network based SSL and TLS traffic decryptor. In: 2nd IEEE CTC-Workshop, pp. 29–33, July 2010
11. Gao, H.: Forensic method analysis involving VoIP crime. In: IEEE Fourth International Symposium on Knowledge Acquisition and Modeling, pp. 241–243, October 2011
12. Manesh, T., Brijith, B., Singh, M.P.: An improved approach towards network forensic investigation of HTTP and FTP protocols. In: PDCTA-11, Springer Heidelberg, pp. 385–392, September 2011
13. Irwin, D., Slay, J.: Extracting evidence related to VoIP calls. In: AICT Conference, pp. 221–228. Springer, Heidelberg, June 2011
14. TLS and SRTP for Skype Connect. June 2011 http://download.skype.com/share/business/guides/skype-connect-technical-datasheet.pdf
15. Azab, A., Watters, P., Layton, R.: Characterizing network traffic for skype forensics. In: IEEE Third Cybercrime and Trustworthy Computing Workshop, pp. 19–27, October 2012
16. Korczynski, M., Duda, A.: Classifying service flows in the encrypted skype traffic. In: IEEE International Conference on Communications, pp. 1064–1068, June 2012
17. Manesh, T., Brijith, B., Bhraguram, T.M., Rajaram, R.: Network forensic investigation of HTTPS protocol. In: IJMER, vol. 3, no. 5, pp. 3096–3106, October 2013
18. Sinam, T., Lamabam, P., Ngasham, N.: An efficient technique for detecting Skype flows in UDP media streams. In: IEEE ICANTS-13, pp. 1–6, June 2013

Probabilistic Graphical Modelling for Semantic Labelling of Crowdsourced Map Data

Musfira Jilani, Padraig Corcoran and Michela Bertolotto

Abstract Concerns regarding the accuracy of crowdsourced information limits its usage for several real world data-driven applications. In this paper we present a novel methodology for automated semantic prediction of street labels in crowdsourced maps. Toward the goal of finding best labels for streets, we use an undirected graphical model to capture three properties: the initial street labels given by the crowd as prior knowledge, the geometrical features of streets, and the inherent spatial relationships existing between streets in a network. Using the structural support vector machine paradigm a potential function is learnt on this model that jointly optimizes over the street labels in the entire network. We evaluate our methodology on the OpenStreetMap data for London and show that our model can predict 8 different street type labels with an accuracy of almost 90 percent. Our approach is more robust and improves upon the previous work where streets were assumed to have an independent and identical distribution.

1 Introduction

Crowdsourcing is increasingly being seen as an effective means of generating useful knowledge [17]. It may be defined as a special case of collective intelligence that emerges from joint efforts made by groups of people connected through the internet. For example, Amazon Mechanical Turk provides a cheap and quick solution for a variety of problems such as obtaining data annotations. Several online gaming portals tap the potential of the crowd for image labelling which otherwise is a challenging task [18]. Google does an implicit crowdsourcing when it uses its search engine database to make seemingly accurate predictions for a range of events, from stock market prediction to detecting epidemics like influenza [16]. And last but not least,

M. Jilani · P. Corcoran(✉) · M. Bertolotto
University College Dublin, Dublin, Ireland
e-mail: musfira.jilani@ucdconnect.ie, {padraig.corcoran,michela.bertolotto}@ucd.ie

S. Berretti et al. (eds.), *Intelligent Systems Technologies and Applications*,
Advances in Intelligent Systems and Computing 385,
DOI: 10.1007/978-3-319-23258-4_19

is the example of Wikipedia where several thousand people from across the world have produced an extensive, high quality encyclopaedia.

Considering the geospatial domain, the OpenStreetMap (OSM) project is one of the most successful examples of crowdsourcing and forms the case study for our research. The project started in London in July 2004 with an aim of creating a free, editable map of the entire world [2]. The major contributors of OSM data are common people; anyone having a GPS equipped mobile device can map their surrounding and upload this to the OSM project via internet. However, as these contributors are not professional cartographers, questions arise regarding the credibility of the OSM database. While the geomtery of map objects mainly comes from GPS trajectories, the associated semantics are almost completely crowdsourced.

Street networks form an integral part of our daily lives and as such they occur as an important entity in the OSM database. A fundamental attribute of any street is its semantic type such as motorway, residential etc. In this paper we present a novel methodology to automatically predict the semantic street type labels of OSM streets. Figure 1 shows a small part of this network consisting of five different street types. Secondary streets are shown in a red colour whereas tertiary are represented in a blue colour. The yellow segments correspond to service streets and residential streets are represented using a green colour. The black segments correspond to unclassified streets (see Table 2 for street type definitions). We derive the following observations from this figure. First, the semantic type information of a street is implicit in its geometrical properties. For example, the residential streets are smaller in length, have several dead ends, and less linear as compared to secondary streets. And second, there are inherent spatial dependencies between various streets. The figure shows several residential streets and no motorway is observable in their vicinity. Previous OSM data quality assessment research has shown that the initial semantic labels given by the crowd can be trusted to a reasonable degree [15, 28]. Our solution toward learning and predicting the street type semantics encodes these observations and statistics. We pose the problem as an undirected graphical model such that the initial labels given by the crowd, observed geometric features of the considered street, and spatial dependencies between various streets can be explicitly defined. The model parameters are learnt using a Structural Support Vector Machine. Finally, Gibbs Sampling Markov Chain Monte Carlo method is used for inference.

Our model considers eight semantic street types. The OSM database for London was used for modelling and testing. The results obtained are positive and improve upon the previous work. Almost 90 percent accuracy was obtained for all the street types considered.

The rest of the paper is structured as follows. In Section 2, we review some of the work describing the quality of OSM data. In Section 3, we first describe the data representation in the OSM database. This is followed by a description of multi-granular representation of the street networks used in this research. The proposed probabilistic graphical modelling framework is presented in Section 4. Section 5 details the experiments performed and the results obtained. Finally, in Section 6 we draw some conclusions and provide an insight for future work.

Fig. 1 A small part of OSM street network showing streets with five different semantic types. Secondary: red; tertiary : blue; residential: green; service: yellow; unclassified: black.

2 OpenStreetMap Data Quality

OpenStreetMap is a crowdsourced spatial dataset. The quality assessment of OSM has recently become a hot topic within the Geographic Information Sciences community. Haklay [20] studied the geometric accuracy and completeness of OSM data for England by comparing it with the official maps provided by the Ordnance Survey of Great Britain. Girres and Touya [15] extended Haklay's work to the French OSM dataset and performed a comprehensive analysis of the data quality by comparing OSM data with BD Topo and other official datasets. Cipeluch et al. [9] compared the Irish OpenStreetMap data with Google Maps and Bing Maps. Helbich et al. [21] studied the positional accuracy for a German city by referencing to TomTom and other official survey datasets. These studies provide some useful insights into the quality of OSM Data. Haklay [20] , for example, showed that the geometric accuracy of OSM Data can be considered fairly accurate with an average positional displacement of about 6 meters, and that there exists almost an 80 percent overlap of motorway objects between the OSM and official datasets. However, this approach of assessing the quality of OSM data by comparing it with reference data suffers from a number of drawbacks. Firstly, owing to the high costs of mapping involved, official datasets are often outdated. Secondly, the very beauty of crowd-sourced geographic information lies in recording even that information which may be logistically impossible to capture by official mapping agencies. This may include mapping minor roads or adding important semantic information which may be very useful for map generalization, data integration, and data analysis. Finally, the task of comparing an official dataset against the one obtained from heterogeneous sources is often daunting and not very efficient [28].

Several attempts have been made to provide an analysis of the quality of OSM data by adopting some conceptual approaches. Goodchild [19], for example, proposed three mechanisms for assuring the quality of geographic information contributed in a volunteered fashion such as OSM. These include crowd-sourcing, social and geographic approaches. Van Exel et al [24] proposed the use of *crowd quality* as an indicator of OpenStreetMap data quality. Mooney and Corcoran [25] and Barron et al [10] explored the OSM history data as a potential indicator of data quality. Kepler and Groot [11] proposed the concept of using *trust* as a proxy measure for determining the quality of OSM data.

Semantic accuracy of OSM data is not a well studied problem as compared to its geometric accuracy and completeness. Ballatore and Bertolotto observed that OSM data is spatially rich but semantically poor [4]. Their research has focused on the extraction of information which is implicit in the OSM data descriptions and on linking it to other online resources [5]. To the best of our knowledge, there exist only a few works [8, 23] where a machine learning approach has been used for assessing the quality or predicting the semantics of map objects. However, all of them assume map objects having an independent and identical distribution (*iid*).

3 Data Preprocessing

In this section, we first provide a description of the OpenStreetMap data. We explain how different geographic entities are stored inside the OSM database and the taxonomy for storing their properties or attributes. Later, we explain the multi-granular representation used in this research for representing the OSM street networks.

3.1 OpenStreetMap Database

OSM data consists of nodes, ways, and relations. All point entities are represented by a node which also stores the latitude and longitude coordinate information. Linear or polygonal entities are represented by means of ways. A way basically consists of a list of ordered nodes. A closed way (a way with the first and last nodes being the same) represents an area (or polygonal) entity. Relations contain information on how two or more entities are related to each other.

Along with the coordinate information of geographic objects, OSM database also stores properties (or attributes) of various entities by means of tags. Tags in OSM are stored as (key,value) pairs. For example, building=terraced. Here building is the key that represents a class of map data entities and terraced is the value for this class. Any number of tags, including user-defined, can be associated to describe an object.

Streets in the OSM database are generally tagged using the key ‘highway’ [3]. Other keys such as ‘name’, ‘ref’, ‘maxspeed’, ‘oneway’ etc. are also present as metadata for a given street. For example, a residential street known as ‘The Palms’ maybe tagged as highway=residential; name=The Palms; ref=R101; maxspeed=30 km/h; oneway=yes.

3.2 Multi-granular Street Network Representation

Appropriate data representation is necessary to conduct meaningful experiments. In this section we will describe the multi-granular representation proposed by Jilani et al. [22] for representing the OSM street networks. We use this representation as a data preprocessing step for our probabilistic graphical modelling framework (Section 4).

A small street network is shown in Figure 2. The names and types for the street segments in the network are given in Table 1. Consider the street 'Hammingway' consisting of four street segments namely A,B,C, and D. The geometric features for this street and its spatial context becomes more meaningful when it is considered as a group of four segments and not as distinct segments. The multi-granular street network representation enables us to define a street as a group of segments having same name and type.

The multi-granular representation is a graph based representation that builds upon the two popular street network representations namely the primal [26] and the dual [27] representation. The corresponding multi-granular representation for Figure 2 is shown in Figure 3. This is a two layered graph structure where every node in the top layer of the graph is a subgraph in itself. Adjacent street segments having same name and type in a given network (Figure 2) are identified as sets of street segments that form a single node in the top layer of the multi-granular graph. By using the multi-granular representation the geometric and topological properties of sets of street segments become explicit and therefore easy to extract.

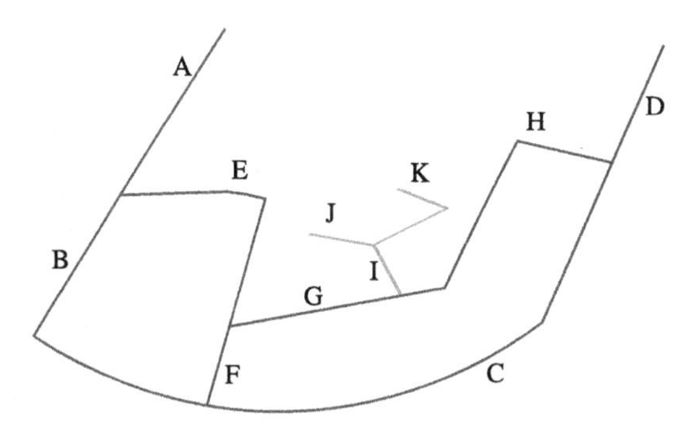

Fig. 2 A section of street network with the names and types of the segments(A-K) as shown in Table 1

4 Probabilistic Graphical Modelling

We now describe our approach for predicting the semantic type of streets in crowd-sourced maps. Using the multi-granular representation described in the previous section, we group street segments on the basis of name and type similarity to form a

Table 1 Names and street types associated with street segments of the network shown in Figure 2

Street Segments	Name	Type
A, B, C, D	Hammingway	Secondary
E, F	Seamount Drive	Tertiary
G, H	Hallows Road	Tertiary
I, J, K	Hallows Road	Residential

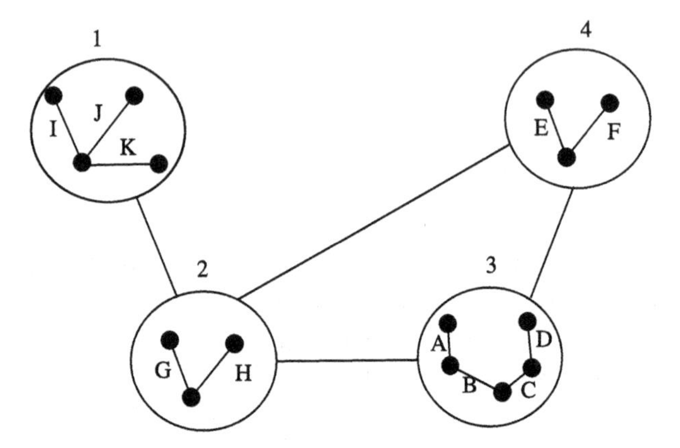

Fig. 3 The multigranular representation of the street network shown in Figure 2. Each vertex of the graph is a subgraph in itself.

street. These streets comprise the basic units in our model that we want to label. Our goal is to develop a model that incorporates the following properties:

Crowdsourced Labels as Priors: The success of various crowdsourcing platforms and previous quality assessment research on OSM data suggests that the initial semantic type labels given by the crowd to streets can be considered reasonably accurate. Hence, we want to capture this information into the model as a prior for the true label value of a street.

Geometrical Features: Different streets have their own characteristics features. For example, a motorway is generally long whereas a residential road is generally short. We compute three geometric features of streets that are suggestive of their semantic type. These include length, number of dead ends, and linearity. Our goal is to be able to use these features for determining street labels.

Spatial Context: Spatial relationships between various streets are inherently present in a street network. Certain streets spatially occur in a specific configuration. For example, motorways tend to be directly connected to trunk roads and not residential roads. We want to encode these spatial relationships between different streets into our model.

4.1 Model Formulation

Suppose we have a street network consisting of n streets $\mathbf{x} = \{x_1, x_2, \ldots, x_n\} \in \mathcal{X}$. Our goal is to predict a street type labelling $\mathbf{y} = \{y_1, y_2, \ldots, y_n\} \in \mathcal{Y}$ for these streets. Figure 4 shows our graphical model. x_{ip}, x_{if} are the observed features of a street x. More specifically, x_{ip} represents the initial crowd sourced labels or priors and x_{if} represents the geometric features. Toward the goal of jointly learning the labels $\mathbf{y}$, our model maximizes the conditional probability of $\mathbf{y}$ given $\mathbf{x}$:

$$\mathbf{y}^* = \arg\max_{\mathbf{y} \in \mathcal{Y}} P(\mathbf{y}|\mathbf{x}; \mathbf{w}) \tag{1}$$

where, using the Hammersley Clifford Theorem [6], $P(\mathbf{y}|\mathbf{x}; \mathbf{w})$ may be expressed as:

$$P(\mathbf{y}|\mathbf{x}; \mathbf{w}) = \frac{exp(\psi(\mathbf{y}, \mathbf{x}; \mathbf{w}))}{\sum_{\mathbf{y}' \in \mathcal{Y}} exp(\psi(\mathbf{y}', \mathbf{x}; \mathbf{w})} \tag{2}$$

where $\psi(\mathbf{y}, \mathbf{x}; \mathbf{w})$ is a potential function measuring the compatibility between the output labels $\mathbf{y}$ given the input $\mathbf{x}$ and $\mathbf{w}$ are the model parameters. This potential function consists of three types of potentials corresponding to the three properties described previously: two unary potentials and one pairwise potential. More specifically, one unary potential measures the compatibility between the label y given the initial crowdsourced prior for the label x_p. Another unary potential measures the compatibility between the label y given the geometric features of the street x_f. Finally, the pairwise potential models the compatibility between neighbouring street labels. The potential $\psi(\mathbf{y}, \mathbf{x}; \mathbf{w})$ is linear in these basis potentials and parameters and can be compactly written as:

$$\psi(\mathbf{x}, \mathbf{y}; \mathbf{w}) = \mathbf{w}^T \Psi(\mathbf{x}, \mathbf{y}) \tag{3}$$

4.2 Learning and Inference

Learning

We use the 1-slack formulation of structural support vector machine (SSVM) [14] to learn the parameter vector $\mathbf{w}$. This involves the minimization of the following quadratic optimization problem:

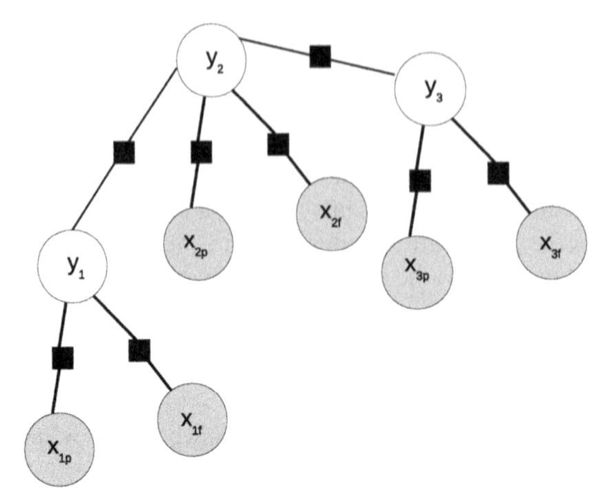

Fig. 4 Graphical modelling of the street network. Label y_1 for the street x_1 depends on its observed features x_{1p} and x_{1f} by two unary potentials and on its neighbouring street's label y_2 by a pairwise potential.

$$
\begin{aligned}
&\min_{\mathbf{w},\xi\geq 0} \frac{1}{2}\mathbf{w}^T\mathbf{w} + C\xi \\
&s.t.\ \forall(\bar{y}_1,\ldots,\bar{y}_n) \in \mathcal{Y}^n : \\
&\frac{1}{n}\mathbf{w}^T\sum_{i=1}^{n}[\Psi(x_i,y_i) - \Psi(x_i,\bar{y}_i)] \geq \Delta(y_i,\bar{y}_i) - \xi
\end{aligned}
\tag{4}
$$

where $\Delta(y_i, \bar{y}_i)$ represents the loss function. We used hamming loss in our experiments. Equation 4 intuitively implies that for a given training example (x_i, y_i), the value of the potential $\psi(\mathbf{x}, \mathbf{y}; \mathbf{w})$ must be greater than its value for any other pair $(x_i, \bar{y}_i)$ by at least the loss $\Delta(y_i, \bar{y}_i)$. We used the PyStruct Library [7] for implementing the SSVM framework.

Inference

Given the model parameters, the task of predicting the best semantic labelling y for the streets involves maximizing the conditional probability in Equation 1. This maximization problem involves a combinatorial search over all possible $\mathbf{y} \in \mathcal{Y}$ and is solved by using the Gibbs Sampling Markov Chain Monte Carlo (MCMC) method [12]. Given a sufficient burn-in period, this inference technique is guaranteed to converge to the optimum solution.

Table 2 Street types definitions

motorway	High capacity highways designed to safely carry fast motor traffic.
footway	Urban paths designated for pedestrians only.
service	Generally for access to a building, motorway service station, beach, campsite, industrial estate, business park, etc.
residential	Road in a residential area
primary	A highway linking large towns.
unclassified	Public access road, non-residential.
trunk	Important roads that are not motorways.
tertiary	A road linking small settlements, or the local centres of a large town or city.

5 Experiments and Results

Our goal is to learn and predict the 8 most common label types used for streets in the OSM database. These include motorway, trunk, primary, secondary, tertiary, service, residential, and unclassified streets. Table 2 provides brief definitions for them. These definitions are taken from [1]. Finding a good ground truth for any crowdsourcing project is a challenging task. The OSM project was born in London more than a decade ago. Due to a strong and active mapping community here the quality of OSM data for this city can be considered fairly accurate [20]. Therefore, we are using the OSM London street network for our training and testing purposes.

The original OSM data was filtered to only consist of streets with the aforementioned 8 label types. A multi-granular representation discussed previously (Section 3) was obtained for this network. Our model uses prior knowledge i.e., the initial crowd sourced labels for streets. In case of OSM London database, our assumption is that the priors are the same as the ground truth. Therefore, it is necessary to avoid overfitting to the priors during learning. This was achieved by randomizing 20 percent prior values. The geometric features used in our experiments include length, number of dead-ends, and linearity.

We trained our model on north-eastern London and evaluated it on the south-western part of the city. The evaluation results with respect to some metrics along with the corresponding confusion matrix are shown in Table 3. The last row of the table corresponds to the total number of instances of the respective street types in the test data. We will now discuss these results in some detail.

Precision: Precision is the ratio of the number of true positives to the sum of true positives and false positives. We have shown class-wise precision of our model evaluation in the Precision row of Table 3. The obtained precision values for certain street types such as footway, residential, and service are quite high. The precision value for motorway is only 0.02. This maybe attributed to high data imbalance in the test set: only 2 motorways compared to 6663 residential streets.

Recall: Recall is the ratio of true positives to the sum of true positives and false negatives. The Recall row in Table 3 shows the class-wise recall for all 8 street types

Table 3 Model evaluation on the test set consisiting of 8 street types. Rows 2-9 represent the corresponding confusion matrix. Precision, Recall, f1-score, and the number of instances of the respective street types are shown in the last four rows of the table.

classified as	motorway	footway	service	residential	primary	unclassified	trunk	tertiary
motorway	2	0	0	0	0	0	0	0
footway	53	1942	37	37	12	43	10	51
service	6	27	1758	30	20	39	14	39
residential	0	103	75	6052	54	120	39	148
primary	0	2	2	3	247	2	4	9
unclassified	2	7	8	9	2	588	3	27
trunk	0	0	0	1	2	2	73	0
tertiary	6	5	0	2	13	3	6	345
Precision	0.02	0.91	0.92	0.98	0.64	0.70	0.43	0.50
Recall	1.00	0.87	0.89	0.91	0.90	0.90	0.89	0.90
f1-score	0.03	0.89	0.90	0.94	0.75	0.79	0.58	0.64
support	2	2223	1977	6663	275	653	82	385

considered in our experiments. An average recall of almost 0.90 was obtained for all street types.

f1-score: f1-score is the harmonic mean of precision and recall and provides a more robust metric than precision or recall considered separately. The f1-score for most street types is greater than 0.75.

Confusion Matrix: A confusion matrix provides a tabular visualization of a classification performance in terms of actual and predicted instances of each class [13]. Rows of the matrix represent instances present in actual classes whereas the columns of the matrix show instances in predicted classes.The diagonal elements of the matrix represent the number of correctly classified instances of each class whereas the off-diagaonal elements represent the misclassified instances. Table 3 (excluding the last four rows) represents the confusion matrix obtained for the test set.

Besides quantifying the evaluation errors, a confusion matrix also helps to identify the nature of these errors. For example, the precision values for motorways and footways only informs that the developed model can predict motorways with a precision of 0.02 and footways with that of 0.91. However, the confusion matrix further reveals that the low precision value for motorways could be due to 43 footways being classified as motorways. This implies that there is a need to use more features that can discriminate between motorways and footways. Similar inferences can be drawn for other streets.

Finally, an average accuracy of 89.76 percent was obtained on all street types considered which improves upon our preliminary work [23] where an average of 58.03 percent was obtained for the corresponding street types.

6 Conclusions and Future Work

While the geometry of map objects in crowdsourced maps mainly comes from GPS, the semantics are entirely contributor dependent and can be misleading. Automatically inferring the semantics using the geometry can improve the map quality. In this paper we proposed and evaluated a novel methodology for predicting street labels in crowdsourced maps. We posed the problem as an undirected graphical model that allowed us to incorporate prior knowledge, geometric features, and the spatial relationships between streets into our model. We used Structural Support Vector Machines to learn the model parameters in an efficient way and predict street labelling. Our model performs considerably better than our previous work where streets were assumed to have an *iid*. The major contributions of this paper are:

1. Proposing and implementing a probabilistic graphical modelling framework for predicting street type labels in crowdsourced maps and jointly inferring over all the labels in a network
2. Using the initial labels given by the crowd as a prior for the true label values.

There is a scope for improving the current results by using more distinguishing geometric features and forms future work of the presented research. Our model is flexible and can be extended for semantic prediction of other map objects such as building types. Future work also involves testing the model on street networks outside UK. Street network infrastructure varies to a certain extent with respect to their geographical location. This problem can be overcome by training the model over a larger dataset and adding a location specific factor into the model.

References

1. https://taginfo.openstreetmap.org/keys/highway#values (accessed April 05, 2015)
2. OSM. http://wiki.openstreetmap.org/wiki (accessed March 18, 2015)
3. OSM: key:highway. http://wiki.openstreetmap.org/wiki/Key:highway (accessed June 18, 2014)
4. Ballatore, A., Bertolotto, M.: Semantically enriching VGI in support of implicit feedback analysis. In: Kim, K.-S. (ed.) W2GIS 2011. LNCS, vol. 6574, pp. 78–93. Springer, Heidelberg (2010)
5. Ballatore, A., Bertolotto, M., Wilson, D.C.: Linking geographic vocabularies through WordNet. Annals of GIS **20** (2), 73–84
6. Blake, A., Kohli, P., Rother, C.: Markov Random Fields for Vision and Image Processing. MIT Press (2011)
7. Müller, A.C., Behnke, S.: Pystruct - Learning Structured Prediction in Python. Journal of Machine Learning Research **15**, 2055–2060 (2014). http://jmlr.org/papers/v15/mueller14a.html
8. Ali, A.L., Schmid, F., Al-Salman, R., Kauppinen, T.: Ambiguity and plausibility: managing classification quality in volunteered geographic information. In: Proceedings of the ACM SIGSPATIAL Conference (2014)

9. Ciepluch, B., Jacob, R., Mooney, P., Winstanley, A.: Comparison of the accuracy of openstreetmap for ireland with google maps and bing maps. In: Proceedings of the Ninth International Symposium on Spatial Accuracy Assessment in Natural Resuorces and Enviromental Sciences, p. 337 (2010)
10. Barron, C., Neis, P., Zipf, A.: Towards intrinsic quality analysis of openstreetmap datasets. In: International Workshop on Action and Interaction in Volunteered Geographic Information (ACTIVITY), Leuven, Belgium, pp. 43–48 (2013)
11. Kepler, C., de Groot, R.T.A.: Trust as a proxy measure for the quality of volunteered geographic information in the case of openstreetmap. In: Geographic Information Science at the Heart of Europe, pp. 21–37 (2013)
12. Koller, D., Friedman, N.: Probabilistic graphical models: principles and techniques. MIT Press (2009)
13. Witten, I.H., Frank, E.: Data Mining: Practical machine learning tools and techniques. Morgan Kaufmann (2005)
14. Tsochantaridis, I., Joachims, T., Hofmann, T., Altun, Y.: Large margin methods for structured and interdependent output variables. Journal of Machine Learning Research, 1453–1484 (2005)
15. Girres, J.F., Touya, G.: Quality assessment of the french openstreetmap dataset. Transactions in GIS **14**(4), 435–459 (2010)
16. Ginsberg, J., Mohebbi, M.H., Patel, R.S., Brammer, L., Smolinski, M.S., Brilliant, L.: Detecting influenza epidemics using search engine query data. Nature **457**(7232), 1012–1014 (2009)
17. Howe, J.: Crowdsourcing: How the power of the crowd is driving the future of business. Random House (2008)
18. Ahn, L.V., Dabbish, L.: Labeling images with a computer game. In: Proceedings of the SIGCHI Conference on Human Factors in Computing Systems, pp. 319–326. ACM (2004)
19. Goodchild, M.F., Li, L.: Assuring the quality of volunteered geographic information. Spatial Statistics **1**, 110–120 (2012)
20. Haklay, M.: How good is volunteered geographical information? a comparative study of openstreetmap and ordnance survey datasets. Environment and Planning **37**(4), 682–703 (2010)
21. Helbich, M., Amelunxen, C., Neis, P., Zipf, A.: Comparative spatial analysis of positional accuracy of openstreetmap and proprietary geodata. In: Proceedings of GI Forum, pp. 24–33 (2012)
22. Jilani, M., Corcoran, P., Bertolotto, M.: Multi-granular street network representation towards quality assessment of openstreetmap data. In: Proceedings of the Sixth ACM SIGSPATIAL International Workshop on Computational Transportation Science. ACM (2013)
23. Jilani, M., Corcoran, P., Bertolotto, M.: Automated highway tag assessment of openstreetmap road networks. In: Proceedings of the ACM SIGSPATIAL Conference (2014)
24. Van Exel, M., Dias, E., Fruijtier, S.: The impact of crowdsourcing on spatial data quality indicators. In: Proceedings of GiScience (2011)
25. Mooney, P., Corcoran, P.: Characteristics of heavily edited objects in openstreetmap. Future Internet **4**(1), 285–305 (2012)
26. Porta, S., Crucitti, P., Latora, V.: The network analysis of urban streets: a primal approach. Environment and Planning B: Planning and Design **33**, 705–725 (2006)
27. Porta, S., Crucitti, P., Latora, V.: The network analysis of urban streets: a dual approach. Physica A: Statistical Mechanics and its Applications **369**(2), 853–866 (2006)
28. Koukoletsos, T., Haklay, M., Ellul, C.: Assessing data completeness of vgi through an automated matching procedure for linear data. Transactions in GIS **16**(4), 477–498 (2012)

An Enhancement of the MapReduce Apriori Algorithm Using Vertical Data Layout and Set Theory Concept of Intersection

S. Dhanya, M. Vysaakan and A.S. Mahesh

Abstract The process of Association Rule Generation is an important task in Data Mining. It is widely used for the Market Basket Analysis. The data that is used for the generation of Association Rules is usually distributed and complex. This can be efficiently implemented using the Hadoop Framework as it can process large datasets with less cost and good performance. We have proposed an efficient algorithm for MapReduce Apriori based on Hadoop- MapReduce model using Vertical Database Layout and Set Theory concept of Intersection.

Keywords: Apriori · Hadoop · MapReduce · Cloud computing · Association rule · Frequent itemset mining

1 Introduction

Data mining is a new powerful technology with great potential which discovers information within the data that queries and reports can't effectively reveal. Current developments are producing enormous amount of data day-by-day resulting in the need for persistent storage, analysis and efficient processing of these complex data. Almost all industries stores huge quantities of their operational data in their local and distributed databases. These data can be used for analyzing customer trends which can be helpful in marketing the products to maximize profit and to efficiently manage the inventory [10].
The mining of Association Rules helps to extract interesting patterns, relationships or associations between the itemsets in a transactional database. The Apriori

S. Dhanya(✉) · M. Vysaakan(✉) · A.S. Mahesh(✉)
Amrita School of Arts and Sciences Amrita Vishwa Vidyapeetham, Brahmasthanam, Edappally, Kochi 682024, Kerala, India
e-mail: {dhanya283,vysaakan,asmaheshofficial}@gmail.com

S. Berretti et al. (eds.), *Intelligent Systems Technologies and Applications*,
Advances in Intelligent Systems and Computing 385,
DOI: 10.1007/978-3-319-23258-4_20

Algorithm can be used for the mining of Association Rules which involves Frequent Itemset Mining and Association Rule Generation over the operational data stored in transactional databases [1]. However, processing of such voluminous data requires greater processing capabilities as the data is distributed in nature.

Cloud computing helps in setting up an infrastructure required for distributed environment and enables companies to consume compute resources by increasing the capability of the shared resources [20]. The Apache Hadoop Framework provides a programming model called MapReduce for processing massive datasets and it provides reliability, scalability and fault tolerance [11]. In our work, we are implementing an efficient MapReduce Apriori algorithm based on Hadoop-MapReduce model using Vertical Database Layout and Set Theory concept of Intersection.

2 Background and Related Work

2.1 Background

Apriori and Association Rule Mining. Association Rules [16] reveals the patterns or associations that have applications in many domains like marketing, decision making and inventory management. It is also used in areas like Market Basket Analysis, Web Mining, studying patterns of Biological databases, analyze Population and Economic Census, Medical diagnosis, and Scientific Data Analysis. Association rules shows itemsets that occur frequently in transactions and the generated rules are probabilistic in nature [1].
The mining or formation of association rules consists of two steps:

- Frequent Itemset Mining: finding the itemsets that occurs frequently in the database based on support.
- Association Rule Generation: generating strong rules from the frequent itemsets based on confidence.

Usually, an Association Rule Mining Algorithm results in the formation of a large number of association rules and it is difficult for users to validate all the rules manually. Therefore, only interesting rules or non-repeating rules are to be generated.

The Apriori Algorithm, which uses a bottom up approach for the generation of candidate itemsets, is the most widely accepted algorithm for Association Rule Mining. This algorithm works based on the property called as Apriori Property which means that “all the subsets of a frequent itemset must also be frequent” [1]. If we know that a particular itemset I is infrequent, then it is not required to count its supersets. If an itemset I do not satisfy the minimum support threshold, then I is not frequent [6].

Hadoop and MapReduce. Hadoop which is a part of the Apache project is an open-source Java-based framework for scalable, reliable, distributed computing. It coordinates the work on the created cluster of nodes. The major parts of Hadoop are the HDFS or the Hadoop Distributed File System and the MapReduce engine. With Hadoop it is possible to run many applications on a number of nodes involving millions of gigabytes of information. Its file system, which is distributed in nature, enables rapid data rates within the nodes and uninterrupted operation is allowed even if a particular node fails. This approach reduces the risk of a complete system failure even if numerous nodes stop functioning [11] [13].

MapReduce is a software framework that helps programmers to write and execute programs that requires considerable amount of unstructured and structured data to be processed in parallel [13]. It works in a cluster of processors which are distributed or in stand-alone computers. It was initially developed for indexing Web pages at Google. Every node in the cluster will report back with status updates and completed work periodically. Hence, the MapReduce framework is fault-tolerant. If a node doesn't respond within an expected interval, the work is re-assigned to other nodes by the master node.

In MapReduce, the algorithms are specified using two functions: map and reduce. The input dataset is split into independent chunks of data and these are processed in a parallel manner by the map function and key and value pairs are generated. The sorted output of the map function is used as the input of the reduce function. The HDFS stores both the inputs and outputs. In this setup, there are slave Task Trackers which are handled by master Job Tracker. The scheduling of a job on the slave node is controlled by the master node. The tasks are executed in the slaves as specified by the master [11] [13] [19].

2.2 *Related Work*

Most of the enhancements done in the parallel Apriori Algorithm are focused on the process of Frequent Itemset Mining. The Vertical Data Format or Horizontal Data Format is used for the representation of transactions. The data representation is in transac_ID versus itemset format in horizontal data layout whereas the data representation is in itemsets versus transac_ID format in the vertical data layout [2].

There are Parallel Algorithms for finding frequent itemsets and for generating rules from frequent itemsets. The algorithms used for finding frequent itemsets are Count Distribution Algorithm, Data Distribution Algorithm and Candidate Distribution Algorithm [1].

The Count Distribution Algorithm reduces the communication cost, but the problem is the replication of the whole hash tree. Data Distribution Algorithm ensures memory efficiency, but cost of communication between the processors is very high. Common Candidate Partitioned Algorithm makes use of complicated hash structures. In the Parallel Partition Algorithm, the coordinator will control the client processors [2].

Apriori Algorithm is an iterative algorithm which can be parallelized and implemented in Cloud. But data Security is a major concern when using cloud computing for Data Mining [3].

The Enhanced Apriori Algorithm reduces the time for generating the frequent item by using the intersection method. It uses Vertical Database Layout. In this approach, to calculate the support, we count the common transactions in each element's candidate set, by using the Intersection Algorithm [7].

3 Existing System

The existing system is Map/Reduce Apriori Algorithm based on the legacy Apriori Algorithm [4]. The majority of the Parallel Apriori Algorithms concentrates on parallelization of the task of creating frequent itemsets which requires large amount of computing resources consumption. Here, it makes use of the Candidate Distribution Algorithm [2].

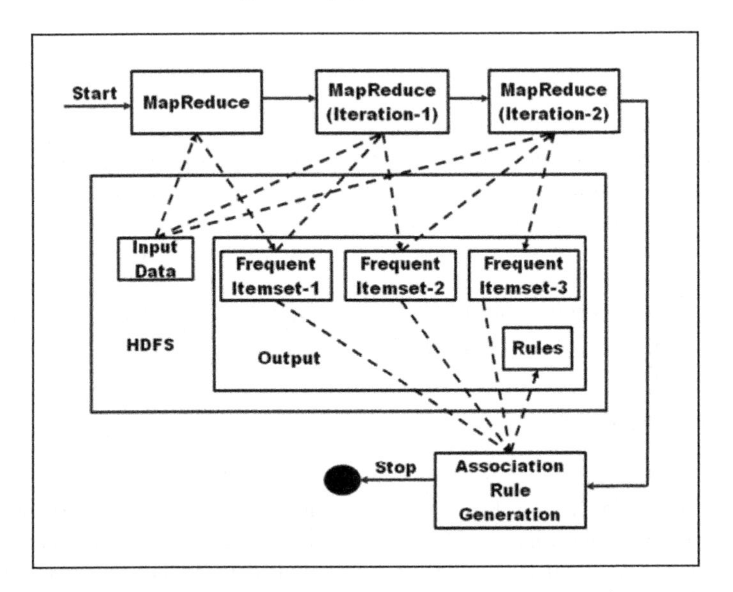

Fig. 1 Data flow diagram showing two iterations

In Candidate Distribution Algorithm, both the data and candidates are partitioned so that every mapper may proceed in parallel. There is no communication of counts or data tuples, but for the prune step involved in candidate generation, a reducer may have to gather the frequent itemsets determined by other mappers.

The input data set is saved on HDFS which has provision for storing considerable amount of data [11] [13]. It help in data localizations for the map/reduce task. The value is written as output if it exceeds the support threshold [1] [6]. The value gets discarded if the criterion of minimum support is not met. This would result in the generation of frequent-1 itemset. The set of one item pattern is found in Frequent -1 itemset. The candidate set is generated using previous iteration's frequent item set [8].

The master node splits the input into smaller divisions in the map function, and allots those to the slave nodes. The algorithm generates the frequent itemset for every map node. Then it gathers all frequent itemsets and removes itemsets that doesn't meet the criteria of minimum support in the reduce nodes. In every map node, the generation of frequent itemsets by addition of an item is performed using join_sort and eliminating the redundant items. At every reduce nodes, frequent itemsets are collected and itemsets that doesn't meet the criteria of minimum support are removed.

The drawback of the existing system is that it suffers from massive communication cost because of the redistribution of the database and the need to repeatedly scan local partitions. When using the horizontal data layout, to get the support count of an itemset, one has to get through all the transactions and check if the transactions contain the itemset.

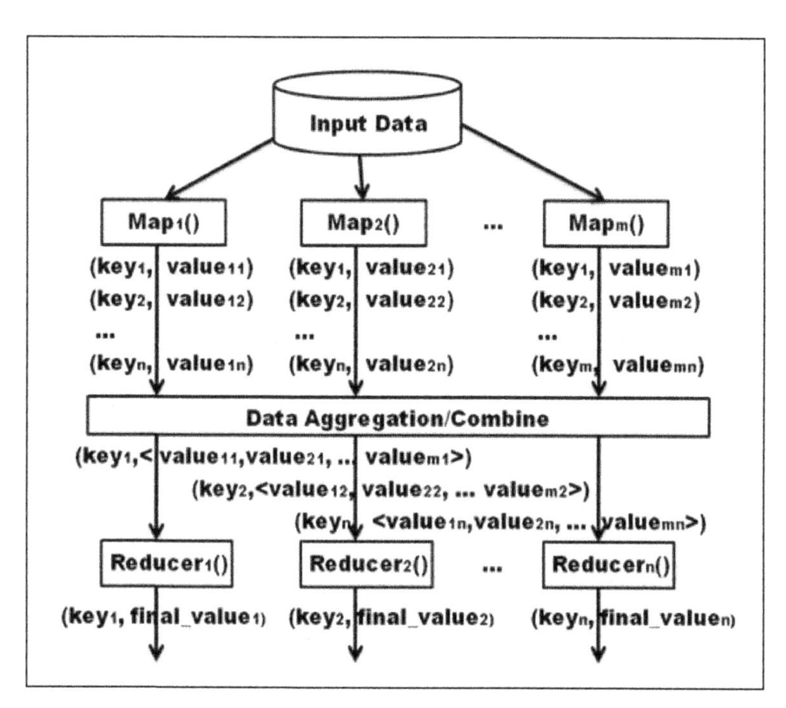

Fig. 2 Map/Reduce Flows

4 Proposed System

In MapReduce Apriori Algorithm, to count the support of candidate set, each record is scanned one by one to check the existence of each candidate. This process is very time consuming, and it requires the iterative scan of the entire database for each candidate set, which is equal to the maximum length of candidate item set.

- Each Mapper scans its local database and constructs partial tid_list for all itemsets.
- Construction of global tid_list
    - Each Mapper sends tid_list of itemsets belonging to other mappers, and receives the tid_list for the itemsets it has, from other Mappers (Local Reduction on Key-Value Pairs)
    - Reducers reduces the key value pairs obtained by total reduction to get the global tid_list.

Fig. 3 Steps for converting Horizontal to Vertical Data Layout in MapReduce

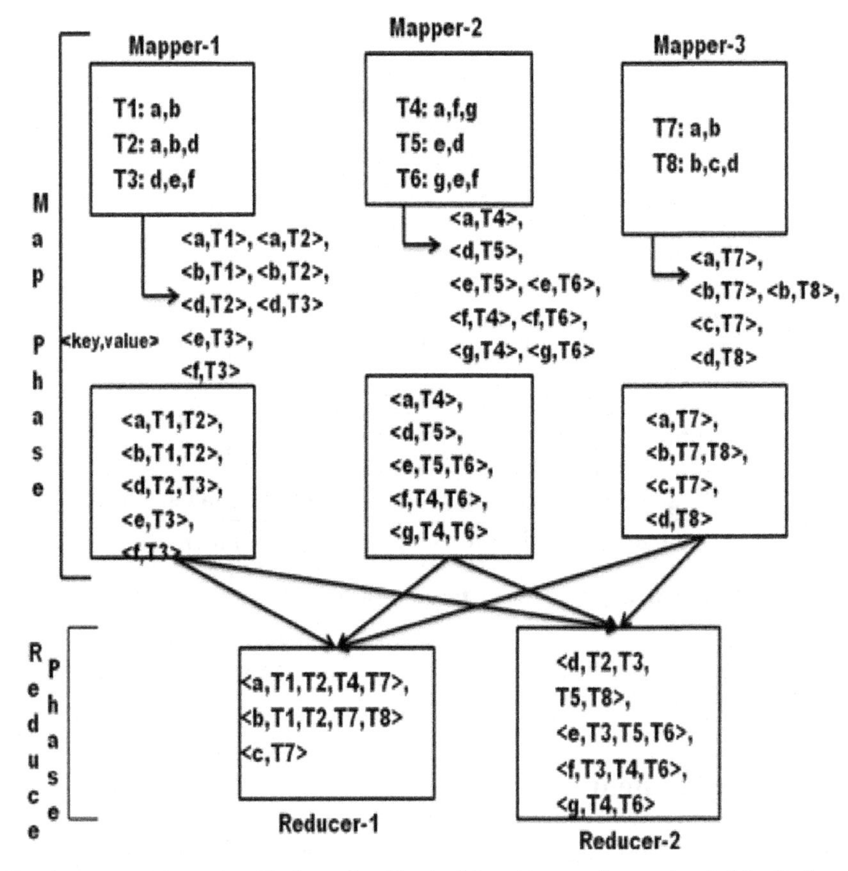

Fig. 4 Diagram illustrating Horizontal to Vertical Data Layout Conversion in MapReduce

The limitation of the existing Apriori algorithm is that it wastes time by scanning the whole database to find the support count by searching the frequent itemsets. The Proposed System can be broken down into the following modules:

1) Transformation from Horizontal to Vertical Layout
2) Finding Frequent Itemsets using Intersection Algorithm
3) Generating Association Rules based on Confidence

We have to change the local database which is in horizontal format to the vertical trans_id-list format. The steps for this conversion are illustrated in Fig. 3 and the diagram illustrating the same is shown in Fig. 4.

The steps for finding frequent itemset are shown in Fig. 5.

The output file contains the list of frequent itemsets and their support counts. These are used for the generation of the association rules. By making use of the frequent-n itemset, association rules are created and the result is saved. This procedure is performed based on Confidence. The confidence threshold reduces the formation of irrelevant rules.

1. Calculate Candidate Itemsets Size-1 for each MapNode (C_{m1})
2. Collect Candidate Itemsets Size-1(C_1)
3. Remove items from the Candidate Itemsets whose support count is less than minimum support in the reduce nodes, and generate Frequent Itemset Size-1 (L_1).
4. Calculate Candidate Itemsets Size-k by addition of an item which is performed using join_sort and eliminating the redundant items in each map node ($C_{m(k+1)} = L_k$ join_sort L_{mk})
5. From $C_{m(k+1)}$, reduce nodes to collect Candidate Itemsets Size-k(C_{k+1})
6. C_{k+1} is mapped to the required Map Nodes as $C_{m(k+1)}$
7. Count the no. of transactions of each item in $C_{m(k+1)}$ by using Set Intersect method to get item frequencies at the map nodes.
8. From $C_{m(k+1)}$, reduce nodes to collect Candidate Itemsets Size-k(C_{k+1}) with support count.
9. Generate Frequent Itemset Size-k(L_k) by removing itemsets that have support count less than the minimum support.

Fig. 5 Steps for finding frequent itemsets

The suggested system uses Vertical Database Layout [17] [18] and Set Theory concept of Intersection [8]. In this approach, to calculate the support, we count the common transactions in each element's candidate set, by using the Intersection Algorithm. It reduces the time spent for scanning the database. The advantage of using the proposed system is that it reduces the number of steps and extraction time. The support of an itemset is the size of its transaction ID set when using the vertical layout in the transaction database. This helps to decrease frequent database scans, and it'll speed up the overall process.

5 Conclusion and Future Work

The limitation of the MapReduce Apriori algorithm is that it wastes time by searching the whole database scanning for the frequent itemsets as it uses horizontal data layout. The proposed system uses Vertical Database Layout and Set Theory concept of Intersection. This can be implemented in a Hadoop Single Node Cluster setup or by making use of the Amazon Web Services like Amazon EC2 and Amazon S3.

In this work, a Hadoop Single Node Cluster Setup was employed to evaluate the performance of the algorithm. Here, the Frequent-1 Itemsets for the given input dataset are generated. Then the proposed solution generates the Frequent-2 Itemsets, Frequent-3 Itemsets and so on without scanning the entire set of transactions. It reduces the number of steps and extraction time. It also decreases frequent database scans, and hence will speed up the overall process.

In the future work, algorithms can be evaluated with a much larger cluster such as Amazon Elastic Compute Cloud (EC2). Also, a hybrid implementation between in-memory and Hadoop implementations can be achieved. This will allow even better execution times.

References

1. Agrawal, R., Shafer, J.C.: IBM Almaden Research Center, "Parallel Mining of Association Rules". IEEE Transactions on Knowledge and Data Engineering (February 1996)
2. Sakhapara, AM., Bharathi, H.N.: Comparative Study of Apriori Algorithms for Parallel Mining of Frequent Itemsets. International Journal of Computer Applications (0975–8887) 90(8) (March 2014)
3. Qureshi, Z., Bansal, J., Bansal, S.: A Survey on Association Rule Mining in Cloud Computing. International Journal of Emerging Technology and Advanced Engineering 3(4) (April 2013); ISSN 2250-2459, ISO 9001:2008 Certified Journal
4. Li, J., Roy, P., Khan, S., Wang, L., Bai, Y.: Data Mining Using Clouds: An Experimental Implementation of Apriori over MapReduce. The 12th IEEE International Conference on Scalable Computing and Communication (ScalCom 2102), Changzhou, China (December 2012)

5. Woo, J.: Apriori-MapReduce Algorithm. In: International Conference on Parallel and Distributed Processing Techniques and Applications, Las Vegas (July 2012)
6. Agrawal, R., Srikant, R.: Fast algorithms for mining association rules. In: Proc. 1994 Int. Conf. Very Large Data Bases, Santiago, Chile, pp. 487–499 (September 1994)
7. Geetha, K., Mohiddin, S.K.: An Efficient Data Mining Technique for Generating Frequent Itemsets. International Journal of Advanced Research in Computer Science and Software Engineering 3(4) (April 2013)
8. Goswami, D.N., Anshu, C., Raghuvanshi, C.S.: An Algorithm for Frequent Pattern Mining Based on Apriori. International Journal on Computer Science and Engineering 2(4) 942–947 (2010)
9. Al-Maolegi, M., Arkok, B.: An Improved Algorithm For Association Rules. International Journal on Natural Language Computing 3(1) (February 2014)
10. Agarwal, R., Imielinski, T., Swami, A.: Mining Association Rules between Sets of Items in Large Databases. ACM SIGMOD Record **22**, 207–216 (1993)
11. Apache Hadoop Project. http://hadoop.apache.org/
12. Market Basket Analysis Example in Hadoop. http://dal-cloudcomputing.blogspot.com/2011/03/market-basket-analysis-example-in.html, Jongwook Woo (March 2011)
13. Saritha, R.C., Usha Rani, M.: Mining Frequent Item Sets Using Map Reduce Paradigam. International Journal of Engineering Sciences Research-IJESR 4(Special Issue) (January 2013)
14. Dhamdhere Jyoti, L., Deshpande Kiran, B.: A Novel Methodology of Frequent Itemset Mining on Hadoop. International Journal of Emerging Technology and Advanced Engineering 4(7) (July 2014)
15. Ziauddin, S.K., Khan, K.Z., Khan, M I.: Research on Association Rule Mining. Advances in Computtational Mathematics and its Applications 2(1), (2012)
16. Umarani, V., Punithavalli, M.: A Study on Effective Mining of Association Rules From Huge Databases. International Journal of Computer Science and Research 1(1) (2010)
17. Zaki, M.J., Parthasarathy, S., Ogihara, M., Li, W.: Parallel Algorithms for Discovery of Association Rules. Data Mining and Knowledge Discovery 1(4), 343–373 (1997)
18. Foster, I., Kesselman, C.: The Grid: Blueprint for a New Computing Infrastructure. Morgan Kaufmann, San Francisco (1999)
19. Zaki, M.J.: Parallel and Distributed Association Mining: A Survey. IEEE Concurrency 7(4), 14–25, (1999)
20. Foster, I., Kesselman, C., Nick, J., Tuecke, S.: The Physiology of the Grid: an Open Grid Services Architecture for Distributed Systems Integration. Technical report, Global Grid Forum (2002)
21. Mridul, M., Khajuria, A., Dutta, S., Kumar, N., Prasad, M.R.: Analysis of Bidgata using Apache Hadoop and MapReduce. International Journal of Advanced Research in Computer Science and Software Engineering 4(5) (May 2014)
22. Ruxandra-Stefania PETRE, Data Mining in Cloud Computing. Database Systems Journal III(3) (2012)

Structuring Reliable Distributed Storages

C.K. Shyamala and N.V. Vidya

Abstract Distributed storage is gaining ample prevalence due to the increase in both data and user base. Design of distributed storage system is not limited to the factors: Fault tolerance, Availability, Consistency and Durability. Potential malicious behavior gives rise to untrusted operating environments. Reliability thus becomes one of the most crucial aspects of the design. Structuring reliable distributed storages require a systematic and detailed understanding of the various challenges and the related limitations. Based on extensive reviews, analysis conducted on prominent storage systems the paper streamlines the vital aspects of distributed storage design. The survey, summarization, streamlining presented in the paper facilitates comprehension of the real time challenges and issues of globally networked Distributed Storage Systems. The paper provides a thorough insight into the design of reliable storage systems with respect to architecture, consistency, redundancy and hybrid failures. Significant recommendations presented in the paper facilitate design trade-offs.

1 Introduction

With exponential increase of digitized information, storage has become one of the most important aspects of computing field. Concept of storage emerged with a purpose to facilitate the recollection of already used data, in the future. Storage in single node emerged to serve the purpose. Privacy is its immanent nature, but in addition to single point of failure it caused inconsistencies, security breach, unpredictability, delay, availability, space requirement issues [2]. In order to address the issues unhandled by single node storage, distributed storage was introduced. Distributed storage, done over independent computer systems is interfaced to the user as a single coherent system and thus transparency during access is ensured. The location of the storage node is immaterial as networks viz.

C.K. Shyamala · N.V. Vidya (✉)
Amrita Vishwa Vidyapeetham University, Ettimadai,
Coimbatore 641112, Tamil Nadu, India
e-mail: ck_shyamala@cb.amrita.edu, nv.vidya@gmail.com

S. Berretti et al. (eds.), *Intelligent Systems Technologies and Applications*,
Advances in Intelligent Systems and Computing 385,
DOI: 10.1007/978-3-319-23258-4_21

LAN, WAN help to connect nodes across the breadth and length of any geographical area [11]. It has to serve request for a global audience. Settling for nominal bandwidth consumption is one of the major challenges faced by the storage. Assurance of consistency is another main concern. An update operation should be reflected immediately in the succeeding data access itself, thus keeping the storage consistent. Unpredictability is yet another challenge faced by distributed storage systems. Availability of a system is the prime concern for a client and hence keeping the data available for a client at the instant of access becomes high priority. A system is said to be available when it responds correctly at any instant of time and is said to be unavailable when it fails to respond positively to pings by a client or a periodic health checking monitor [3]. All these factors contribute to the reliability of a system. Another major concern is the trust of the operating environment. LAN provides trusted operating environment where as WAN cannot assume the usual trust of LAN; it provides partially trusted/untrusted operating environments. The predictability and controlled nature of a trusted environment makes client server architecture well suited for distributed storage while un-trusted and partially trusted environments work efficiently in peer to peer architecture which is dynamic [1], [11]. Reparability is yet another issue to address. There are non-homogeneous systems which makes use of multiple storage schemes like maximum distance separable (MDS) and non-maximum distance separable codes which has smaller repair bandwidth and can operate at a smaller field size [30]. Certain non-homogeneous systems have different storage sizes and is closer to real time systems. They provide solution to the minimum bandwidth regenerating problem as well. The two rack design of DSS considers the repair bandwidth of each rack while fixing the storage size [29]. The challenges discussed here need to be dealt appropriately while building an efficient storage system. There has been a significant line of research [1], [5], [6], [9] investigating and proposing architectures, protocols and models to overcome these challenges.

Outline of this paper is as follows; Section 2 describes various features of distributed storage which include the architecture, consistency, protocols for replication and and failure model. Section 3 presents survey of various storage systems. Section 4 discusses insights into DSS design based on the features with section 5 the paper concludes.

2 Features of Distributed Storage

2.1 Architecture

Arrangement of nodes in a storage system with respect to its centralization and its behavior owes to the architecture. The most prominent architectures are peer to peer architecture and client server architecture. In a client server architecture, a node acts either as a client or a server for ever, where as in a peer to peer network nodes can take any of the two roles according to the context [1]. As illustrated in

Fig. 2.1 the client server architecture is either globally centralized or locally centralized. In globally centralized architecture there is a single central authority for control. On the other hand, locally centralized architecture distributes the role of central server among several servers and avoids single point of failure.

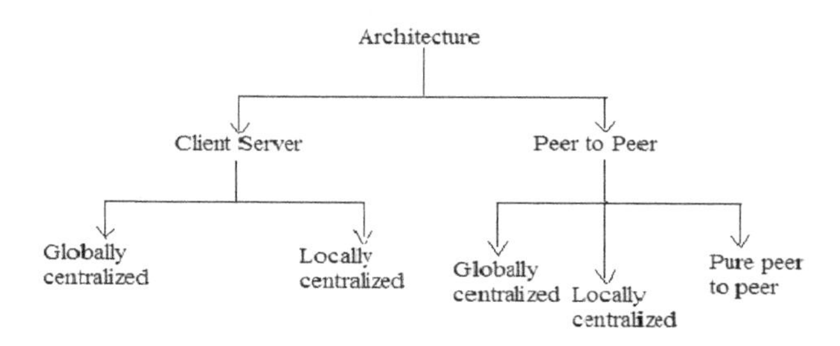

Fig. 1 Architecture classification

Peer to peer architecture has three levels of centralization namely globally centralized, locally centralized and pure peer to peer as shown in the fig. 1. In a globally centralized architecture, peers contact a central server which contains the information regarding other peers in the network. To overcome the demerit of a single point of failure, locally centralized architecture came into existence. Here, certain nodes with reliable characteristics are treated as super nodes and details regarding other nodes are obtained from these super nodes. In pure peer to peer architecture all the nodes are symmetrical, thus provides better scalability [1]. Every architecture has its merits and demerits. Hence it is always reasonable to make use of hybrid architecture for large storage systems [23].

2.2 *Consistency*

A system is said to be in a consistent state if it continues to generate correct data even after the occurrence of concurrent events [1]. In static storages where updates seldom take place consistency assurance is not a big problem. Whereas, the active storages (eg storage for shopping site data) are subjected to dynamic updates and the changes have to be propagated to all the servers holding redundant data. The issue of consistency pops up here. There are wide varieties of consistency models viz. optimistic consistency, strong consistency, eventual consistency, sequential consistency (Depicted in Fig. 2). Strong consistency (Pessimistic consistency) ensures that the data beholds ACID properties (Atomicity, Consistency, Isolation and Durability) and hence is consistent. But the level of concurrency achieved and availability of data gained are very low. Locking and leasing techniques [28] are used to achieve strong consistency. Weak consistency (optimistic consistency) model, allows a system to function despite inconsistencies. It allows multiple reads and writes without a central locking mechanism. Sequential consistency projects as if all memory storage accesses are executed in a single order by all the processes [11]. In Eventual consistency all the replicas eventually become identical without losing any of the writes. Here file synchronization is gradually performed at a later stage [11].

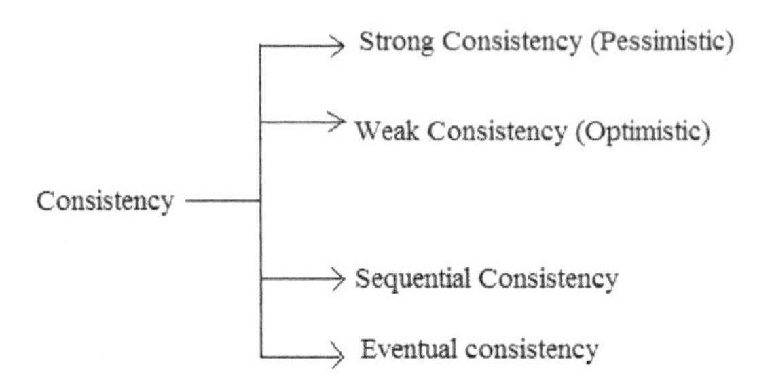

Fig. 2 Types of consistency

2.3 Protocols For Data Redundancy

Redundant copies of data enable a storage system to be fault tolerant. Public nature of distributed system makes it vulnerable to numerous faults and hence a reliable storage approach should embrace fault tolerance. Replication and dispersal are the two major approaches used to achieve redundancy in a distributed environment. A practical replication or dispersal mechanism must be capable of optimizing end to end client perceived performance metrics such as system throughput, request processing latency, recovery time after failure etc. Primary/Backup, Paxos, Quorum system, Dual quorum, Carbonite, Error correcting codes and Information dispersal algorithm are the most popular ones [4], [14],[2].

2.3.1 Replication Protocols

2.3.1.1 Primary Backup

Primary backup, a replication protocol for obtaining consistent data, handles all queries locally using primary servers and resort to secondary only when updates are processed. If no updates happen secondary servers do not have a role and the protocol resembles closely the non-replicated case [2][28]. Update operation commits only when it is *prepared* on all the replicas. In case a replica fails, the protocol is prevented from taking a new step until the failed replica is reconfigured.

2.3.1.2 Paxos

Paxos achieves consistency of data with less overhead than the Primary backup protocol. In Paxos, requests processing is similar to that of primary/backup except that the requests are committed when a majority of the replicas are *prepared.* The protocol does not wait for the entire set of replicas to be *prepared.* At the same time, a request cannot be processed if majority of the replicas become unavailable [2].

2.3.1.3 Carbonite

Carbonite is a replication algorithm predominantly used to achieve durability of data. A durable system assures that the data is preserved even during disk failures. The independence of disk failures is assumed. It inserts a set of replicas initially and new replicas are generated in case of transient failures [9]. Over a period of time the system acquaints itself to ignore transient failures and produce replicas only in case of permanent failures.

2.3.1.4 ROWA

Read One Write All (ROWA) protocol is predominantly used for read dominant environment. It is optimized and provides good availability. A transaction *reads* from a single copy usually the local one, to minimize the communication cost. All the servers that hold data performs *write* operations [8]. The availability of all the replicas is a must to perform a write operation. A variation of ROWA, Read One/Write All –Asynchronous protocol, performs *write* operations on the available ones. Writes are propagated asynchronously thereby increasing availability.

2.3.1.5 Quorum

Quorum offers a solution to issues related to network partitioning. It provides efficient fault tolerance and consistency, with minimum number of nodes [8]. Quorum system is a collection of two sets of servers, read servers and write servers, and the intersection between these two is non empty [7].The intersection enables the quorum to take decision on behalf of the entire distributed system while maintaining consistency. There are several variations to this quorum protocol like Majority, Tree, Grid protocols. Tree quorums are most suited for read dominant environments whereas Grid and Majority quorums are used of *write* dominant environment.

2.3.1.6. Dual Quorum

Dual quorum (DQ) protocol meets the needs of multi- reader multi- writer objects and serves edge services better. It provides ideal read performance and availability as that of ROWA. Apart from this, DQ strengthens consistency and staleness guarantees. DQ has two separate quorums namely input quorum and output quorum to handle write and read requests respectively. In DQ, intersection is not mandatory, regular semantics is maintained via communication between the input and output quorums. With DQ, read quorum and write quorum are able to provide higher availability for the requests along with low latency. The concept of leases is also incorporated in order to allow writes to complete even during network partitions [14].

2.3.2 Dispersal Protocols

2.3.2.1 Error Correcting Codes

Error correcting codes (ECC) are used for the detection and correction of errors in data. The basic idea is to add redundancy to the stored data thus enabling retrieval of data even after the occurrence of an error or an erasure. ECC is expected to have good error correction capability which is measured in terms of the number of errors it can correct. It also has fast and efficient encoding and decoding capabilities. Binary ECC and non-binary ECC codes are the two classifications of ECC. Binary ECC performs encoding and decoding on bit basis (Convolution codes) and non binary ECC performs encoding and decoding on byte basis or symbol basis (Block codes) [6], [26]. One of the very common techniques of erasure coding is Reed Solomon code [19], [20], [21] [22]. At times in order reduce repair cost in the design of erasure codes for distributed storage systems, locality of symbols are also considered which thereby helps in reducing the repair cost [31].

2.3.2.2 Information Dispersal Algorithm

IDA, a dispersal technique [25] is a special case of ECC. It distributes k information symbols among n $(n>k)$ nodes. It enables lossless reconstruction of data with the help of any k nodes among the n nodes which brings security and confidentiality to the situation. The major advantage of IDA is its high storage efficacy which in turn saves network bandwidth [4]. The compute intensive nature increases the overhead of the processing unit which is a disadvantage [12].

2.4 *Failure*

A deviation from the expected flow of a system is a Failure and this affects the reliability of the system. Failures are predominantly two kinds: crash failures and non crash failures. Crash failure happens when a system halts unexpectedly, due to errors, before its completion time. Here the system stops functioning when it fails and the situation is referred to as fail stop failures. Non crash failures are classified into two- omission failures, timing failures and byzantine failures. During Omission failures server fails to respond to incoming requests. It can be a receive omission server fails to receive the intended message or a send omission where it fails to send a message. Timing failure occurs when a server's response is not received within the stipulated time. During Byzantine failures system behaves arbitrarily at different points of time. This error is usually caused when servers are under the control of an adversary, either a mobile adversary or a static adversary, which brings out the worst performance [11].

2.4.1 Adversary

Adversaries pose threat to a system in terms of security. The different modes of adversary attacks are intelligent guesswork, dictionary attacks, crowd-sourcing etc [16].

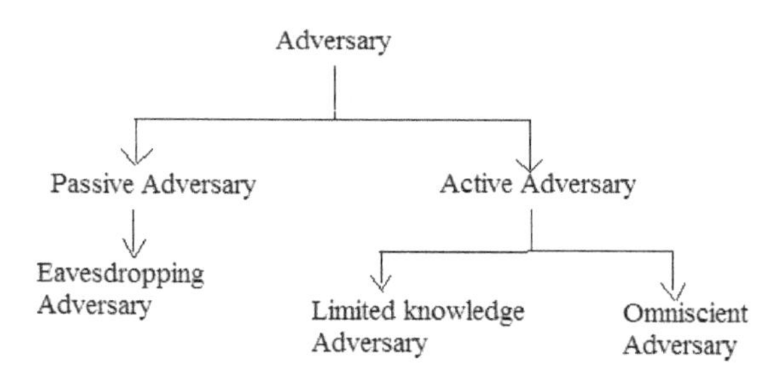

Fig. 3 Types of adversary

As shown in fig. 3 adversaries can be broadly classified into two types - passive adversary and active adversary. A passive adversary sniffs to obtain metadata or data itself. It does not interrupt communication and hence it is difficult to recognize its presence. On the other hand, an active adversary interrupts the data flow by instantiation of an evident attack in the likes of corruption of data, generation of DoS attack etc. An active adversary becomes more malicious when it is mobile as it propagates its activity from one node to another. An active adversary of Omniscient type has a complete knowledge of the system; it can control and change the behavior of nodes, whereas a Limited Knowledge kind active adversary knows only the nodes it controls. Both the adversaries evidently affect confidentiality of data thereby compromising security. Cryptographic methods help in the maintenance of integrity and confidentiality of data [15]. Numerous methodologies are developed to resist adversaries, for instance sending random hashes of stored data to verifiers, usage of polytype codes [18]. In order to measure the resistance of the system to passive eavesdroppers, the metric used is Secrecy capacity and resiliency capacity [17].

3 Survey of Storage Systems

There are ample number of storage systems of which the predominant ones (Ocean store, FARSITE, Google file system, HAIL, Coda, and Ivy) are discussed here.

FARSITE is Federated Available Reliable Storage for partially trusted environment. It has a centralized server which controls the functionalities distributed among a set of untrusted systems (servers). Servers involved are loosely coupled, insecure and unreliable in nature. Erasure coding as well as replication is used for storing redundant data. FARSITE also uses cryptographic measures and byzantine fault tolerance to enhance security [10].

Ocean store is an architecture for persistent storage of active data as well as static data at a global scale. Ocean store is built in an un-trusted environment. Replication of data is undertaken to make data available during network partitioning. Fault tolerance is achieved using additional neighbor links and Denial of Service attack is withstood using redundancy which is created by hashing node identifiers with salt values. It also provides consistency, security and privacy via cryptographic techniques and redundancy [5].

GFS is a scalable distributed file system for large, distributed, data intensive application. GFS comprises of a single master server, chunk servers and client servers. A file is divided into chunks and stored in the chunk servers where each chunk is identified using chunk handle assigned by the master server. Client servers access the data via the master server. Constant monitoring, replication and automatic recovery enables GFS to be fault tolerant [13].

HAIL (High availability and Integrity Layer) is a distributed cryptographic system. In HAIL a client distributes a file with redundancy across several servers. Client uses Proof of Retrievability (POR), which is a strategy of Test and redistribute, to check if any form of corruption has occurred [6]. HAIL is reactive against adversaries and it ensures availability even in their presence.

Coda(Constant data availability), a general purpose file system, works in a peer to peer trusted environment to provide optimistic consistency and high availability along with the ability to withstand network outages. The architecture of Coda has two components mainly Vice (server) and Venus (client). The architecture is heavily client oriented. It provides benefits of conventional file server by the utilization of decentralized architecture.

Ivy is a distributed read/write network file system set in a peer to peer architecture. Each node consists of mainly two parts, Chord/Dhash and the Ivy server. Chord provides distributed storage and enables Dhash to send and receive data from other nodes through the server interfaces. Ivy maintains a log for every participant and the logs are stored in distributed hash table [1], [24]. It maintains meta-data consistency and resolves conflicts in case of concurrent reads or writes.

4 DSS Design Recommendations

A distributed storage should be tailored to meet the requirements of specific applications and workloads it should handle. Table 1 illustrates the most suitable choice of protocol, consistency, architecture for various DSS serving purposes like general purpose, publish/share, performance oriented systems or customized applications. DSS designs should efficiently handle storage and bandwidth requirements, failures and network partitions, and security requirements. In terms of centralization, nodes should be decentralized to the possible extent so as to provide availability and to avoid single point of failures. The architecture may adopt either a linear or hierarchical arrangement for assignment of roles and responsibilities. Consistency model need not necessarily be strong, but should be strong enough to provide availability with consistency. For optimum read/write performances consistency can be liberalized (eventual consistency). To obtain redundancy at storage mere replication is not the most optimal solution. Data should be dispersed with just the required amount of replication (Dispersals, combination of dispersal and naïve replication). Failure modeling should be done and incorporated in the design to enable DSSs resistant to not only failures but resistant to adversaries and faults caused by malicious nodes.

Table .1 Insight to DSS design

Functions Features	Archival	General purpose file system	Performance	Publish/ Share	Custom
Architecture	Centralized or decentralized client server	Client server architecture/ Peer to peer architecture	Centralized Client server architecture	Peer to peer	Peer to peer/Locally centralized
Consistency	Weak consistency model	Consistency as required by the file system- Eventual consistency	Strong consistency (Locking mechanism)	Sequential consistency	Eventual consistency
Protocol	Carbonite, ROWA, Primary back up	Paxos/ primary backup	IDA/ECC Dual quorum	Quorum/ Dual quorum/ Paxos	Customized Protocol
Storage mechanism	Replication	Replication and(or)Dispersal	Dispersal/Replication	Replication	Replication or dispersal

5 Conclusion

Structuring of reliable storage is a real challenge in terms of architecture, centralization, consistency, space, bandwidth optimization, fault tolerance and security. The design should consider suitable trade-offs for the above listed.

Design of DSS should take into consideration a suitable architecture along with appropriate approach to consistency and redundancy that is well suited for the specific application at hand. Read write workloads influence the design choices. Approach to consistency is not only driven by the geographical spread and number of nodes but also by the frequency in which updates are performed on the stored data. Data redundancy techniques invariably present a tradeoff between the storage space, computation complexity and security. Spatial requirements override the concern for security in many designs. Performance and scalability of the environment is influenced by varying degree of centralization. Challenges of distributed storage system along with their practical relevance are discussed in the paper. The paper presents issues, approaches and recommendations to enable appropriate trade-offs for reliable and secure DSS.

References

1. Placek, M., Bhuyya, R.K.: A Taxonomy of Distributed Storage Systems, The university of Melbourne revision 1.148, Reporte técnico, Universidad de Melbourne, Laboratorio de sistemas distribuidos y cómputo grid (2006)
2. Lin, W., Yang, M., Zhang, L., Zhou, L.: PacificA: Replication in Log-Based Distributed Storage Systems, Technical Report MSR-TR-2008-25, Microsoft Research (2008)
3. Ford, D., Labelle, F., Popovici, F.I., Stokely, M., Truong, V.-A., Barroso, L., Grimes, C., Quinlan, S.: Availability in Globally Distributed Storage Systems. In: OSDI, pp. 61–74 (2010)
4. Zhao, D., Burlingame, K., Debains, C., Alvarez-Tabio, P., Raicu, I.: Towards High-Performance and Cost-Effective Distributed Storage Systems with Information Dispersal Algorithms. In: 2013 IEEE International Conference on Cluster Computing (CLUSTER). IEEE (2013)
5. Kubiatowicz, J., Bindel, D., Chen, Y., Czerwinski, S., Eaton, P., Geels, D., Gummadi, R., Rhea, S., Weatherspoon, H., Weimer, W., Wells, C., Ben: OceanStore: An Architecture for Global-Scale Persistent Storage, Technical report UCB/CSD-00-11-1102, pp. 8–10 (2000)
6. Bowers, K.D., Juels, A., Oprea, A.: HAIL: A High-Availability and Integrity Layer for Cloud Storage. In: Proceedings of the 16th ACM Conference on Computer and Communications Securit. ACM (2009)
7. Peleg, D., Wooly, A.: The Availability of Quorum Systems. Information and Computation **123**(2), 210–223 (1995)
8. Jimnez-Peris, R., Patio-Martnez, M., Alonso, G., Kemme, B.: How to Select a Replication Protocol According to Scalability, Availability and Communication Overhead. In: 20th IEEE Symposium on Reliable Distributed Systems Proceedings. IEEE (2001)
9. Chun, B.-G., Dabek, F., Haeberlen, A., Sit, E., Weatherspoon, H., FransKaashoek, M., Kubiatowicz, J., Morris, R.: Efficient Replica Maintenance for Distributed Storage Systems. In: NSDI, vol. 6, p. 4 (2006)
10. Adya, A., Bolosky, W.J., Castro, M., Cermak, G., Chaiken, R., Douceur, J.R., Howell, J., Lorch, J.R., Theimer, M., Wattenhofer, RP.: FARSITE: Federated, Available, and Reliable Storage for an Incompletely Trusted Environment. In: 5th Symposium on Operating Systems Design and Implementation (OSDI 2002), Boston, MA (December 2002)

11. Tanenbaum, A.S., Van Steen, M.: Distributed Systems Principles and paradigms, 2nd edn. Pearson and Princeton Hall
12. Debains, C., et al.: IStore: Towards High Efficiency, Performance, and Reliability in Distributed Data Storage with Information Dispersal Algorithms. In: IEEE MSST, 2013 (2013)(under review)
13. Ghemawat, S., Gobioff, H., Leung, S.-T.: The Google File System. ACM SIGOPS Operating Systems Review **37**(5). ACM (2003).
14. Gao, L., Dahlin, M., Zheng, J., Iyengar, A: Dual-Quorum: A Highly Available and Consistent Replication System for Edge Services. IEEE Transactions on Dependable and Secure Computing 7(2) (April-June 2010)
15. Ozan Koyluoglu, O., Rawat, A.S., Vishwanath, S.: Secure Cooperative Regenerating Codes for Distributed Storage Systems, 1-1 (2012)
16. Shah, N.B., Rashmi, K.V., Ramchandran, K., Vijay Kumar, P.: Privacy-preserving and Secure Distributed Storage Codes. In: Globecom 2011 and ISIT 2012
17. Pawar, S., Rouayheb, S.E., Ramchandran, K.: Securing Dynamic Distributed Storage Systems Against Eavesdropping and Adversarial Attacks. IEEE Transactions on Information Theory 57(10) (October 2011)
18. Kosut, O.: Polytope Codes for Distributed Storage in the Presence of an Active Omniscient Adversary. In: 2013 IEEE International Symposium on Information Theory Proceedings (ISIT). IEEE (2013)
19. Ernvall, T., Rouayheb, S.E., Hollanti, C., Vincent Poor, H.: Capacity and Security of Heterogeneous Distributed Storage Systems. IEEE Journal on Selected Areas in Communications 31(12) (December 2013)
20. Rashmi, K.V., Shah, N.B., Ramchandran, K., Vijay Kumar, P.: Regenerating Codes for Errors and Erasures in Distributed Storage. In: IEEE International Symposium on Information Theory Proceedings (2012)
21. Seroussi, G., Roth, R.M.: On MDS Extensions of Generalized Reed-Solomon Codes. IEEE Transactions on Information Theory IT 32(3) (May 1986)
22. Reed, I., Solomon, G.: Polynomial Codes Over Certain Finite Fields. Journal of the Society for Industrial and Applied Mathematics **8**(2), 300–304 (1960)
23. Meye, P., Raipin, P., Tronel, F.: Emmanuelle Anceaume, Toward a distributed storage system leveraging the DSL infrastructure of an ISP. In: Proceedings of the 11th IEEE Consumer Communications and Networking Conference (2014)
24. Chen, B., Gil, TM., Morris, R., Muthitacharoen, A.: Ivy: Read-Write Peer-to-Peer Filesystem. ACM SIGOPS Operating Systems Review 36.SI, 31–44 (2002)
25. Rabin, M.O.: Efficient Dispersal of Information for Security, Load Balancing, and Fault Tolerance. Journal of the Association for Computing Machinery 36(2) (1989)
26. Bose, R.: Information Theory Coding and cryptography. Tata McGraw-Hill Publishing Company Limited, New Delhi (2002)
27. Thekkath, C.A., Mann, T., Lee, E.K.: Frangipani: A Scalable Distributed File System 31(5). ACM (1997)
28. Gray, C.G., Cheriton, DR.: Leases: An Efficient Fault-Tolerant Mechanism for Distributed File Cache Consistency 23(5). ACM (1989)
29. Pernas, J., Gaston, B., Yuen, C., Pujol, J.: Non-homogeneous two-rack model for distributed storage systems. In: ISIT 2013 (2013)
30. Van, V.T., Yuen, C., Li, J.: Non-homogeneous distributed storage systems. Allerton 2012 (2012)
31. Song, W., Dau, S.H., Yuen, C., Li, T.J.: Optimal Locally Repairable Linear Codes. IEEE JSAC-Special Issue on Distributed Storage, pp. 1019–1036 (May 2014)

Distributed Node Fault Detection and Tolerance Algorithm for Controller Area Networks

Nithish N. Nath, V. Radhamani Pillay and G. Saisuriyaa

Abstract The major concern in a Controller Area Network based safety critical system are node failures, its early detection and failure tolerance. Aiming at the improvement of node fault detection timing response and reliability of a CAN based distributed system, this paper mainly focuses on a new fault detection and tolerance algorithm named Distributed Node Fault Detection and Tolerance [**DNFDT**] Algorithm for Controller Area Network. The purpose of the algorithm is to have an effective fault detection method for (n-1) node faults for n node system by reducing the fault detection cycle timings and by limited checking on fault free nodes. A four node experimental hardware platform was implemented and performance evaluation of the time elapsed for one complete fault detection cycle under varying CAN bus loads have been obtained.

1 Introduction

Fault diagnosis in a distributed system is of major concern. The introduction of CAN by Bosch in 1980s, revolutionized the in-vehicular communication network design using multiprocessor systems. Being a safety critical system, monitoring the communication reliability of in-vehicular network requires special attention. Once the network is perfectly functioning using CAN, the next main challenge is to overcome fault. Faults in the network can be basically classified as node failures and link failures [3]. The difficulty of Fault detection is because of the huge possibilities of fault prone areas. However, power supply failure, link (wiring) failure, entire node failure are some of the factors that determines the efficiency and reliability of the communication network. Node failures arise mainly due to processor errors, memory faults, degraded cables on the nodes, over tightened cable connection to the node on movable parts, unprotected power

N.N. Nath(✉) · V.R. Pillay · G. Saisuriyaa
Amrita Vishwa Vidyapeetham, Coimbatore, India
e-mail: nithishnnath@gmail.com, {vr_pillay,g_saisuriyaa}@cb.amrita.edu

S. Berretti et al. (eds.), *Intelligent Systems Technologies and Applications*,
Advances in Intelligent Systems and Computing 385,
DOI: 10.1007/978-3-319-23258-4_22

supply to the nodes, and electromagnetic interference problems due to closely placed sensors and actuators to the node ECU. When node fault occur the message transmitted may be corrupted or the nodes may not respond at all. In worse cases these non-responses can cause entire system shut down or may lead other nodes to bus-off state, which may invoke multiple node failures. More over if the node fault is intermitting, then considerable time may be elapsed to detect the fault node and numerous detection test trials may be required. Even though there are many fault algorithm, most are not related to CAN, or it may consume more fault detection cycle time and more CPU cycles of the node. Not only in automotive systems, but in many of the applications where a distributed network is involved, implementing an efficient fieldbus protocol would be helpful to achieve deterministic response time. In this paper a CAN fieldbus is used for a 4 node system and a new distributed node fault detection and tolerance algorithm is proposed. The proposed algorithm is implemented on an experimental layout consisting of ARM7 and ARM Cortex-M architecture boards. The next section deals with the understanding of CAN based system and fault detection and tolerance methods.

2 Earlier Work

Each module connected to the CAN bus has unique identifiers for their messages which denoted its priority. The CAN protocol focuses on the Data link Layer and the Physical Layer in the ISO/OSI model. The bus configuration of CAN, allows ECUs connected on the bus to receive and transmit in-vehicle signals digitally encoded in specified CAN message format almost simultaneously which significantly enhances its real-time capabilities. The identifiers in a CAN message format may be of 11 bit or 29bit. The basic data frame format is shown in Fig. 1.

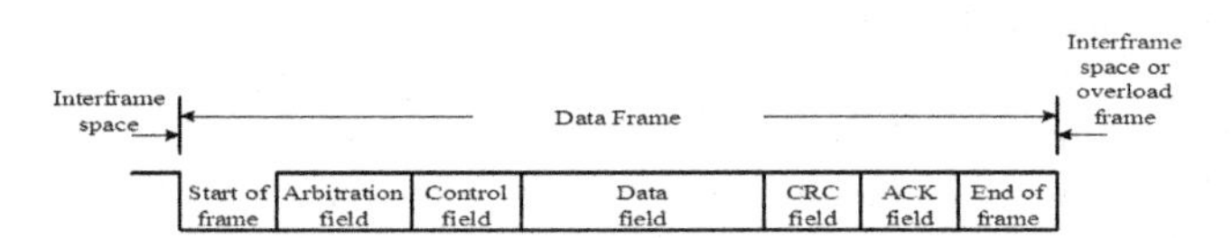

Fig. 1 CAN data frame format

CAN arbitration feature controls the bus access by ECUs in order to avoid transmission collision which causes communication errors. Security systems has been implemented using CAN [5]. Two types of communication such as Distributed control system and Field bus control system are widely used for safety critical distributed system. Scheduling in CAN includes algorithms such as Deadline monotonic (DM), Earliest Deadline First (EDF), and even by using Mixed traffic scheduler [6]. Guaranteeing fault tolerance through scheduling in CAN bus is a major aspect for safety critical systems [7]. High level of fault-tolerance in the CAN bus has been achieved by incorporating time redundancy and task schedulability tests concurrently with processor redundancy [7]. A reliability analysis method has been implemented in [8] considering the effect of faults on schedulability analysis and its impact on the reliability estimation of the system. The errors like Bit Error, Bit Stuffing Error, CRC

Error, Form Error and Acknowledgement Error can be handled by error management feature in the communication controller by notifying ECUs. Proper message delivery and deterministic behavior of the distributed network can be only achieved if accurate response time is known. A probabilistic distribution of CAN message response time can be computed using statistical analysis [9]. The major challenges in engine ignition system diagnosis is that, multiple faults may appear simultaneously, such problem refers to simultaneous-fault diagnosis. It was resolved in [2] by extraction, probabilistic classification, and decision threshold optimization. The timely detection of fault nodes in fly by wire and drive by wire technologies are being achieved in a distributed manner by developing distributed failure diagnosis algorithms under deadline and resource constraints scenarios. The use of CAN in safety-critical applications has been controversial due to dependability limitations. In particular, in a CAN bus, there are multiple components such that if any of them is affected by fault, a general failure of the entire system may happen. The algorithm proposed in this paper is designed specifically for distributed systems implemented using CAN protocol. In this section the comparison of DNFDT with other algorithms [11]-[15] for distributive fault diagnosis is done.

Algorithms in [12]-[15] are used in computer systems connected with Ethernet, whereas DNFDT and [11] are used in CAN based distributed systems. All the algorithms considered here are adaptive in nature, so do DNFDT algorithm. The main advantage of DNFDT algorithm compared to any other CAN node fault detection algorithm is that, it is able to detect all the node faults even if only one node of the entire system is perfectively functioning. The execution of DNFDT algorithm is periodic as well as sequential, whereas the algorithms [12]-[15] are being executed in parallel. The added advantages of DNFDT algorithm compared to [11] is that all diagnostic messages transmitted contains information about the status of all nodes, which will reduce the response time of nodes to get the complete system status. A node will not be checked in the next detection cycle, if it was found fault in the previous detection cycle, thus an improvement in response time and detection time are gained, which is a proposed modification in [11]. As in [11] and [15], DNFDT algorithm also considers the fault node as a fail stop model. In [11] the status of all the nodes are available only when the final node broadcast the information. This is a serious drawback, because the failure of the last node can cause entire system to fail. But in DNFDT algorithm each node broadcasts the status of all other verified nodes in a detection cycle.

3 Methodology and Approach

3.1 System Model

The proposed algorithm has been implemented on a four node test bed, which uses CAN protocol at its data-link and physical layer. Basic system design consist of four nodes N1, N2, N3 and N4 functioning as embedded hardware (ECUs), as shown in Fig. 2. The networked multiprocessor system comprises, three ARM7TDMI LPC2129 microcontroller, two ARM cortex-M3 NXP LPC1768 microcontroller and CAN bus for interconnection. Specific tasks are scheduled in each of these nodes to replicate the node functionalities. The complete system consist of one critical node and three non-critical nodes. The critical nodes of the system N1 has cold standby redundancy features incorporated into it, as an example N1' is the redundant unit of N1 which together forms the N1 module. The switching of N1 to N1|| is initiated by the fault detection algorithm. The hardware used, its clock timing parameters are explained in detail in the implementation section.

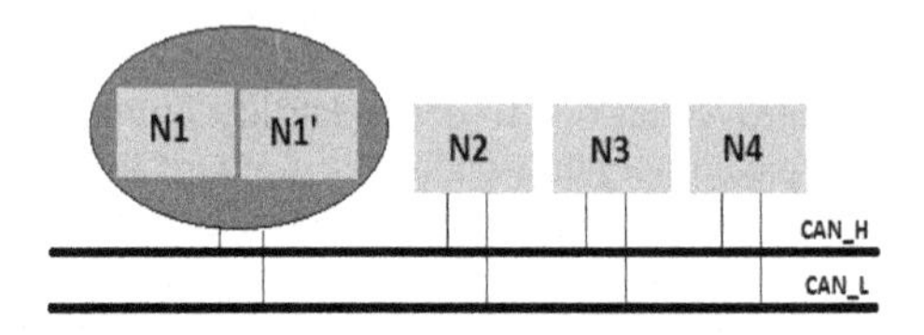

Fig. 2 Basic model of system

3.2 *Objective of DNFDT Algorithm*

- Detection of (n-1) node failures in 'n' node distributed system.
- Initiation of the fault detection algorithm irrespective of the primary node failure

3.3 *Assumption Considered for DNFDT Algorithm*

A Node is considered as fault node, if it is not responding to a message within a specified time.

3.4 *Features*

- If there are' n' nodes in the system, then the algorithm can detect (n-1) node failures.
- In a fault detection cycle a node is only checked once if it is perfectly responding to the first check, otherwise it is checked again.
- If the status of the node varies during a detection cycle once it has been checked, then the new status is not updated in the ongoing detection cycle. However the status is updated during the next detection cycle.
- If a new node is added during the execution of a detection cycle, it may not be included in the network till the end of the detection cycle. So it will be only checked during the next detection cycle.
- Initiation of the fault detection algorithm irrespective of the primary node failure.

3.5 DNFDT Algorithm

Detection of a fault node requires a minimum of single fault free node in the system and all the fault free nodes take part in finding the fault nodes. A node can have more than one type of messages with different identifiers called message identifiers (MID). Each node in the network has a buffer known as status buffer (SBUF), which consist of identifiers and status of all the nodes in the system. The size of the status buffer varies during the addition of new nodes during hot plugging. SBUF takes binary values, '0' indicates the corresponding node is fault and '1'denotes properly working nodes. The highest priority node present in the system at a given point of time initiates fault detection process for the current cycle. The indefinite waiting by working nodes for the detection message from a fault node N1with the highest priority will lead to a system failure. This algorithm overcomes such a situation by onset of timers in each nodes. Timeout interrupt generation cause the second higher priority node to start the detection process. The algorithm waits for a predefined response from another node within a specified time. The diagnostic message reserves 7 bytes for the node status which will be obtained from the buffer (SBUF) of each nodes. As mentioned earlier a node is considered as fault if it is not responding within a specified time called Good Node Time (GNT).

3.6 Typical Failure Pattern and Algorithm Behavior

Case 1: All nodes fault free - In a functional system, consider N1 has the highest priority, N2 has the second highest priority, N3 and N4 are having third and fourth priority respectively. The fault detection algorithm is initiated by the highest priority node (N1). It broadcasts a message informing other nodes about the reception of a new message from node (N2) within a short time (GNT). After obtaining the message from N1 all nodes will revise the status of N1 to good in their SBUF and will be waiting for the message from N2 within the GNT. In perfectly working condition, N2 will broadcast within the GNT and it will be apprising other nodes about the reception of a message from the node N3 within GNT.

The message from N2 encompasses the status information about N2 and earlier inspected N1. After the acceptance of the message from N2 all nodes will update the status of N1 and N2 in their SBUF and waits for a message from N3. Thus N3 and N4 also repeats the same procedure and finally the last node (i.e. N4) broadcasts a message which consist of the status information regarding all nodes in the system. This information has been used by rest of the nodes for status updating in their corresponding SBUF, which makes one detection cycle complete. The occurrence of detection cycle is mostly periodic in nature but highly depends on bus availability.

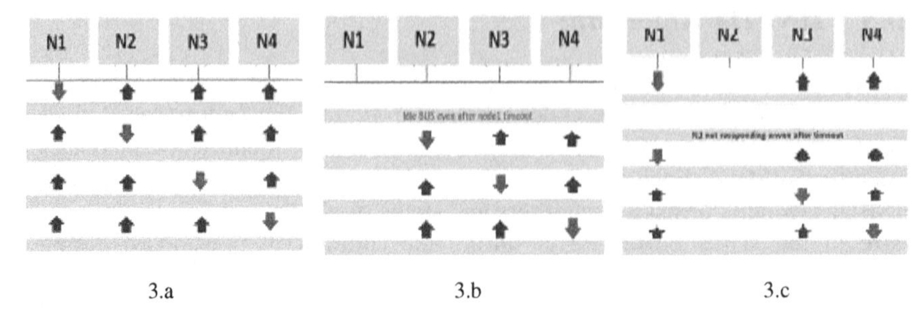

3.a 3.b 3.c

Fig. 3 Complete detection cycle- a: when all nodes are fault free, b: when highest priority node fault, c: when node 2 is fault.

Case 2: One node fault

Highest priority node fault–This can be considered as the first node fault, since we had already assumed that the first node has the highest priority. Significant importance is given to this case because fault detection algorithm is triggered by the highest priority node. The absence of the highest priority node, assigns the duty of initiation of algorithm to the second highest priority node by default. The process of overriding the function of the highest priority node is done when the timer value of the second highest priority node exceeds a threshold value known as MSTAI (Maximum Silent Time without Algorithm Initiation). Once the algorithm is initiated then the detection cycle executes normally.

Any node fault other than highest priority node – Nodes responding within the good node time are considered to be good otherwise as a fault. Let N2 be the fault node. N1 starts the detection algorithm and all other nodes will be waiting for a message from N2 within GNT. Once the time has been elapsed, N1 re-broadcasts a message, and if this message is received by N3 then it can be inferred that either the bus connected to N2 or N2 itself is fault. This information will be updated in SBUF of all the nodes. Thus in a similar manner if any node is not responding within GNT, the previous transmission is repeated to conform failure of the non-responding node.

Case 3 : New rode entry and reentry of repaired nodes

The physical connection of a node to the CAN bus and its power up results in the entry of a node to the network. All nodes entering the system are having the DNFDT algorithm implemented in it along with the code for its specific node functionality. Once the node has entered, it waits for the bus to be idle and transmits its first frame indicating its presence. In the case of new node arrival, all other nodes in the system will increase their buffer size for accommodating the new node status, while the re-entry of repaired node will retain its original location and becomes a part of the system.

4 Implementation

4.1 DNFDT Algorithm

This section covers the fundamental design of proposed DNFDT algorithm and various sub-modules in the algorithm. The algorithm implemented on each node, the algorithm during new node entry and the algorithm to increase the buffer size in each node is given in the following sections.

- Algorithm implemented on each node.

1) Start the detection cycle at the specific time period
2) Initialization
3) Wait for the Detection Message Transmission timeout (DMTT)
   DMTT >= (.5*n)
4) Obtain the recent status from the SBUF(status buffer)
5) Broadcast the detection message
   If pre-defined detection message is received
   Else if MSTAI timeout occurs
6) If the detection message is not received within the GNT
   Wait for the specific timeout condition and the retransmit the detection message
   Or
   Wait till the next instance timeout condition.
7) If a detection message is received within the GNT
   Revise the status of that node as good (1) in the SBUF
   And
   Acknowledge it by broadcasting next detection message within the GNT
   8) If detection message is not received
   Update the status of the expected message buffer as bad (0) in the SBUF
   And
   Wait for timeout and go to step 1 or 3 depending on detection message received.
9) End.

- Algorithm for broadcasting detection message during a new entry or a re-entry of a repaired node.

1) Start and initialization.
2) Wait for the bus to be idle.
3) Broadcast the detection message with the new Detection Message Identifier Byte (DMIB), for new node entry.
4) Wait for the timeout condition.
   TP(i+1)= (TP(i) + ((50ns*n)+(5000ns))) , where TP(i) is the time period of 'i'th instance.
5) If no messages are received within the above time period, go to step 2.
   If any message is received go to step 3 of Algorithm A.

- Algorithm of increasing the buffer size for a new node or a re-entry node

1) Start and initialization.
2) If a new DMIB is received, then
   a) Increment the SBUF array size by 1.
      s+=1; where 's' is the size of the array.
   b) Save the content of the received DMIB as the address location of SBUF[s].
3) Go to step 7 of Algorithm A.

4.2 *Hardware Implementation*

The four node experimental hardware layout is shown in the fig.4. Nodes N1and N2 uses ARM7TDMI LPC2129 microcontroller and N3 and N4 has ARM Cortex-M3 NXP LPC1768 microcontroller. The LPC2129 has a clock frequency of 10 MHz and LPC1768 has 96 MHz.

Fig. 4 Experimental hardware layout

Periodic fault detection cycles are automatically initiated between the control algorithms of individual node tasks. The complete system health are displayed on the display LCD, connected to node N4. The results and discussions of the fully functioning DNFDT algorithm are shown in the next session.

5 Results and Discussion

The actual implementation and evaluation of DNFDT algorithm is done using an experimental hardware layout as discussed in the hardware implementation section, shown in Fig. 4. A number of fault conditions pointed in the section 3.6 and its detection cycle test results of the different fault condition were displayed on the node 4 LCD display. In the condition where all the nodes are perfectively functioning, all the broadcasted test messages are correctly received by the corresponding nodes and finally all the node status reaches the display node where it's indicated as "All nodes good". When the power supply to node N2 is turned off, the DNFDT algorithm detects the node 2 as fault and indicated as "node 2 fault". Similarly when node 3 power is cut off, the failure is displayed as "node 3 fault". The multiple node failure condition is simulated by turning off the power to the nodes 2 and 3, DNFDT algorithm indicates the failure by "nodes 2 and 3 are fault".

The failure of the node N1 is considered as a special case, due to the hardware redundancy associated with it. As indicated in the case 2, the failure of node N1 causes its redundant unit to switch ON. The display unit indicates the bad health of N1 module as "node 1 fault, redundant ON". The red indicator on N1 indicates it's not functioning and the green indicator on N1' signifies it's active.

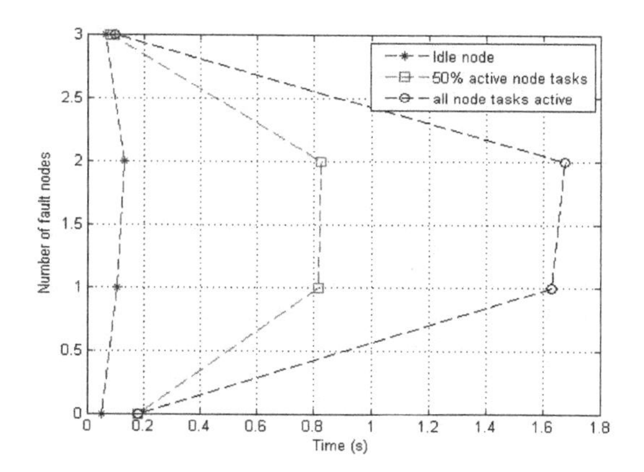

Fig. 5 One detection cycle time under varying task execution.

Fig. 5: shows the plot obtained using MATLAB software by Hardware-In-Loop simulation. The graph denotes the relation between number of fault nodes and the time required to complete one fault detection cycle. The graph is plotted for various bus loads like when the nodes are idle, 50% of node tasks are active and all node tasks are active.

6 Conclusion

The proposed algorithm detects all fault nodes within a detection cycle without any overhead of additional hardware. Analysis of the results shows the proposed algorithm has a powerful failure detection mechanism which allows it to find all the fault nodes and has valuable practical significance. A single node algorithm initiation failure is avoided which surges the algorithm reliability. DNFDT uses a definite number of detection messages in a detection cycle, depending on the number of node failures, and uses a definite bandwidth. For the best of authors knowledge, the number of detection messages per detection cycle is least for DNFDT algorithm compared to any other fault detection algorithm in a CAN based distributed embedded system. The fault detection cycle of DNFDT are mostly periodic, but highly depends on the activation of node functionality messages. The critical module switches to its redundant part in the very next detection cycle, replicating the fault node functionality within the least time. The analysis of the proposed algorithm using the MATLAB environment also adds validation for the detection cycle timings. However this paper mainly concentrates on the development of the fault detection algorithm. The implementation of the application layer and the power management layer should also be considered during the development of user friendly practical applications.

References

1. Oliveira, M.P., Fernandes, A.O., Campos, S.V.A., Zuquim, A.L.A.P.J., Mata, M.: Guaranteeing Fault Tolerance through Scheduling on a CAN bus. In: CAN in Automation, International CAN Conference (2003)
2. Vong, C.-M., Wong, P.-K., Lp, W.-F., Chiu, C.-C.: Simultaneous-Fault Diagnosis of Automotive Engine Ignition Systems Using Prior Domain Knowledge and Relevance Vector Machine. Mathematical Problems in Engineering 2013, Article ID 974862, 19 pages (2013). doi:10.1155/2013/974862.
3. Rangarajan, S., Dahbura, A.T., Ziegler, E.A.: A distributed systemlevel diagnosis algorithm for arbitrary network topologies. IEEE Transactions on Computers **44**(2), 312–334 (1995)
4. Duarte, E.P.: A hierarchical adaptive distributed system-level diagnosis algorithm. IEEE Transactions on Computers **47**(1), 34–45 (1998)
5. Mazran, E., Redzuan, A.M., Badrul, H.A., Adie, M.K., Amat, A.B.: Security System using CAN bus. Journal of Telecommunication Electronic and Computer Engineering 1 (July-December 2009)
6. Khawar, M.Z., Shin, K.G.: Design and Implementation of Efficient Message Scheduling for Controller Area Network. IEEE Transactions on Computers **49**(2), 182–188 (2000)
7. Rodriguez-Navas, G., Roca, S., Proenza, J.: Orthogonal, Fault-Tolerant, and High Precision Clock Synchronization for the Controller Area Network. IEEE Transactions on Industrial Informatics **4**(2) 92–101 (2008)
8. Hansson, H.A., Nolte, T., Norstrom, C., Punnekkat, S.: Integrating Reliability and Timing Analysis of CAN-Based Systems. IEEE Transactions on Industrial Electronics **49**(6), 165–172 (2002)
9. Zeng, H., Di Natale, M., Giusto, P., Sangiovanni-Vincentelli, A.: Using Statistical Methods to Compute the Probability Distribution of Message Response Time in Controller Area Network. IEEE Transactions on Industrial Informatics **6**(4), 678–691 (2010)
10. Bianchini, R.P., Buskens, R.W.: Implementation of on-line distributed system-level diagnosis theory. IEEE Transactions on Computers **41**(5), 616–626 (1992)
11. Kelkar, S., Kamal, R.: Adaptive Fault Diagnosis Algorithm for Controller Area Network. IEEE Transaction on Industrial Electronics **61**(10), 5524–5537 (2014)
12. Bagchi, A., Hakimi, S.L.: An optimal algorithm for distributed system level diagnosis. In: 21st IEEE International Symposium on Fault-Tolerant Computer, Montreal, QC, Canada, pp. 214–221 (1991)

13. Bianchini, R.P., Buskens, R.W.: Implementation of on-line distributed system-level diagnosis theory. IEEE Transactions on Computers **41**(5), 616–626 (1992)
14. Rangarajan, S., Dahbura, A.T., Ziegler, E.A.: A distributed systemlevel diagnosis algorithm for arbitrary network topologies. IEEE Transactions on Computers **44**(2), 312–334 (1995)
15. Duarte, E.P., Nanya, T.: A hierarchical adaptive distributed system-level diagnosis algorithm. IEEE Transactions on Computers **47**(1), 34–45 (1998)
16. Albini, L.C.P., Brawerman, A.: An algorithm for distributed hierarchical diagnosis of dynamic fault and repair events. In: Proceedings Seventh International Conference on Parallel and Distributed Systems, pp. 299–306 (July 2000)
17. Choi, K., Luo, J., Pattipati, K., Namburu, S.M., Qiao, L., Chigusa, S.: Data reduction techniques for intelligent fault diagnosis in automotive systems. In: IEEE International Conference Autotestcon, pp. 66–72 (September 2006)
18. Barranco, M.: An Active Star Topology for Improving Fault Confinement in CAN Networks. IEEE Transactions on Industrial Informatics **2**(2), 78–85 (2006)
19. Suwatthiku, J.: Fault detection and diagnosis for in-vehicle networks. In: Zhang, W. (ed.) Fault Detection (2010)
20. Zeng, H., Di Natale, M., Giusto, P., Sangiovanni-Vincentelli, A.: Using Statistical Methods to Compute the Probability Distribution of Message Response Time in Controller Area Network. IEEE Transactions on Industrial Informatics **6**(4), 678–961 (2010)
21. Abd-El-Barr, M.: System Level Diagnosis-I, Design and Analysis of Reliable and Fault- Tolerant Computer Systems (2006)
22. CAN in Automation. http://www.canopen.org

Part III
Business Intelligence and Big Data Analytics

Ensemble Prefetching Through Classification Using Support Vector Machine

Chithra D. Gracia and Sudha

Abstract Owing to the steadfast growth of the Internet web objects and its multiple types, the latency incurred by the clients to retrieve a web document is perceived to be higher. Web prefetching is a challenging yet achievable technique to reduce the thus perceived latency. It anticipates the objects that may be requested in future based on certain features and fetches them into cache before actual request is made. Therefore, to achieve higher cache hit rate group prefetching is better. According to this, classification of web objects as groups using features like relative popularity and time of request is intended. Classification is aimed using Support Vector Machine learning approach and its higher classification rate reveals effective grouping. Once classified, prefetching is performed. Experiments are carried out to study the prefetching performance through Markov model, ART1, linear SVM and multiclass SVM approach. Compared to other techniques, a maximum hit rate of 93.39% and 94.11% with OAO and OAA SVM multiclass approach is attained respectively. Higher hit rate exhibited by the multiclass Support Vector Machine demonstrates the efficacy of the proposal.

Keywords Prefetching · Classification · Machine learning · SVM · Hit rate · ART1

1 Introduction

Since, the World Wide Web (WWW) has grown to be well popular, services over the web and requirement for web objects by users has grown vividly [1]. Therefore, the latency incurred by the users in accessing the web objects over the web

C.D. Gracia
Department of Computer Science and Engineering,
National Institute of Technology, Tiruchirapalli, India
e-mail: chithragracia@gmail.com

Sudha(✉)
Department of Electrical and Electronics Engineering,
National Institute of Technology, Tiruchirapalli, India
e-mail: sudha@nitt.edu

S. Berretti et al. (eds.), *Intelligent Systems Technologies and Applications*,
Advances in Intelligent Systems and Computing 385,
DOI: 10.1007/978-3-319-23258-4_23

has also increased to a predominant issue that is to be met. Web prefetching is a promising and outstanding technique to reduce this latency. Prefetching as groups will produce more precise and accurate results. The challenge is to find a way to categorize this massive data in some meaningful structure. Prefetching in groups based on similarity measures can be obtained through data mining techniques like classification and clustering [2]. Classification is a supervised machine learning technique that groups related objects based on similarity of features or attributes. Support Vector Machine is a multivariate machine learning algorithm which supports classification.

Current literature gives subjective and descriptive ideas on prefetching and classification using SVM's. Most of the prefetching techniques relay on Markov model approaches since they are effective in predicting the next to be accessed page [3]. However, it not fast in adapting to the new patterns because the predictions are solely history based that is accessed during the previous time period. The effectiveness of using prefetching to resolve the problems in handling dynamic web pages is studied [4]. Yet, only the temporal properties of the dynamic web pages are explored.

Metadata prefetching based on relationship graphs with a significantly lesser hit rate of 70% is presented [5]. Semantic based prefetching relies on predicting the future requests based on semantics. The semantic preferences are exploited by analyzing the keywords in the URL anchor text or the documents that are previously accessed [6].

Clustering is an unsupervised data mining technique that groups data according to similarity such that intra data instances within a cluster are similar to each other and the inter data instances between clusters are much dissimilar. As told by Vakali.A. et. al, a wide range of web data clustering schemes present in the literature just focuses on clustering of inter-site and intra-site web pages[7]. Clustering can be based on statistical and evolutionary algorithms.

Classification, is assigning a class label to a set of unclassified cases. The main idea of classification and clustering lies in finding the hidden patterns in data. Support Vector Machines (SVM's) are a new generation learning system based on recent advances in statistical learning theory and are accurate with lesser training samples [8]. The tuning of the parameters of binary support vector machines in multiclass decomposition using genetic algorithm is performed [9]. SVM is used to predict the to be visited objects in order to optimize cache usage [10]. A Multiclass Support Vector Machine classifier approach in hypothyroid detection is performed [11]. However, all the aforementioned works focus on various applications other than prefetching. To our knowledge, no work is reported in literature that classifies the web objects through multiclass classification for prefetching.

SVM being a case-based classifier (non parametric) does not require any prior knowledge other than the training samples [12]. Moreover, training using SVM is faster with high generalization ability and avoids premature convergence. Taking these positive factors of SVM into consideration, an attempt is made to classify the web objects into classes to prefetch the related web objects of the class that the request corresponds to. To study the performance of classification, both binary and multi-class classification is intended. The following section explains in brief the classification techniques.

2 Overview of Binary and Multiclass SVM

The primary intuition behind SVM is to maximize the margin 'm' between the hyperplanes. Fig. 1 shows the hyperplanes and the margin between them. The idea is to find the function of the optimizing hyperplane between the classes.

2.1 *Binary SVM*

The purpose of SVM is to find the hyperplane that best separates the classes. If the number of classes is two, then binary classification is performed.

The hyperplane is characterized by the decision function [14]

$$f(x) = \mathrm{sgn}(< w, \Phi(x) > +b)$$

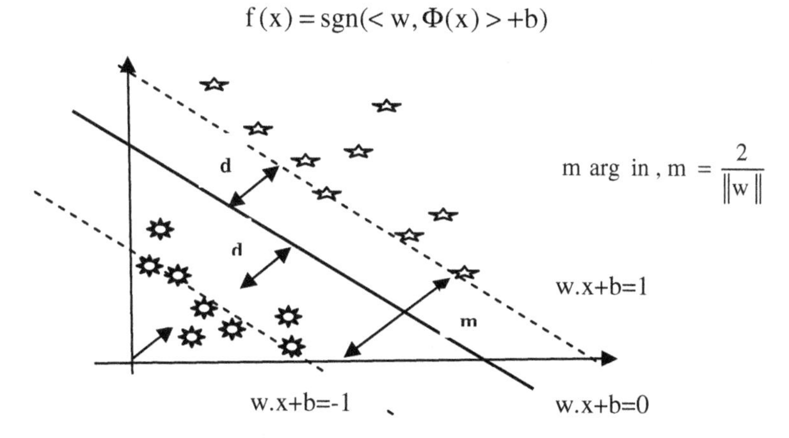

Fig. 1 Optimal separating hyperplane with maximum margin

where

w	- weight vector orthogonal to the hyperplane
x	- current input
$\Phi(x)$	- kernel transforming input data into feature space
b	- scalar that represent the margin of the hyperplane
$<,>$	- dot product
sgn	- signum function which extracts the sign of a number (i.e.,) returns 1 if value > 0 and -1 otherwise

If 'w' has unit length (norm=1), then $(< w, \Phi(x) >)$ is the length of $\Phi(x)$ along the direction of 'w'.'w' leads to largest 'b'. 'w' is scaled by $\|w\|$ to obtain a unit vector where, $\|w\| =< w, w >$.The normal vector 'w' that leads to largest 'b' of the hyperplane is to be found.

2.1.1 Linearly Separable Points

These are the points that can be linearly separated by a hyperplane in the input space itself. Let $x, y \in R^n$ where $x = (x_1, x_2, ..., x_n)$ and $y = (y_1, y_2, ..., y_n)$ The optimal separating hyperplane is found by [12]

$$\underset{w \in H, b \in R}{\text{maximize}} \min\{\|x - x_i\| \mid x \in H, < w, x > +b = 0, i = 1, ..., m\}$$

Margin $\frac{2}{\|w\|}$ is to be maximized. It is obtained by solving the objective function,

$$\underset{w \in H, b \in R}{\text{min imize}} \quad \tau(w) = \frac{1}{2}\|w\|^2$$

subject to $y_i(< w, x_i > +b) \geq 1 \quad \text{for all} \quad i = 1, ..., m$

where ' τ ' is an objective function. The constraints ensure

$$f(x_i) = +1 \quad \text{for} \quad y_i = +1$$

$$f(x_i) = -1 \quad \text{for} \quad y_i = -1$$

To solve the optimization which is in primal form, it is converted to its dual form by applying the Lagrangian multipliers $\alpha_i \geq 0$, that leads to the dual optimization problem.

$$L(w, b, \alpha) = \frac{1}{2}\|w\|^2 - \sum_{i=1}^{m} \alpha_i (y_i(< x_i, w > +b) - 1)$$

The Lagrangian 'L' must be maximized with respect to ' α_i ' and minimized with respect to primal variables 'w' and 'b'. At the saddle point, the partial derivatives of L with respect to primal variables must be 0. Hence,

$$w = \sum_{i=0}^{m} \alpha_i y_i x_i \quad , \quad \sum_{i=0}^{m} \alpha_i y_i = 0$$

The training points with non zero ' α_i ' are called support vectors. As per Karush-Kuhn-Tucker condition (KKT), ' α_i ' that are non zero at the saddle point corresponds to the constraints that are precisely met. Hyperplane in dual optimization problem is

$$f(x) = \text{sgn}\left(\sum_{i=1}^{m} y_i \alpha_i < x_i, x > +b\right)$$

Where 'b' is computed by KKT conditions. When training errors are encountered, slack variables ' ξ_i ' is introduced and the objective function to be solved is

$$\tau(w,\xi) = \frac{1}{2}\|w\|^2 + C\sum_{i=1}^{m}\xi_i$$

subject to $y_i(< w, x_i > +b) \geq 1 - \xi$ for all $i = 1,...,m$

where $0 \leq \alpha_i \leq C$. Finally, $\sum_{i=0}^{m}\alpha_i y_i = 0$

2.1.2 Non Linearly Separable Points

The points that are not separable linearly by a hyperplane in the input space are mapped to a higher dimensional feature space as in Fig. 2. The kernel function $'\Phi'$ implicitly maps the training data in the input space to a higher dimensional feature space and constructs a separating hyperplane in the feature space. It is given by the kernel trick $K(x,x') =< x, x' >$. The decision function obtained [14] is

$$f(x) = \text{sgn}(\sum_{i=1}^{m} y_i\alpha_i K(x,x') + b)$$

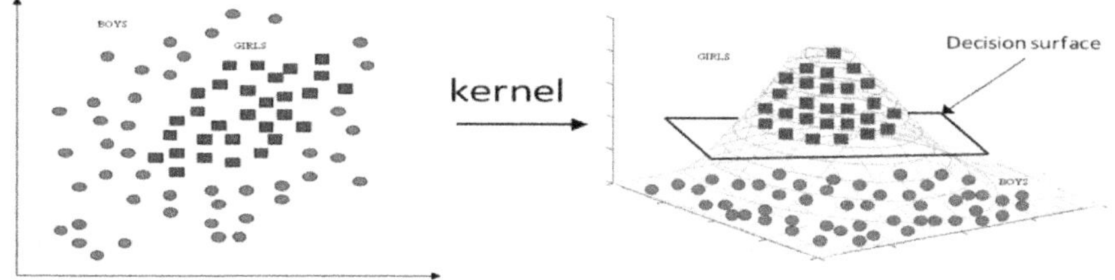

Fig. 2 Kernel mapping from input space to feature space

2.2 *Multiclass SVM*

If the number of labels is greater than two, then multiclass classification is performed. In the One Against All (OAA) approach, to have a classification of 'M' classes, a set of binary classifiers is constructed where each training separates one class from the rest. The output obtained will be the maximum of all the decision functions that is obtained [14] by

$$\underset{j=1,...,M}{\text{argmax}}\ g^j(x) \quad \text{where } g^j(x) = \sum_{i=1}^{m} y_i\alpha_i^j K(x,x_i) + b^j$$

and the decision function is $f^j(x) = \text{sgn}(g^j(x))$. This approach computes 'M' hyperplanes.

In the One Against One (OAO) approach, pairwise classification is performed. Two classes are chosen and a hyperplane is constructed between them. This process is repeated for each pair of classes in the training set. The output is the decision function of the hyperplane with lesser number of support vectors.

3 Proposed Work

The objective of the proposal is to minimize the user perceived latency (client) while accessing the web server. Prefetching using lists tend to consume more memory with low hit rates, because the list is large in size. Especially, with the current Internet supporting multiple types of web objects like audio, video, images and text files creating lists would worsen this further. So, it will be ideal if the web objects are classified into groups which would minimize the group size. Minimizing the group size accounts to higher hit rate as only the objects related to a specific group are prefetched. As grouping is possible through classification, classification of multiple web objects is attempted through Support Vector Machine. The proposal comprises of three phases namely; preprocessing of web logs, classification and prefetching. Fig. 3 depicts the various tasks carried out in each of the phases and the co-ordination between them. Each of these phases is dealt briefly in the succeeding section.

Phase I: Data Preprocessing

The access logs consists of ten fields namely; timestamp, elapsed time, client's IP address, log tag with HTTP code, type of request(method), URI, user identification, hierarchy data and host name and content type. The log is pre-processed to extract the necessary fields. Preprocessing of data involves data formatting, filtering and extracting required fields in a format suitable for classification. As the log files are unstructured, identification of the various fields is carried out through data formatting.Data filtering involves elimination of entries that have unsuccessful HTTP status codes. Only URL's with status code 200 are considered. Requests with question marks in URL's, cgi-bin is discarded. As the data extracted is free of labels, it is to be presented in a form suitable for labelling. So, it is displayed in the form <*a1 a2 a3*> where *a1* denotes the requested URL (web object), *a2* the time of request and *a3* the access count. An example of preprocessed data is shown in Table 1.

Table 1 Preprocessed log file

URL_Request	*Timestamp*	*Access count*
/academics/departments/ece/programmes/btech/curriculum/	0.125000	13
/home/students/facilitiesnservices/hostelsnmess/hostels/	0.541666	2

Phase II: Classification and Training

The SVM uses (i) training data to generate the learned model (classifier) and (ii) when test data is given as input to the learned model, it classifies the data. Labelling is performed based on the relative popularity of URL's and their timestamp.The relative popularity (RP_i) of the 'ith' URL is calculated using Eqn. 1. Then the weighted average, w_{avg} of all URL's is found using Eqn. 2. It is the weighted average of the relative popularity of all unique URL's. This measure is used as a threshold to decide the URL's that are to be prefetched. This avoids prefetching of all objects corresponding to the entire class. By this only the URL's whose access count is above the threshold (those URL's in the class that are more frequently accessed) are prefetched to achieve high hit rate.

$$\text{Relative Popularity of a URL}(RP_1) = \frac{\text{Number of accesses to the 'i}^{th}\text{' URL}}{\text{Highest access count in the log}} \tag{1}$$

$$\text{Weighted Average, } w_{avg} = \frac{\sum_{i=1}^{N} RP_i * w_i}{\sum_{i=1}^{N} w_i} \tag{2}$$

Binary classification is done by taking the access count of the URLs into consideration and rules are framed to assign labels. The rules framed are as in Eqn.3.

$$\begin{aligned} &Rule\ 1: && IF\ (RP_i > w_{avg})\ THEN\ \ URL_i \leftarrow label\ 1 \\ &Rule\ 2: && IF\ (RP_i < w_{avg})\ THEN\ \ URL_i \leftarrow label\ 2 \end{aligned} \tag{3}$$

Based on these rules the URL's are labeled. To train the SVM the data is organized in the form <*x1 la* > where '*x1*' represent the relative popularity of the web object and 'la' denotes the label of the URL.

Further, multi classification is intended for which more labels are to be included. Hence, the feature set is extended to accommodate more rules and thereby more labels. So, timestamp at which the URL is accessed is also taken in addition to the Relative popularity. The rules framed are given in Eqn.4.

$$\begin{aligned} &\text{Rule 1:} && \text{IF}(RP_i > w_{avg}) \text{ and time}_i = \text{recent time}_i \pm 2 \text{ THEN URL}_i \leftarrow \text{label 1} \\ &\text{Rule 2:} && \text{IF}(RP_i > w_{avg}) \text{ and time}_i <> \text{recent time}_i \pm 2 \text{ THEN URL}_i \leftarrow \text{label 2} \\ &\text{Rule 3:} && \text{IF}(RP_i < w_{avg}) \text{ and time}_i = \text{recent time}_i \pm 2 \text{ THEN URL}_i \leftarrow \text{label 3} \\ &\text{Rule 4:} && \text{IF}(RP_i < w_{avg}) \text{ and time}_i <> \text{recent time}_i \pm 2 \text{ THEN URL}_i \leftarrow \text{label4} \end{aligned} \tag{4}$$

where recent $time_i$ is the recently accessed time of the 'i'th URL and the constant '2' refers to 2 hours. Based on the rules framed above, the URL's are labeled. To train the SVM, the data is converted to the pattern < *la x1 x2* > where 'la' denotes the label of 'i'th URL, '*x1*' represents the relative popularity of 'i'th URL, *x2* denotes the time of request of 'i'th URL. A snippet of the data set which is to be trained is displayed in Fig. 4.

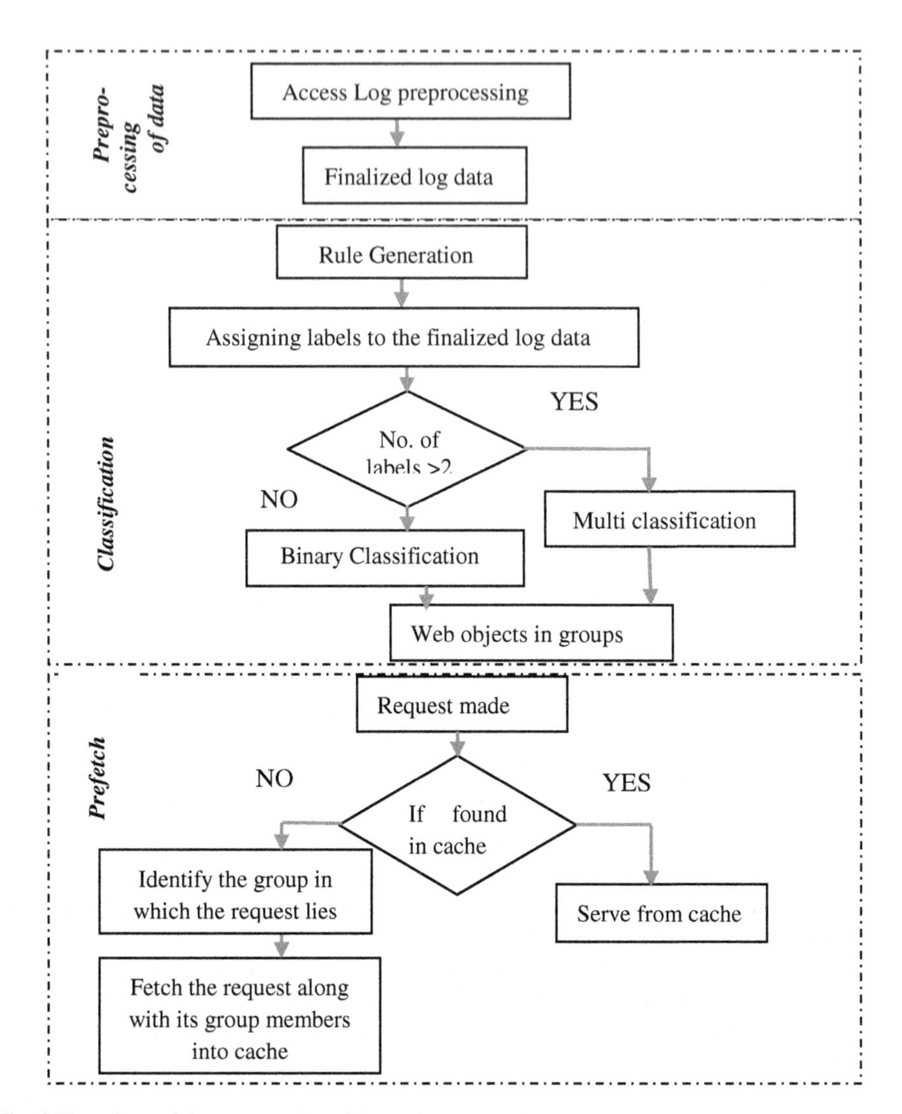

Fig. 3 Flowchart of the proposed model

1	0.250000	0.936218678815
1	0.250000	1.000000000000
2	0.125000	0.173120728929
2	0.125000	0.002277904328
2	0.125000	0.002277904328
2	0.125000	0.002277904328
2	0.125000	0.002277904328
2	0.125000	0.002277904328
2	0.125000	0.002277904328

Fig. 4 Sample of the log file features and their labels

Once the dataset is labeled, classification focusing both on binary and multiclass SVM's is performed.

Training with Binary SVM and Multi Class SVM

The binary classification is performed by separating the two classes with a linear optimal separating hyperplane that maximizes the margin. Further, when a new data is given, the data is classified based on the sign value of the hyperplane function which is discussed in section 2.

For the non-linear cases, the Gaussian RBF kernel (Radial Basis Function) kernel is employed since it is considered better than other kernel functions as it is localized and has finite response across the range of real x-axis [13]. The RBF kernel on two samples x and y is represented as feature vectors in some input space which is given by $K(x, y) = \exp\left(-\frac{\|x - y\|^2}{2\sigma^2}\right)$ where $\|x - y\|^2$ is recognized as the squared Euclidean distance between the two feature vectors. 'σ' is a free parameter. Two parameters namely the kernel parameter 'γ' where $\gamma = -\frac{1}{2\sigma^2}$ and cost parameter 'C' is to be set. By altering various values for C and 'γ', the generalization capability of the SVM is controlled and the training is done. It is found that choice of kernel function and suitable value for the parameters of the kernel is essential for a given data to perform better classification. A combination of binary SVMs is involved in the multi-class classification using one-against-all (OAA) and compared against one-against-one (OAO) approach. In the OAA approach, the members that belong to a class are differentiated from the rest of the classes. To perform classification of 'M' classes, each training constructs a set of binary classifiers where each class separates one class from the others. Finally, all the classifiers are combined by multiclass approach. The output of this approach corresponds to the maximum obtained from all the decision functions.

Phase III: Prefetching

When the browser is idle, the classified data is subjected to prefetch in the cache. On making a request, the URL is searched in cache and if found, the URL is served from the cache. If not, the web object corresponding to the URL is obtained from the server along with its corresponding class members and placed in cache.

4 Results and Analysis

The hardware platform used for the implementation is a high end workstation with i7 processor configured as server running @3.40GHz and 16GB RAM. The access logs of NITT web server (nitt.edu) is taken for classification and a snapshot is displayed in Fig. 5.

A mirror of the NITT web server is obtained to carry out the experimentation. 30 log files (corresponding to 30 days) collected from 24th August 2014 to 23rd September 2014 is used for our experimentation. The cache size is 1024MB.

```
117.193.32.150 - - [24/Aug/2014:09:42:47 +0530] "GET
/home/academics/departments/civil/ HTTP/1.1" 200 14126 "-"
"Mozilla/5.0 (Windows NT 6.1) AppleWebKit/537.36 (KHTML, like
Gecko) Chrome/27.0.1453.110 Safari/537.36"
117.193.32.150 - - [24/Aug/2014:09:42:48 +0530] "GET
http://www.nitt.edu/home/academics/departments/civil/MTTP/1.1"
304 -
"http://www.nitt.edu/home/students/facilitiesnservices/tp/"
"Mozilla/5.0 (Windows NT 6.1) AppleWebKit/537.36 (KHTML, like
Gecko) Chrome/27.0.1453.110 Safari/537.36"
117.193.32.150 - - [24/Aug/2014:09:42:48 +0530] "GET
http://www.nitt.edu/home/academics/departments/cse/services/"
"Mozilla/5.0 (Windows NT 6.1) AppleWebKit/537.36 (KHTML, like
Gecko) Chrome/27.0.1453.110 Safari/537.36"
117.199.127.189 - - [24/Aug/2014:09:42:47 +0530] "GET
/home/students/facilitiesnservices/hostelsnmess/hostels/
HTTP/1.1" 200 22655 "http://www.nitt.edu/home/" "Mozilla/5.0
(Windows NT 6.1) AppleWebKit/537.36 (KHTML, like Gecko)
Chrome/27.0.1453.116 Safari/537.36"
117.193.32.150 - - [24/Aug/2014:09:42:48 +0530] "GET
/cms/templates/corporate-final/images/tooltip/phoneicon1.gif
HTTP/1.1" 304 -
"http://www.nitt.edu/home/students/facilitiesnservices/tp/"
"Mozilla/5.0 (Windows NT 6.1) AppleWebKit/537.36 (KHTML, like
Gecko) Chrome/27.0.1453.110 Safari/537.36"
117.193.32.150 - - [24/Aug/2014:09:42:48 +0530] "GET
/cms/templates/corporate-final/images/tooltip/emailicon1.gif
HTTP/1.1" 304 -
"http://www.nitt.edu/home/students/facilitiesnservices/tp/"
"Mozilla/5.0 (Windows NT 6.1) AppleWebKit/537.36 (KHTML, like
Gecko) Chrome/27.0.1453.110 Safari/537.36"
```

Fig. 5 Snippet of the log file for training

4.1 Classification Results

These 30 log files are divided into 2 log files namely log1 and log2 each corresponding to 15 days. Log1 is used for training the classifier. Log2 is further divided into dataset1 of size 650MB with 3891 URL accesses and dataset2 of 458MB with 2904 URL accesses to test the classifier. Classification is performed using svmtoy module of the LIBSVM 3.20 software implemented in MATLAB 2013a. Experiments are carried out to study the performance of ART1, binary and multiclass SVM's. Their results are summarized below.

Test Case 1: Binary Classification

Log1 is labeled using Eqn.3 and the labeled log file is used to train both the linear and non-linear SVM binary classifier. Dataset1 and Dataset2 is used to test the accuracy of the trained classifier. Fig. 6(a) & (b) shows the binary classification output of dataset1 for both linear and non linear SVM.

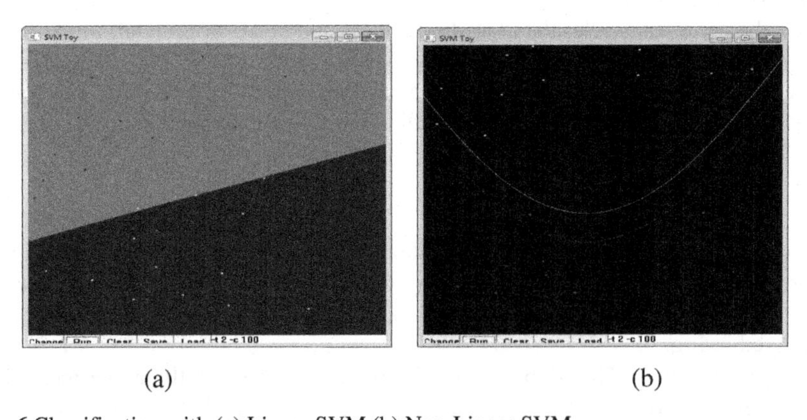

Fig. 6 Classification with (a) Linear SVM (b) Non-Linear SVM

From the above results, the misclassification rate with linear SVM and non linear SVM for dataset1 is found to be 13.28% and 12.01% respectively.

Test Case 2: Multiclass Classification

Log1 is labeled using Eqn.4 and the labeled log file is used to train both the OAA and OAO SVM multiclass classifiers. Dataset1 and Dataset2 is used to test the accuracy of the trained multiclass classifiers. The output obtained for dataset1 through OAA and OAO classifiers is displayed in Fig. 7(a) & (b) respectively.

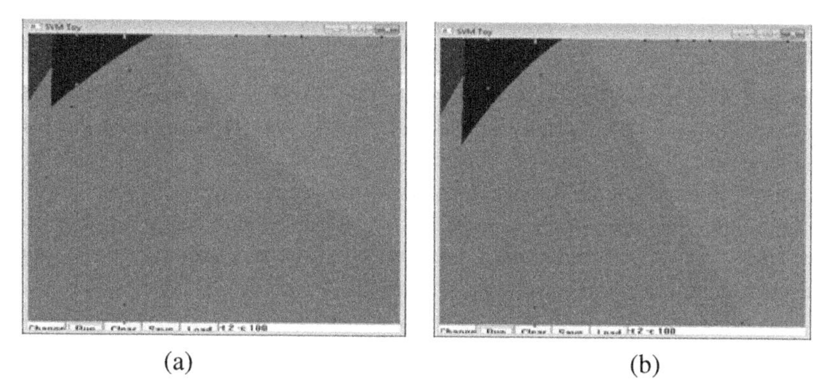

(a) (b)

Fig. 7 Classification with (a) OAO SVM (b) OAO SVM

From the above results, the misclassification rate is found to be around 10.81% and 11.27% through OAA and OAO multiclass SVM models respectively. Similarly, classification with ART1 is also carried out. The classification rate obtained through various classification techniques with both datasets is tabulated in Table2.

Table 2 Classification rate through various techniques

Techniques		Dataset1 (%)	Dataset2 (%)
ART1 Network		85.45	86.25
Binary class	Linear SVM	86.23	86.21
	Non linear SVM	87.97	88.9
Multiclass	OAA	90.23	91.35
	OAO	89.23	90.00

From Table 2, the classification rate through multiclass SVM is inferred to be higher than the binary SVM. Further, of the multiclass classifiers OAA is found to be even much better than OAO confirming the efficacy of multiclass OAA SVM.

Further, to study the prefetching performance based on the requested web object prefetching by different methods such as Popularity Based Markov model, binary and multiclass SVM is performed. The hit rate obtained through these techniques using dataset1 and dataset2 is displayed in Table 3.

Table 3 Hit rate obtained by different techniques

Techniques		Dataset1 (%)	Dataset2 (%)
Popularity Based Markov Model		88.94	89.01
ART1 Network		85.11	86.29
Binary Class	Linear SVM	89.67	88.39
	Non linear SVM	92.22	92.91
Multiclass	OAA	94.07	94.11
	OAO	93.39	93.13

From Table 3, prefetching through multiclass SVM is found to yield higher hit rate than Binary SVM, Markov model and ART1. This is because of the history based and unclassified data of Markov, large group size of ART1 and limited classes of binary SVM. Whereas, OAA SVM multiclass approach gives higher hit rate than others because of (i) its high classification rate and (ii) group size limitation. Hence, it is concluded that prefetching through ensembled web objects achieve high hit rates and especially, with multi class OAA SVM.

5 Conclusion

Web prefetching minimizes latency. Literatures present various methods of prefetching including binary classification. But none of the work is reported using multiclass classification for ensemble prefetching. Hence the proposed work focuses on group prefetching of web objects through multi-class classification and subsequent prefetching. As the log files contain unsupervised data, they are initially converted to supervised data by assigning labels. Further, training is carried out to compare the classification efficiency using ART1 neural network, binary and multiclass SVM. Further, web object prefetching is performed using the classification and Markov models. Experimental results using OAA multi-class classification yielded higher classification rate than the other techniques. Moreover, higher hit rate is observed through prefetching based on OAA multi-class classification than ART1 neural network and Markov model approach. Thus, the proposed multi-class SVM classification is found to support prefetching effectively and efficiently.

References

1. Domenec, J., Gil, J.A., Sahuquillo, J., Pont, A.: Using current web page structure to improve prefetching performance. Computer Networks **54**, 1404–1417 (2010)
2. Liao, S.-H., Chu, P.-H., Hsiao, P.-Y.: Data mining techniques and applications – A decade review from 2000 to 2011. Expert Systems with Applications **39**(12), 11303–11311 (2012)

3. Chithra, D.G., Sudha, S.: MePPM- Memory efficient Prediction by Partial Match Model for Web Prefetching. In: 3rd IEEE International Advance Computing, Conference (IACC), pp. 736–740 (2013)
4. Lam, K.Y., Ngan, C.H.: Temporal Pre-Fetching of Dynamic Web Pages. Info. Systems Journal, Elsevier, 31149–31169 (2006)
5. Gu, P., Wang, J., Jiang, H.: A novel weighted-graph based grouping algorithm for metadata prefetching. IEEE Transactions on Computers 59(1), 1–14 (2010)
6. Alexander, P.P.: Semantic Prefetching Objects of Slower Web Site Pages. The Journal of System. and Software **79**, 1715–1724 (2006)
7. Vakali, A.I., Pokorný, J., Dalamagas, T.: An Overview of Web Data Clustering Practices. In: Lindner, W., Fischer, F., Türker, C., Tzitzikas, Y., Vakali, A.I. (eds.) EDBT 2004. LNCS, vol. 3268, pp. 597–606. Springer, Heidelberg (2004)
8. Mathur, A., Foody, G.M.: Multiclass and binary SVM classification: Implications for training and classification users. IEEE Geos. and Rem. Sens. Letters **5**(2), 241–245 (2008)
9. Lorena, A.C., de Carvalho, A.C.P.L.F.: Evolutionary tuning of SVM parameter values in multiclass problems. Neurocomputing **71**, 3326–3334 (2008)
10. Ali, W., Shamsuddin, S.M., Ismail, A.S.: Web Proxy Cache Content Classification based on Support Vector Machine. Journal of Artificial Intelligence **4**, 100–109 (2011)
11. Chamasemani, F.F., Singh, Y.P.: Multi-class Support Vector Machine (SVM) Classifiers–An Application in Hypothyroid Detection and Classification. In: 2011 Sixth International Conference on Bio-Inspired Computing: Theories and Applications (BIC-TA). IEEE (2011)
12. Denœux, T., Smets, P.: Classification using belief functions: relationship between case-based and model-based approaches. IEEE Trans. System Man Cybernetic. Part B: Cybernetic. **36**(6), 1395–1406 (2006)
13. Arora, M., Kanjilal, U., Varshney, D.: Efficient and Intelligent Information Retrieval. International Journal of Soft Computing and Engineering (IJSCE) **1**(6), 39–43 (2012)
14. Morariu, D.: Classification and Clustering using SVM, Ph.D Report, University of Sibiu (2005)

An Effective Stock Price Prediction Technique Using Hybrid Adaptive Neuro Fuzzy Inference System Based on Grid Partitioning

Atanu Chakraborty, Debojoyti Mukherjee, Amit Dutta, Aruna Chakraborty and Dipak Kumar Kole

Abstract Prediction of forthcoming stock price is an important area of research. A large number of data is used by the system to predict the possible upcoming events in future. The stock prediction work is done primarily for overnight as it gets more volatile in a longer span. However in this work an effective effort is made to extend the duration of prediction to 14 days. A fuzzy logic approach based on grid partition is adopted in the paper to deal with the uncertainty factors while predicting the stock price of any company. The *premise* and *consequent* parameters of the learning rules are optimized in an adaptive fashion using a hybrid neural learning mechanism. This Adaptive Neuro-Fuzzy Inference System (ANFIS) using grid partition is undertaken to deal with the problem of stock price prediction, which leads to an accuracy of 94-95%.

Keywords Stock prediction · Grid partitioning · ANFIS · Normalization

1 Introduction

Future stock price controls many a thing of business including company's vision, business policy and projects to be undertaken. So, there is always a craze to predict the future of stock market price to gain the maximum amount of profit and

A. Chakraborty(✉) · D. Mukherjee · A. Chakraborty
Department of CSE, St. Thomas' College of Engineering & Technology, Kolkata, India
e-mail: {atanu.here2011,debmuk1993,aruna.stcet}@gmail.com

A. Dutta
Department of IT, St. Thomas' College of Engineering & Technology, Kolkata, India
e-mail: to.dutta@gmail.com

D.K. Kole
Department of CSE, Jalpaiguri Government Engineering College,
Jalpaiguri, West Bengal, India
e-mail: dipakk.cse@jgec.in

S. Berretti et al. (eds.), *Intelligent Systems Technologies and Applications*,
Advances in Intelligent Systems and Computing 385,
DOI: 10.1007/978-3-319-23258-4_24

adapt to the situation beforehand. But to predict the stock price in advance is very difficult due to the volatility of the market and the dependency of large number of influencing factors. Prediction on a day to day basis is somehow simpler than to predict the same in large span of time as the outcome gets more uncertain and vague. There are several works on share price prediction for short span. Mahdi *et al.,* implemented a multi-layer perceptron and Elman recurrent network to predict future stock price of a company [1]. Cheng *et al.*, had done the prediction of daily stock prices using neural network and time series analysis over 60 trading days [2]. By using ANN based system Devadoss *et al.,* had predicted the daily stock price and estimated the percentage error, Mean Absolute Deviation and Root Mean Square Error to indicate the performance of the networks [3]. In [4], the data is trained using general approach of back propagation to predict the daily stock price. As the process needs a large amount of data to be analyzed the normalization is done to make the input within the range of 0 and 1. C. Saranya and G. Manikandan had proved that the min max normalization technique have minimum misclassification error in their work [5]. In [6], Chauhan *et al.,* proposed an approach to predict the daily stock price by using three tire structure of neural network and supervised learning. Mayank *et al.*, implemented their method on stock price data of Axis bank, BOB and Hindalco etc. to predict the daily stock using Multi-Layer Perception (MLP) [7]. A hybridized market indicators is used for stock prediction purpose in [8]. In [9], Stock data is predicted based on clustering and fuzzy set. Data mining techniques can also be utilized in stock prediction [10]. Four-layer fuzzy multi agent system (FMAS) architecture was developed to make a hybrid artificial intelligent model as proposed in [11]. Some work had been done on the basis of Recursive feature elimination (RFE) which is based on Support vector machine (SVM) as in [12]. A technique called hybrid pruning algorithm was developed to predict the future of stock price with the help of backpropagation algorithm in [13]. The structure of the fuzzy model was utilized based on the log-likelihood value of each data vector generated by a trained Hidden Markov Model by Hassan *et al.*, in [14]. Manjul Saini and A.K. Singh proposed an advanced backpropagation based stock and weather prediction in [15].In the paper of Z. Bashir and M. El-Hawary PSO based neural logic is used to forecast short term load [16]. Stock price prediction can also be done by using hybrid machine learning technique [17]. Regression methodology is also another approach to predict stock price [18].

Pre-processed historical stock data is used for grid partitioning and rule formation of initial Sugeno-Takagi Fuzzy Inference System (FIS). The initial FIS is further trained using adaptive hybrid learning technique. The final structure is tested with stock price data (Adjusted closing price) and results are obtained. In this paper, Section 2, 3, 4 and 5 represent preliminaries, proposed method along with the necessary algorithms, experimental results and conclusion respectively.

2 Preliminaries

The stock data are collected from authentic website. These extracted raw data must be preprocessed for creating the proper fuzzy inference structure. It mainly involves gap filling and normalization. The preprocessed data is used to create Sugeno Fuzzy Inference Structure which is finally adjusted using adaptive hybrid neural learning.

Gap Filling: The data extracted does not always correspond to the exact duration of 14 days. In order to train and test the system a proper sequential data is essential. So the necessary gap that is missing data points is calculated by an average of the preceding and following data points.

Normalization: The artificial neural network is trained better when the input is within the range of 0 to 1 i.e. normalized. In this work the data is normalized by Min-Max normalization technique.

If a is the input and A is the set where a belongs then the normalized form $\bar{a}$ is calculated as in equation (1).

$$\bar{a}= (a-min\ (A))/\ (max\ (A)-min\ (A)) \tag{1}$$

where min (A) and max (A) are the minimum and maximum element in the set A.

Sugeno Fuzzy Inference Structure: Sugeno-Takagi is a special type of Fuzzy Inference Structure (FIS). It is basically a rule based architecture that is based on fuzzified inputs. Fuzzification consists of calculating the membership grades of different inputs in different fuzzy sets. A FIS consists of set of fuzzy *If-Then* rules that describes the behavior of fuzzy sets formed from the input training data. Sugeno-Takagi type FIS always has a linear or constant output function. A typical first order Sugeno fuzzy model rule structure is like,

$$If\ input_1=x\ and\ input_2=y\ then\ output\ z=ax+by+c$$

Neural Network Learning: The parameters, weights and biases of the Fuzzy Inference Structure are updated using learning algorithm based on artificial neural network. Hybrid learning mechanisms is a combination of two or more standard techniques.

3 Proposed Method

The extracted historical stock data first undergoes the preprocessing. Grid partitioning is then used to eliminate the noise and identify the rule structure to create the Sugeno type FIS. The initial FIS structure is trained using hybrid learning to get the final ANFIS structure. Thus the proposed method consists of three basic modules data preprocessing, Fuzzy Inference Structure formation and adaptive hybrid neural training. A structural representation of the proposed method is given in Fig. 1.

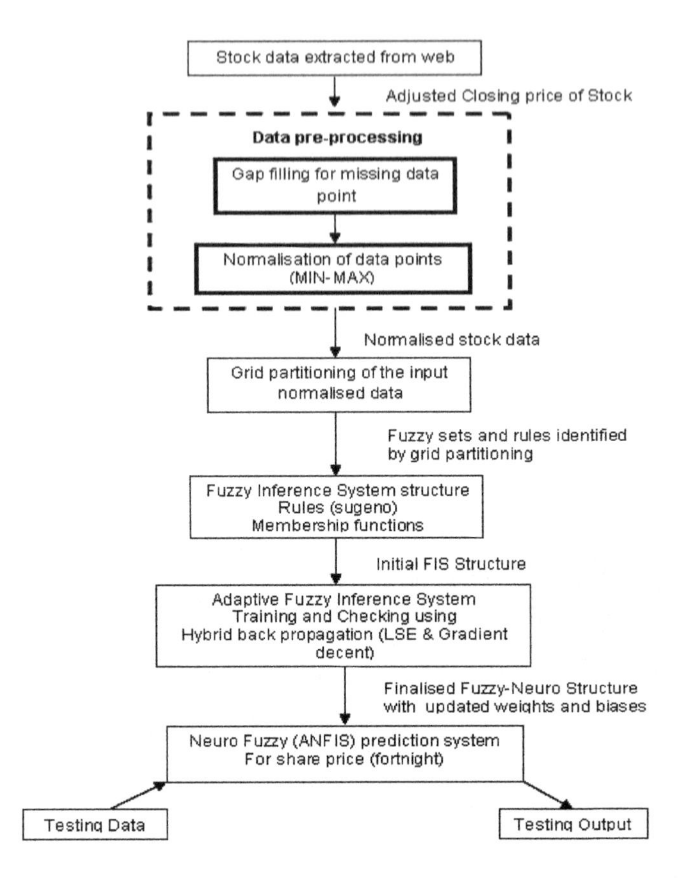

Fig. 1 Structural representation of proposed method

Data preprocessing includes gap filling and normalization as described in section 2. According to the study Min-Max normalization technique gives better efficiency in the field of large data processing [5].

The fuzzy inference structure is created using the preprocessed data. The Sugeno-Takagi fuzzy inference structure is used as an initial condition for ANFIS training. It uses grid partitioning to determine the number of rules and detect most significant data by eliminating the noisy data points.

The output of each rule is weighted by firing strength W_i where, $W_i = AND\ (F_1\ (x),\ F2\ (y))$ 'AND' is defined by the *product implication*. It is the arithmetic product of the two membership grades that is $F_1(x).F_2\ (y)$. $F_1(x)$, $F_2\ (y)$ are the member ship functions associated with the input *x*, *y* respectively. The final output is the weighted average of all rule outputs.

ANFIS uses the previous FIS structure for updating its parameters. In this study the architecture of the ANFIS is made of four layers given in Fig. 2. Among those layers both 1st and 3rd consists of adaptive neurons. The adapted neurons get updated in each of the subsequent iterations while the fixed neurons do not include any parameters.

Rule 1: *If* $x= A_1$ *then* $f_1=p_1\,x+r_1$;
Rule 2: *If* $x=A_2$ *then* $f_2=p_2\,x+r_2$; similarly for other rules.

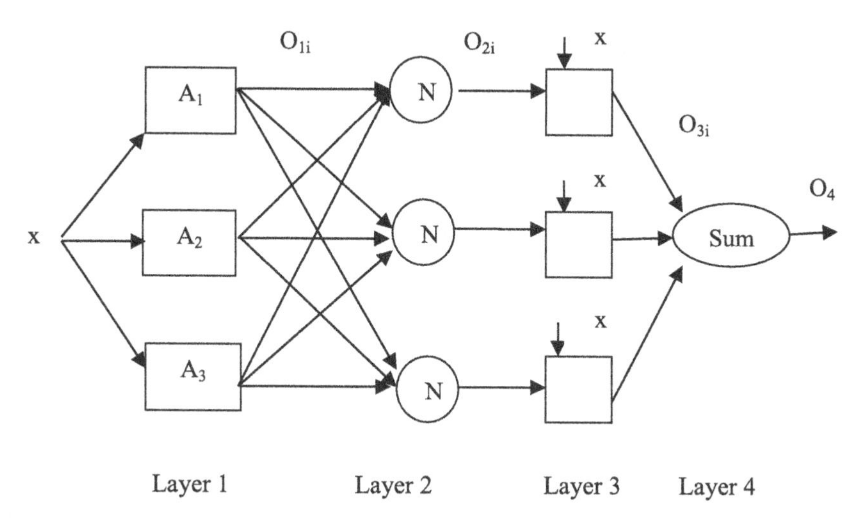

Fig. 2 ANFIS Structure

Layer 1: These are adaptive neurons. The output of these layers is fuzzy membership grade of the inputs (Fuzzification).

$O_{1i}=\mu\, A_i(x)$ *for* $i=1, 2\ldots n$; O_{1i} is membership grade of fuzzy set (A_1,A_2,A_3). This work is implemented with different membership functions (mf) i.e., generalized Bell shaped mf, Gaussian mf, Gaussian combination mf. The membership function for Bell shaped generalized function is given in equation (2).

$$\mu_i A(x) = \frac{1}{1+\left|\frac{(x-c_i)}{a_i}\right|^{2b_i}} \tag{2}$$

Where a_i, b_i and c_i are referred as premise parameters.

The comparison of using different membership function is shown in result and analysis portion. As only one input is concerned over here the membership grades are itself treated as weights and fed into next normalization layer (3).

$$O_{1i} = W_{1i} \tag{3}$$

Layer 2: In this layer the firing strengths are normalized (4) .These are fixed neurons.

$$O_{2i} = \overline{W}_i = \frac{w_i}{\sum_{i=1}^{3} w_i} \tag{4}$$

Layer 3: It gives output as the product of the normalized firing strength and a first order polynomial given in equation (5).

$$O_{3i}=\overline{W}_i f_i=\overline{W}_i\,(p_i x+r_i) \tag{5}$$

Where p_i *and* r_i are linear consequent parameters.

Layer 4: It calculates the output of the weighted sum of the outputs (defuzzifications) using equation (6).

$$O_4 = \frac{\sum_{i=1}^{3} w_i . F_i}{\sum_{i=1}^{3} w_i} \quad (6)$$

In this work hybrid learning technique is used to train the ANFIS structure. This is based on hybrid back propagation which includes least square estimation and gradient descent. In the forward pass the consequent parameters are updated using least square estimation and in the backward pass the premise parameters are optimized using gradient descent algorithm. The convergence of the hybrid learning is faster than standard back propagation learning technique [8].

When this trained ANFIS structure is provided with the stock price of a particular company as input it can predict the probable stock price of that company after duration of the 14 days. The input and output are the adjusted closing stock price of a company.

Proposed Algorithm*: Hybrid Adaptive Neuro Fuzzy algorithm (HANF Algorithm).*
The proposed algorithm represents the entire process undertaken to predict the future stock price. Algorithm mainly consists of three different phases each of which includes multiple steps. These are input preprocessing, construction of fuzzy inference structure and training the system using hybrid adaptive learning. The first phase of the HANF algorithm is to preprocess the stock data. The processed input then undergoes grid partitioning to identify the rules for Sugeno-Takagi fuzzy inference structure. The final ANFIS structure is obtained using a hybrid adaptive training of the initial fuzzy structure. The ANFIS output is denormalised to produce the predicted stock.

Input and Output: The input to the system is stock data (adjusted closing price) of a particular company and the output is the predicted closing after 14 days.

Step 1: The Stock market data (adjusted closing price of a particular company) is *extracted from web*.

Step 2: The raw data is *pre-processed* which includes the following.

- ***2.1***: In the first stage of preprocessing the input data undergoes gap filling. The missing data points are calculated as an approximate average of the preceding and following day's data. As a result the data gets arranged in sequence.
- ***2.2***: The next and final phase of preprocessing includes normalization the normalization technique adapted here is Max-Min.

Step 3: The normalized data is fed into the *Sugeno-Takagi Fuzzy inference system* as the initial training data to form the FIS structure.

- ***3.1***: The input data undergoes Grid Partitioning to identify the initial rule structure and eliminate the noise.
- ***3.2***: The *Fuzzy If-Then rules* are identified with predefined membership functions and their premise parameters and consequent parameters.

Step 4: This initial FIS and the training data is fed into *adaptive hybrid neuro learning system* for updating the premise and consequent parameters of the initial FIS.

4.1: The input is passed through the initial FIS structure to calculate the initial predicted output. The difference of the predicted and original output is calculated as *error.*

4.2: In the forward pass the consequent parameters are updated using *least square estimation (LSE).* In the backward pass the premise parameters are updated using *gradient descent.*

Step 5: The final optimized structure is tested with testing data and the output is *denormalised.*

4 Experimental Results

The experiment is made on the basis of stock data collected from three different types of sectors namely companies like “Infosys” from IT Sector, “Colgate” from FMGC Sector and “HDFC Bank” from banking sector. The stocks are listed in New York Stock Exchange and the data for the experiment are extracted from “https://in.finance.yahoo.com”. The stock price is in American Currency (USD) and the experiment is conducted based upon the data in the period of 2010-2015.For each company a sample of size 126(i.e. stock data of 126 days) is considered.

The experiment is conducted using MATLAB (2010), in 32-bit Windows XP Professional with Service Pack 3. The processor used is Core 2 Duo, 3.06GHz and 2.00GB RAM.

75% of the collected data is used for training and 15% is for checking and rest for testing. Three different types of membership functions namely generalized Bell shaped mf, Gaussian mf and Gaussian combination mf are used. A comparative study of average error percentages are shown in the Table 1.

Table 1 Average percentage error in prediction for different companies.

Stock	Bell shaped mf	Gaussian mf	Gaussian combination mf
Infosys	4.27	4.24	4.32
Colgate	4.80	4.87	2.86
HDFC Bank	3.77	4.30	4.14

From the above table (Table 1) it is clear that “Bell shaped mf” gives better result than the other in some cases where as “Gaussian combination mf” gives even better result for some particular cases. The experimental result for all the three companies using “Bell shaped mf” are shown in the following figure (Fig. 3).

The seven different dates D1 to D7 for which the stock price of the above mentioned three companies are predicted during the experiment is listed in Table 2. The gaps between the dates are approximately 14 days.

Table 2 Testing dates used in the experiment

Symbol	Date	Symbol	Date	Symbol	Date
D1	02/01/15	D4	20/11/14	D6	23/10/14
D2	19/12/14	D5	06/11/14	D7	09/10/14
D3	05/12/14				

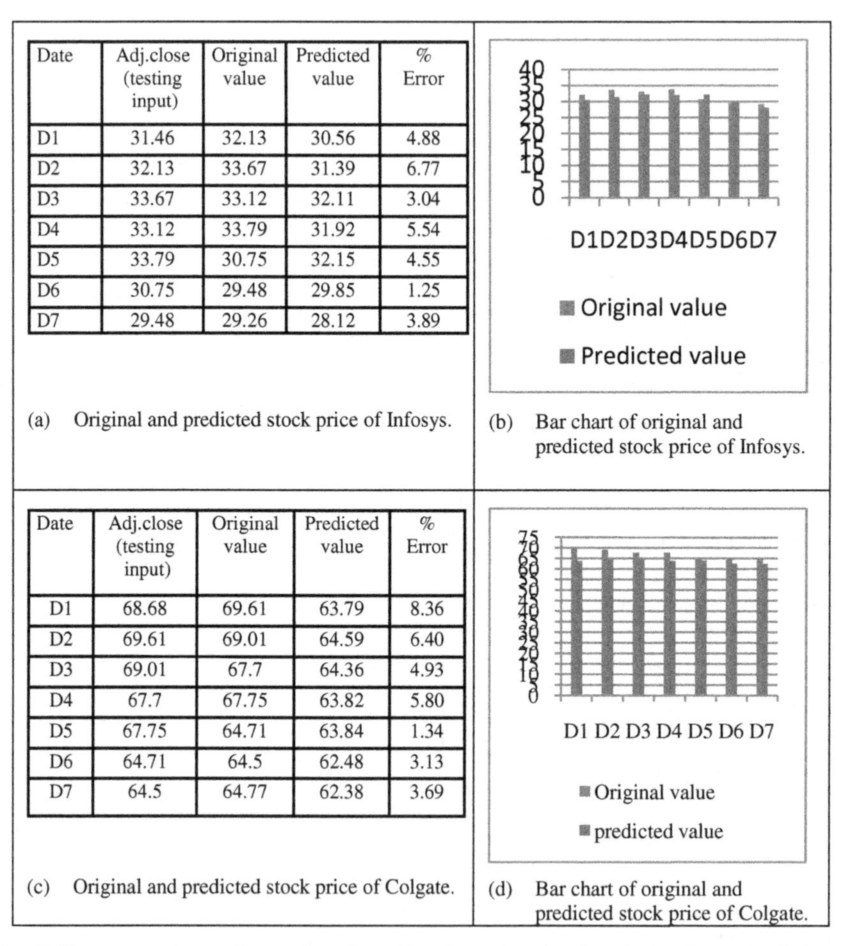

Date	Adj.close (testing input)	Original value	Predicted value	% Error
D1	31.46	32.13	30.56	4.88
D2	32.13	33.67	31.39	6.77
D3	33.67	33.12	32.11	3.04
D4	33.12	33.79	31.92	5.54
D5	33.79	30.75	32.15	4.55
D6	30.75	29.48	29.85	1.25
D7	29.48	29.26	28.12	3.89

(a) Original and predicted stock price of Infosys.

(b) Bar chart of original and predicted stock price of Infosys.

Date	Adj.close (testing input)	Original value	Predicted value	% Error
D1	68.68	69.61	63.79	8.36
D2	69.61	69.01	64.59	6.40
D3	69.01	67.7	64.36	4.93
D4	67.7	67.75	63.82	5.80
D5	67.75	64.71	63.84	1.34
D6	64.71	64.5	62.48	3.13
D7	64.5	64.77	62.38	3.69

(c) Original and predicted stock price of Colgate.

(d) Bar chart of original and predicted stock price of Colgate.

Fig. 3 The comparison of actual and predicted stock price in tabular form for Infosys, Colgate and HDFC bank are represented in (a), (c) and (e) respectively. The same using vertical bar chart diagram is presented in (b), (d) and (f) respectively.

Date	Adj.Close (testing input)	Original value	Predicted value	% Error
D1	51.98	50.69	51.29	1.17
D2	50.69	51.11	50.77	0.67
D3	51.11	52.72	51.04	3.29
D4	52.72	52.89	51.28	3.14
D5	52.89	49.57	51.84	4.39
D6	49.57	48.17	51.81	7.02
D7	48.17	48.07	51.54	6.73

(e) Original and predicted stock price of HDFC Bank.

D1 D2 D3 D4 D5 D6 D7

Original Value

Predicted Value

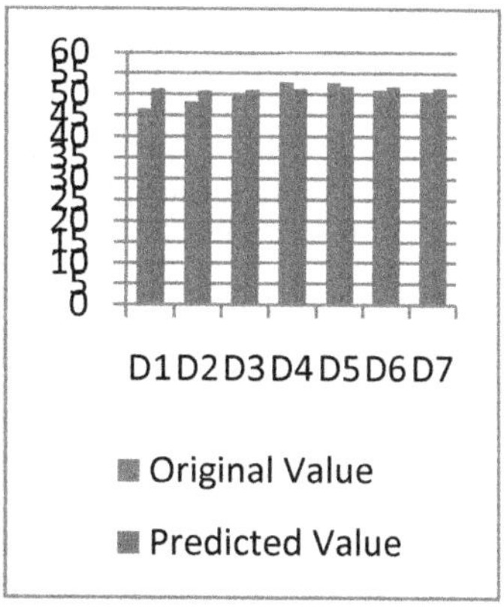

(f) Bar chart of original and predicted stock price of HDFC Bank.

Fig. 3 (*Continued*)

Our proposed technique outperforms the existing methods as proposed by Tsai *et. al,* and Gharehchopogh *et. al.* where the accuracy achieved was 77 % [17] and 61.35 % [18] respectively.

5 Conclusion

The future stock result is of immense importance in the world of business. Majority of stock prediction works done mainly for the duration of overnight or a little more. The proposed method is able to predict the future closing stock price of a particular company in 14 to 15 days duration. This work proposes an adaptive neuro fuzzy system (ANFIS) using grid partition for the stock prediction. The model is a combination of fuzzy logic and hybrid neural learning technique. The efficiency of the prediction system is quite satisfactory with the results correct up to 94-95 percentages. The paper also represents a comparative study of the different membership functions associated with the Sugeno-Takagi fuzzy inference system. In general the Bell shaped membership function gives better results. However, in some cases Gaussian membership function can also outperform others. The overall result shows that hybrid model performs better than standard neural learning techniques. The future work can be extended to a more long term prediction by introducing factors like occurrence of special events such as budgets, new turns in the business or political world, government policy etc. to the learning system.

References

1. Pakdaman, N.M., Hamidreza, H., Baradaran, H.: Stock market value prediction using neural networks. In: International Conference on Computer Information Systems and Industrial Management Applications (CISIM). IEEE publications (2010)
2. Kuo-Cheng, T., Ojoung, K., Tjung, L.C.: Time series and neural network forecast of daily stock prices. Investment Management and Financial Innovations **9**, 32–54 (2012)
3. Victor, D., Antony, T., Ligori, A.: Stock prediction using artificial neural networks. International Journal of Data Mining Techniques and Applications **2**, 283–291 (2013)
4. Haider, K.Z., Sharmin, A.T., Akter, H.Md.: Price Prediction of Share Market using Artificial Neural Network (ANN). International Journal of Computer Applications **22** (2011)
5. Saranya, C., Manikandan, G.: A Study on Normalization Techniques for Privacy Preserving Data Mining. International Journal of Engineering and Technology (IJET) **5**, 14–18 (2013)
6. Bhagwant, C., Umesh, B., Ajit, G., Sachin, K.: Stock Market Prediction Using Artificial Neural Networks. International Journal of Computer Science and Information (IJCSIT) Technologies **5**, 904–907 (2014)
7. Mayank, K., Patel, B., Yalamalle, S.R.: Stock Price Prediction Using Artificial Neural Network. International Journal of Innovative Research in Science **3**, 13755–13776 (2014)
8. Adebiyi, A.A., Ayo, C.K., Adebiyi, M.O., Otokiti, S.O.: Stock Price Prediction using Neural Network with Hybridized Market Indicators. Journal of Emerging Trends in Computing and Information Sciences **3**, 1–9 (2012)
9. Das, D., Uddin, M.S.: Data mining and neural network techniques in stock market prediction: A Methodological review. International Journal of Artificial Intelligence & Applications (IJAIA) **4**, 117–127 (2013)
10. Radaideh, Q.A.Al, Emanalnagi A.A.A: Predicting stock prices using data mining techniques. In: The International Arab Conference on Information Technology (ACIT 2013), pp. 1–8 (2013)
11. Zarandi, M.H., Fazel, H.,Esmaeiland, T.I.B.: A hybrid fuzzy intelligent agent-based system for stock price prediction International Journal of Intelligent Systems, 1–23 (2012)
12. Xu, Y., Li, Z., Luo, L.: A study on feature selection for trend prediction of stock trading price. In: Computational and Information Sciences (ICCIS), pp. 579–582. IEEE publications (2013)
13. Xingcheng, Pu, Pengfei, Sun: A New Hybrid Pruning Neural Network Algorithm Based on Sensitivity Analysis for Stock Market Forecast. Journal of Information & Computational Science **10**, 883–892 (2013)
14. Rafiul, H.Md., Kotagiri, R., Kamruzzaman, R.J., Mustafizur, R., Maruf, H.M.: A HMM-based adaptive fuzzy inference system for stock market forecasting, vol. 104, pp. 10–25. Elsevier Science Publishers (2013)
15. Saini, M., Singh, A.K.: Forecasting Stock Exchange Market and Weather Using Soft Computing. International, Journal of Advanced Research in Computer Science and Software Engineering **4**, 877–882 (2014)
16. Bashir, Z., El-Hawary, M.: Applying wavelets to short-term load forecasting using PSO-based neural networks. IEEE Trans. Power Syst. **24**, 20–27 (2009)
17. Tsai, C.F., Wang, S.P.: Stock price forecasting by hybrid machine learning techniques. In: Proceedings of the International Multi Conference of Engineers and Computer Scientists, vol. 1 (2009)
18. Soleimanian, G.F., Haddadi, B.T., Reza, K.S.: A linear regression approach to prediction of stock market trading volume: A case study. International Journal of Managing Value and Supply Chains (IJMVSC) **4**, 25–31 (2013)

Intelligent Distributed Economic Dispatch in Smart Grids

Meheli Basu, Raghuraman Mudumbai and Soura Dasgupta

Abstract This paper considers the optimal economic dispatch of power generators in a smart electric grid for allocating power between generators to meet load requirements at a minimum total cost. We present a decentralized algorithm where, each generator independently adjusts its power output using only a measurement of the frequency deviation of the grid and minimal information exchange with its neighbors. Existing algorithms assume that frequency deviation is proportional to the load imbalance. In practice this is seldom exactly correct. We assume here that the only thing known about this relationship is that it is an unknown, odd, strictly increasing function. We provide a proof of convergence and simulations verifying the efficacy of the algorithm.

1 Introduction

This paper presents a distributed algorithm for optimal economic dispatch, [1], of power generators in a smart grid. The goal is meet specified power generation requirements at minimum total cost. Like our earlier papers [4], [3] and [2] we assume that each generator can measure the frequency deviation of the grid. In the tradition of [4], we also assume the existence of a local internet that permits generators to exchange their marginal costs with their neighbors. Using such information the generators must autonomously adjust their power output to asymptotically erase the load balance at a minimum cost.

Supported in part by US NSF grants CCF-0830747 and EPS-1101284 and a grant from the Roy J. Carver Charitable Trust.

M. Basu(✉) · R. Mudumbai · S. Dasgupta
Department of Electrical and Computer Engineering, University of Iowa, Iowa City, IA 52242, USA
e-mail: meheli-basu@uiowa.edu, {rmudumbai,dasgupta}@engineering.uiowa.edu

S. Berretti et al. (eds.), *Intelligent Systems Technologies and Applications*,
Advances in Intelligent Systems and Computing 385,
DOI: 10.1007/978-3-319-23258-4_25

The major difference between this paper and [4] is in the fact that, [4] assumes that the frequency deviation is proportional to the power imbalance. Thus its measurement is tantamount to knowing the load imbalance to within a positive constant of proportionality. Such a proportional relationship is only approximate and assumes small load imbalances. In practice the precise relationship between the imbalance and frequency deviation is unknown. Thus in this paper we extend [4] by relaxing the assumption of proportionality. Instead, we assume that all that is known about the frequency deviation is that it is an unknown odd increasing function of the load imbalance.

This work anticipates the needs of future smart grids that will have smart consumer end-nodes [5] and a high penetration of alternative energy generators. As alternative energy sources are intermittent in time and dispersed in geography [6], the electric grid must dynamically adjust generation and consumption. This stands in stark contrast to the traditional grid, where only a small number of large generation units are dispatchable. The future smart grid will likely have a plethora of small distributed generation (DG) [7], storage and demand-response units that will all contribute in varying measures. A centralized control approach will simply not scale and will lack the required agility. Thus a decentralized approach with a limited use of communication infrastructure and using local message exchanges is needed.

Accordingly, we leverage recent advances in the distributed consensus theory to develop such control schemes. We believe that our broad approach extends well beyond the distributed dispatch problem to other control loops in the electric grid such as reactive power control and voltage regulation. We use the terminology of the traditional economic dispatch in a broader sense than usual. Thus "generators" represent all dispatchable units that have primary controllers that follow a power-frequency droop characteristic with negative slope, just like traditional generators. Though developed to ensure stable interconnection of synchronous generators [30], recent studies [31] have shown that the droop curve is useful and effective and it is advantageous to retain this mechanism even for modern microgrids. It is the droop curve that defines the aforementioned relationship between the load imbalance and the frequency deviation.

We observe that traditionally the dispatch problem is treated as a constrained optimization solved at a centralized controller, [1]. Techniques such as "lambda iteration" [14], genetic algorithms, particle swarm optimization or Monte-Carlo methods [15, 16] are typically used. These centralized tehniques are fundamentally different from the distributed approaches of this paper and [2, 3, 4], and do not conform to the vision of distributed control set out in the foregoing.

The rest of this paper is organized as follows. We first introduce the dispatch problem considered in this paper in Section 2. Our distributed dispatch algorithm is presented in Section 3. The stability proof is in Section 4. Section 5 has simulations. Section 6 concludes.

2 The Dispatch Problem

Assume there are N generators that must supply the power P_L which is assumed to be constant. The active power set point for generator i at the rated system frequency by $P_i(k),\ i \in 1 \dots N$. The power imbalance in the system is thus

$$\Delta(k) = P_L - \sum_{i=1}^{N} P_i(k) \tag{1}$$

We neglect the effects of reactive power flows, voltage deviations and transients as is standard for economic dispatch problems.

The actual active power produced by each generator is determined by its primary controller which uses $P_i(k)$ as a reference. More precisely, the primary controller on each generator responds to a power imbalance by adjusting its generated power relative to its generation set-point $P_i(k)$ until the imbalance is erased. This has the side effect of introducing a small frequency deviation $g(\Delta(k))$. We assume that each controller measures $g(\Delta(k))$, akin to the Area Control Error (ACE) signal observed by the secondary controller in a traditional Load Frequency Control (LFC) implementation [28], though it does not know its precise dependence on $\Delta(k)$. Instead all it knows is that $g(\cdot)$ conforms to the assumption below.

Assumption 2.1. The function $g(\Delta)$ is an analytic, strictly increasing and odd memoryless function in Δ. Further

$$\lim_{\Delta \to \infty} g(\Delta) = \infty.$$

Suppose $J_i(P_i)$ is the cost function for generator i. With $P = [P_1, \cdots, P_N]^\top$, define P_i^* to be power allocations that minimize the total cost:

$$\sum_{i \in V} J_i(P_i) \tag{2}$$

$$\text{subject to} \sum_{i \in V} P_i = P_L \tag{3}$$

The goal of the dispatch algorithm is to choose the $P_i(k)$ achieve

$$\lim_{k \to \infty} P_i(k) = P_i^*. \tag{4}$$

The optimization (2,3) requires global communication between all generators. To circumvent this problem, [4] proposed an alternative cost function whose minimum coincides with the minimum of (2,3) but whose gradient descent minimization requires only local information exchange and the measurements the frequency deviation. However, [4] assumes that for some possibly unknown positive β, $g(\Delta) = \beta\Delta$.

Thus effectively, it assumes that a quantity proportional to Δ is available. The knowledge of such a quantity, proves crucial to the generation of the gradient used in [4]. As in this paper we do not assume that $g(\Delta)$ can yield $\beta\Delta$, the algorithm of [4] cannot be implemented in the settings of this paper.

3 The Algorithm

As in [4], we assume that the network of generators and the communications infrastructure form a *possibly directed graph* $G = (V, E)$, where $V = \{1, \cdots, N\}$ is the vertex set indexing the generators. The directed edge $\{i, j\} \in E$ if generator i has access to generator j's marginal cost $J_j'(P_j)$. Define $\mathcal{N}(i)$ as the set of neighbors of i, i.e.

$$\mathcal{N}(i) = \{j \,|\{i, j\} \in E\}. \tag{5}$$

In the sequel we assume a constant load P_L, and a load deficit:

$$\Delta = P_L - \sum_{i=1}^{N} P_i. \tag{6}$$

We make the following assumption.

Assumption 3.1. The load P_L is constant. For all $i \in V$, the cost $J_i(\cdot) : \mathbb{R} \to \mathbb{R}$ is analytic everywhere. Further, there exists a $\gamma > 0$, such that for all $x \in \mathbb{R}$, and $i \in V$, there holds,

$$J_i''(x) \geq \gamma. \tag{7}$$

Finally, for every $i \in V$

$$\lim_{P_i \to \infty} J_i'(P_i) = \infty. \tag{8}$$

Observe, (4) is a convexity assumption that is standard for most cost functions used in the power systems literature. Sometimes, these are obtained by interpolating tabulated data. These data are of a form that allows a convex interpolant, [37]. Convexity reflects the appealing reality that the marginal cost increases with production. We assume the marginal costs to be always positive, which is again a reality. For technical reasons we have not restricted the P_i to be nonnegative, though, in reality they would be.

Optimality of (3) subject to (2) necessitates that the marginal costs be equal subject to (2). Assumption 3.1 ensures that there is in fact a unique operating point meeting this requirement. To see this suppose a second operating point $\bar{P} \neq P^*$ has equal marginal costs and induces $\Delta = 0$. Call the i-th element of $\bar{P}$ and P^*, $\bar{P}_i$ and P_i^* respectively. Since $\Delta = 0$ in both case

$$P_L = \sum_{i \in V} \bar{P}_i = \sum_{i \in V} P_i^*$$

and $\bar{P} \neq P^*$, there must be one element of $\bar{P}$ that is greater than the corresponding element of P^* and another that is less than the corresponding element of P^*. Thus for some i, $\bar{P}_i > P_i^*$ and for some j $\bar{P}_j < P_j^*$. As $J'(P_i^*) = J'(P_j^*)$, convexity ensures that $J'(\bar{P}_i) > J'(\bar{P}_j)$, establishing a contradiction.

Thus, the equality of the marginal costs subject to (2) is both necessary and sufficient for optimality. Thus, we must find P_i that equalize the marginals subject to (2). The equalization of the marginals through their local exchange has similarities to the goals of consensus algorithms, [23]-[27]. An important difference is in the additional requirement of (2).

Define $P(k) = [P_1(k), \cdots, P_N(k)]^\top$. The algorithm we propose is as follows:

$$P(k+1) = P(k) - \mu z(k) \tag{9}$$

where μ is a suitably small adaptation gain and for some scalar $\alpha > 0$, the i-th element of $z(k)$ obeys

$$z_i(k) = -\alpha g(\Delta(k)) + J_i''(P_i) \sum_{j \in \mathcal{N}(i)} \left(J_i'(P_i) - J_j'(P_j) \right). \tag{10}$$

When $g(\Delta) = \beta\Delta$ then (10) is just the gradient of the cost function

$$S(P) = \frac{\alpha\Delta^2}{2\beta} + \frac{1}{2} \sum_{\{i,j\} \in E} \left(J_i'(P_i) - J_j'(P_j) \right)^2. \tag{11}$$

Indeed, [4] critically exploits this fact as under a connectedness assumption on G, the minimization of $S(P)$ is equivalent to the equalization of marginals and imbalance erasure. Of course $z(k)$ is no longer the gradient of $S(P)$ in our more complicated, albeit realistic, model for frequency deviation.

4 Stability

It is a well established fact in averaging theory, [39] that under suitably small $\mu > 0$ the asymptotic stability of the discrete time algorithm

$$x(k+1) = x(k) - \mu f(x(k))$$

can be concluded from the asymptotic stability of its continuous time counterpart:

$$\dot{x}(t) = -f(x(t)).$$

Thus, instead of (9, 10) we will analyze

$$\dot{P}_i = \alpha g(\Delta) - J_i''(P_i) \sum_{j \in \mathcal{N}(i)} \left(J_i'(P_i) - J_j'(P_j) \right) \tag{12}$$

We first prove the following.

Lemma 4.1. Consider (12) under assumptions 2.1 and 3.1, Δ is bounded from below.

Proof. Suppose $\Delta < 0$. Then $g(\Delta) < 0$. It suffices to show that in such a case P is bounded from above. Indeed without loss of generality assume at a given time t, $m \in V$ is such that

$$J_m'(P_m(t)) = \max_{i \in V} \{ J_i'(P_i(t)) \}.$$

Then as by Assumption 3.1, $J_m''(P_m(t)) > 0$, $\dot{P}_m < 0$. Consequently, P_m decreases in value. Thus by convexity at any given time, the largest marginal cost always declines. Thus the P_i are bounded.

Suppose now the lower bound on Δ is Δ_-. Define the cost function

$$J(P) = \alpha \int_{\Delta_-}^{\Delta} g(x)dx + \frac{1}{2} \sum_{\{i,j\} \in E} \left(J_i'(P_i) - J_j'(P_j) \right)^2. \tag{13}$$

Clearly the integral in (13) is well defined and due to Assumption 2.1 nonnegative. Then we have the following Lemma.

Lemma 4.2. Suppose the graph G is connected. Then under the conditions of Lemma 4.1

$$\lim_{t \to \infty} \dot{P}(t) = 0. \tag{14}$$

Proof. Observe $J(P) \geq 0$. Further because of (1)

$$\begin{aligned}
\dot{J} &= \sum_{i=1}^{N} \frac{\partial J(P)}{\partial P_i} \dot{P}_i \\
&= \sum_{i=1}^{N} \alpha g(\Delta) \frac{\partial \Delta}{\partial P_i} \dot{P}_i \\
&+ \sum_{i=1}^{N} \left\{ J_i''(P_i) \sum_{j \in \mathcal{N}(i)} \left(J_i'(P_i) - J_j'(P_j) \right) \right\} \dot{P}_i \\
&= -\alpha g(\Delta) \sum_{i=1}^{N} \dot{P}_i
\end{aligned}$$

$$
\begin{aligned}
&+\sum_{i=1}^{N}\left\{J_i''(P_i)\sum_{j\in\mathcal{N}(i)}\left(J_i'(P_i)-J_j'(P_j)\right)\right\}\dot{P}_i\\
&=-\sum_{i=1}^{N}(\alpha g(\Delta)-J_i''(P_i)\sum_{j\in\mathcal{N}(i)}\left(J_i'(P_i)-J_j'(P_j)\right))\\
&\quad\dot{P}_i\\
&=-\sum_{i=1}^{N}(\dot{P}_i)^2\\
&=-\|\dot{P}\|^2\\
&\leq 0. \qquad (15)
\end{aligned}
$$

Thus $J(P)$ in (13) is bounded from above as well. As each summand in (13) is nonnegative, each must be bounded. The first ensures that $g(\Delta)$ and hence Δ is bounded. The second together with the connectedness of V and the convexity of the J_i ensures that $P_i - P_j$ is bounded for all $\{i, j\} \subset V$. Thus from (1) all P_i are bounded. Thus, as (12) has no explicit dependence on t, from Lassalle's Theorem, [40] P converges to the trajectory where $\dot{P} \equiv 0$.

We can now prove the main result.

Theorem 4.1. Under the conditions of Lemma 4.2 with P_i^* the values of P_i that optimize (2,3), (4) holds for all $i \in V$.

Proof. From Lemma 4.2 for all $i \in V$

$$\lim_{t\to\infty} \dot{P}_i(t) = 0.$$

Thus all variables including P_i and hence Δ have limit points. Consider two cases.

Case I: Limit point of Δ is nonnegative. Thus at this Δ, $g(\Delta) \geq 0$. Consider l so that $J_l'(P_l) \leq J_i'(P_i))$ for all $i \in V$. Then from (12) and Assumption 3.1 all summands in the expression of $\dot{P}_l$ are nonegative and must be zero. Thus in the limit $\Delta = 0$, for all $i \in \mathcal{N}(l)$, $J_l'(P_l) = J_i'(P_i))$ and for all $i \in V$

$$\sum_{j\in\mathcal{N}(i)}\left(J_i'(P_i)-J_j'(P_j)\right)=0 \qquad (16)$$

Then the marginal costs of all neighbors of elements of $\mathcal{N}(l)$ must also equal $J_l'(P_l)$. Continuing in this vein as G is connected all marginal costs are equal. As the point at which $\Delta = 0$ and the marginal costs are equal is unique, the result follows.

Case II: Limit point of Δ is nonpositive. The proof of this case is very similar to Case I. All that is needed is to choose l so that $J_l'(P_l) \geq J_i'(P_i))$ for all $i \in V$.

5 Simulations

This section presents simulations that demonstrate the performance of the approach described in this paper.

We assume that $g(\Delta) = \Delta + \beta\Delta^3$. Fig. 2 shows the evolution of the power imbalance and total generation cost of a system with 6 generators. The cost curves and the total load for this simulation are the same as in the Example 2 in [34]. Specifically, the cost functions are of the form $J_i(P_i) = c_i P_i^2 + b_i P_i + a_i$, with the parameters c_i, b_i, a_i as listed in Table 1.

Table 1 Parameter values for simulations.

Unit	1	2	3	4	5	6
a_i	1122	620	156	950	580.5	560.5
b_i	15.84	15.7	15.94	13.414	14.174	14.147
c_i	312E-5	388E-5	964E-5	264.1E-5	349.6E-5	349.6E-5

The generators are connected by a communication network represented by the undirected graph shown in Fig. 1. Observe that this is a connected graph that satisfies the assumptions of Theorem 4.1.

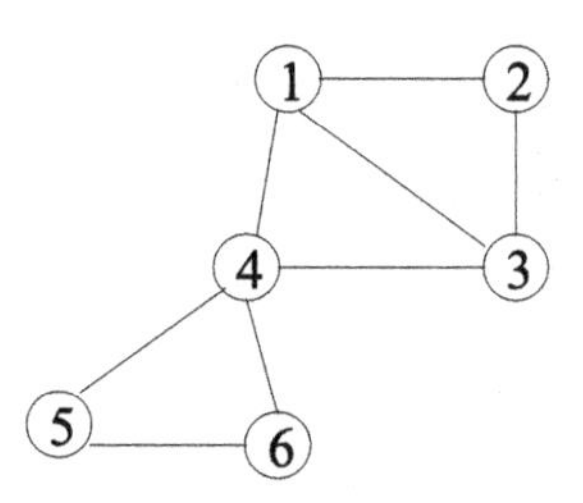

Fig. 1 Connectivity graph for the simulated dispatch problem.

The cost functions are of the form $J_i(P_i) = c_i(P_i)^2 + b_i P_i + a_i$ with the parameters a, b and c as specified in the earlier paper. We initialized the simulation by setting all the six generators to generate equal power with a small randomly chosen initial power imbalance. We set $\alpha = 410^{-4}$, $\mu = 2$ and $\beta = 0.5$

When we initially start with the case where total generated power is greater than the load power, the power imbalance starts out as negative which is increased to zero and also during that time the total cost of generation steadily decreases within the first 500 iterations as shown in Figure 2 .

On the other hand, when we initially start with the case where total generated power is less than the load power, the power imbalance starts out as positive which is reduced to zero and also during that time the total cost of generation steadily decreases within the first 500 iterations as shown in Figure 3.

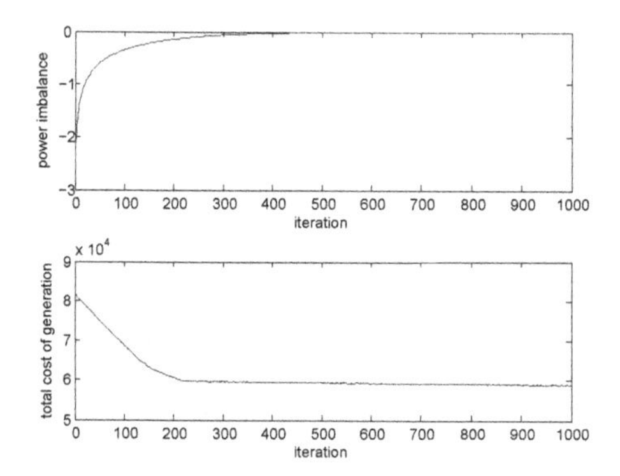

Fig. 2 Connectivity graph for the simulated dispatch problem.eps

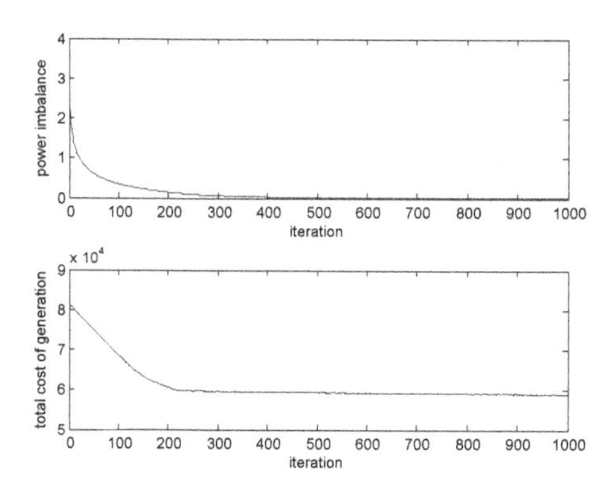

Fig. 3 Connectivity graph for the simulated dispatch problem.

6 Conclusion

We have considered the optimal economic dispatch of power generators in a smart electric grid for allocating power between generators to meet load requirements at a minimum total cost. The algorithm is decentralized. Each generator independently adjusts its power output using only a measurement of the frequency deviation on the grid and minimal information exchange between generators. Existing algorithms assume that frequency deviation is proportional to the load imbalance. In practice this proportional relationship seldom holds. Accordingly our algorithm assumes that the only thing known about this relationship is that it is an unknown, odd, strictly increasing function. We have shown that the algorithm is globally convergent.

An important future area of research is to tune this algorithm to grid dynamics to avoid instabilities, though it is safe to conjecture that sufficiently small μ and large enough sampling intervals in (9) should prevent grid instabilities.

References

1. Chowdhury, B., Rahman, S.: A review of recent advances in economic dispatch. IEEE Transactions on Power Systems **5**, 1248–1259 (1990)
2. Mudumbai, R., Dasgupta, S., Cho, B.: Distributed control for optimal economic dispatch of power generators: the heterogenous case. In: Proc. of the IEEE CDC (2011)
3. Mudumbai, R., Dasgupta, S., Cho, B.: Distributed control for optimal economic dispatch of a network of heterogeneous power generators. IEEE Trans. on Power Systems, 1750–1760 (2012)
4. Mudumbai, R., Dasgupta, S., Mahboob, R.: A distributed consensus based algorithm for optimal dispatch in smart power grids. In: Proceedings of the 32nd IASTED International Conference on Modeling, Identification and Control (MIC), February 2013
5. Marvin, S., Chappells, H., Guy, S.: Pathways of smart metering development: shaping environmental innovation. Computers, Environment and Urban Systems **23**(2), 109–126 (1999)
6. Milborrow, D.: Penalties for intermittent sources of energy, p. 17. Cabinet Office, London (2001)
7. Dugan, R., McDermott, T.: Distributed generation. IEEE Industry Applications Magazine **8**, 19–25 (2002)
8. Rebours, Y.G., Kirschen, D.S., Trotignon, M., Rossignol, S.: A survey of frequency and voltage control ancillary services part i: Technical features. IEEE Transactions on Power Systems **22**, 350–357 (2007)
9. Wu, F., Moslehi, K., Bose, A.: Power system control centers: past, present, and future. Proceedings of the IEEE **93**(11), 1890–1908 (2005)
10. U.S.D. of Energy. Economic dispatch of electric generation capacity. A Report to Congress and the States (2007)
11. Vargas, L., Quintana, V., Vannelli, A.: A tutorial description of an interior point method and its applications to security-constrained economic dispatch. IEEE Transactions on Power Systems **8**(3), 1315–1324 (2002)
12. Lopes, J., Moreira, C., Madureira, A., Resende, F., Wu, X., Jayawarna, N., Zhang, Y., Jenkins, N., Kanellos, F., Hatziargyriou, N.: Control strategies for microgrids emergency operation. In: International Conference on Future Power Systems, Amsterdam, Netherlands (2005)
13. Amin, M., Wollenberg, B.: Toward a smart grid: power delivery for the 21st century. IEEE Power and Energy Magazine **3**(5), 34–41 (2005)
14. Saadat, H.: Power system analysis. WCB/McGraw-Hill, Boston (1999)
15. Mohammadi, A., Varahram, M., Kheirizad, I.: Online solving of economic dispatch problem using neural network approach and comparing it with classical method. In: International Conference on Emerging Technologies. ICET 2006, pp. 581–586. IEEE (2007)
16. Chen, C.: Economic dispatch using simplified personal best oriented particle swarm optimizer. In: Third International Conference on Electric Utility Deregulation and Restructuring and Power Technologies. DRPT 2008, pp. 572–576. IEEE (2008)
17. Yang, S., Tan, S., Xu, J.-X.: Consensus based approach for economic dispatch problem in a smart grid. IEEE Transactions on Power Systems (2013)
18. Bidram, A., Davoudi, A., Lewis, F., Qu, Z.: Secondary control of microgrids based on distributed cooperative control of multi-agent systems. IET Generation, Transmission Distribution **7**(8) (2013)
19. Fathi, M., Bevrani, H.: Adaptive energy consumption scheduling for connected microgrids under demand uncertainty. IEEE Transactions on Power Delivery **28**(3), 1576–1583 (2013)

20. Dall'Anese, E., Zhu, H., Giannakis, G.: Distributed optimal power flow for smart microgrids. IEEE Transactions on Smart Grid **4**(3), 1464–1475 (2013)
21. Kraning, M., Chu, E., Lavaei, J., Boyd, S.: Dynamic network energy management via proximal message passing. Optimization **1**(2), 1–54 (2013)
22. Aganagic, M., Mokhtari, S.: Security constrained economic dispatch using nonlinear dantzig-wolfe decomposition. IEEE Transactions on Power Systems **12**, 105–112 (1997)
23. Olfati-Saber, R., Murray, R.M.: Consensus problems in networks of agents with switching topology and time-delays. IEEE Trans. Autom. Control **49**(9), 1520–1533 (2004)
24. Ren, W., Beard, R.W.: Consensus seeking in multi-agent systems under dynamically changing interaction topologies. IEEE Trans. Autom. Control **50**(5), 655–661 (2005)
25. Rahman, M.M., Mudumbai, R., Dasgupta, S.: Consensus based carrier synchronization in a two node network. In: Proceedings of IFAC World Congress 2011, Milan, Italy (2011)
26. Khan, U.A., Kar, S., Moura, J.M.F.: Distributed Sensor Localization in Random Environments Using Minimal Number of Anchor Nodes. IEEE Transactions on Signal Processing, 2000–2016 (2009)
27. Moreau, L.: Stability of multi-agent systems with time-dependent communication links. IEEE Trans. Autom. Control **50**(2), 169–182 (2005)
28. Christie, R.D., Bose, A.: Load frequency control issues in power system operations after deregulation. IEEE Transactions on Power Systems, 1191–1200 (1996)
29. Jaleeli, N., VanSlyck, L., Ewart, D., Fink, L., Hoffmann, A.: Understanding automatic generation control. IEEE Transactions on Power Systems **7**(3), 1106–1122 (1992)
30. Clough, F.: Stability of large power systems. Journal of the Institution of Electrical Engineers **65**(367), 653–659 (1927)
31. Dobakhshari, A., Azizi, S., Ranjbar, A.: Control of microgrids: aspects and prospects. In: IEEE International Conference on Networking, Sensing and Control (ICNSC), pp. 38–43, April 2011
32. Sundarapandian, V.: An invariance principle of discrete-time nonlinear systems. Applied Mathematics Letters, 85–91 (2003)
33. Hahn, W.: Stability of motion. Springer (1967)
34. Mohammadi, A., Varahram, M., Kheirizad, I.: Online solving of economic dispatch problem using neural network approach and comparing it with classical method. In: International Conference on Emerging Technologies. ICET 2006, pp. 581–586. IEEE (2007). in Proc. 2003 Am. Control Conf., 2003, pp. 951956
35. Jadbabaie, A., Lin, J., Morse, A.S.: Coordination of groups of mobile autonomous agents using nearest neighbor rules. IEEE Trans. Autom. Control **48**(6), 988–1001 (2003)
36. Lin, W., Zhixin, L., Guo, L.: Robust consensus of multi-agent systems with noise. In: Proceedings of the Chinese Control Conference (2007)
37. Aganagic, M., Mokhtari, S.: Security constrained economic dispatch using nonlinear Dantzig-Wolfe decomposition. IEEE Transactions on Power Systems, 105–112 (1997)
38. Happ, H.H.: Optimal power dispatch: A comprehensive survey. IEEE Transactions on Power Apparatus and Systems, 841–854, May 1977
39. Anderson, B.D.O., Bitmead, R.R., Johnson Jr., C.R., Kokotovic, P.V., Kosut, R.L., Mareels, I.M.Y., Praly, L., Riedle, B.D.: Stability of Adaptive Systems: Passivity and Averaging Analysis. M.I.T. Press, Cambridge (1986)
40. Khalil, H.K.: Nonlinear Systems, 3rd edn. Prentice Hall (2002)

Classification of Software Project Risk Factors Using Machine Learning Approach

Prerna Chaudhary, Deepali Singh and Ashish Sharma

Abstract Software project risk can be defined as a various future harms that could be possible on the software due to some non-noticeable mistakes done during the development of software project. Analyzing the risk is required in order to reduce the risk before it can harm the quality of the project. This paper interprets an idea of software project risk factors classification which involves the use of support vector machines (SVM) i.e., machine learning approach to improve the accuracy of the results. Risk assessment is a crucial task as the projects are facing increased complexity with higher uncertainties. In order to make the risk assessment easier, it is necessary for the developers to identify the hardbound and less hardbound risk factors. Classifying the risk factors will help the developers to identify the most effective risk which will ultimately become easy for the software developer to take some mitigation actions as early as possible. Hence the proposed approach reduces the developer's effort and increases the accuracy in identifying the harmful risk factors.

Keywords Analytical Hierarchy Process (AHP) · Support Vector Machines (SVM) · Risk Factors · WEKA Tool

1 Introduction

Software development appears worse in quality from project delays, cost overruns, unutilized system and unfulfilled user needs. Understanding software project risk can help in reducing the failure of the software. Chittister and Haimes [1] define the risk as the probability and impact of an adverse event. Boehm [2] describes the risk similarly and discusses the top ten risks in software development. Any risk factor identified in a particular software project, if ignored, will enhance the

P. Chaudhary(✉) · D. Singh · A. Sharma
GLA University, Mathura, India
e-mail: priyachy1990@gmail.com, {deepali.panwar,ashish.sharma}@gla.ac.in

S. Berretti et al. (eds.), *Intelligent Systems Technologies and Applications*,
Advances in Intelligent Systems and Computing 385,
DOI: 10.1007/978-3-319-23258-4_26

probability of failure of that project. Hence, in order to mitigate the coming risk, there is a need to prioritize those risks.

With the changing requirements of the software projects, corresponding risk factors will also change. Hence, there is a need to achieve such model which relies with the changing risk factors so we have used the machine learning approach. Machine learning deals with programs that learn from experience in order to achieve accurate results.

Support Vector Machines (SVMs) is a popular machine learning method for classification, regression and other learning tasks. SVM was introduced by Vapnik et al. [3][4]. SVM successfully bring out substantial improvement over recent best performing methods and behave robustly over the other machine learning algorithms. Furthermore, they are fully automatic, excluding the need of manual efforts. Major aim of SVM is to construct a hyperplane with maximized margin, in order to clearly separate the data of one class to the data of other class.

This paper proposes a method of prioritizing the risk factors using AHP, at the same time achieving the scalability issue using SVM. Scalability is the ability of the system, to handle the growing amount of work in a capable manner.

The objective of this paper is to present the idea for analyzing the software project risk using machine learning technique. First, we introduce the description of software project risk. Second, we discuss the work done by different authors to formulate the research problem. Third, we proposes a research methodology of classifying the risk factors using SVM [5]. Fourth, the weka tool will give the analysis of classification [6] results. And fifth, on the basis of classified risk factors final harmful risk factors will be provided to the developer.

2 Literature Review

Various researchers consider the probability and impact of the risk factors in order to identify the risk exposure and the combination of which will prioritize the risk.

Table 1 depicts various risk analysis methods that includes the criteria of analysing the risk in the software projects and the technique used to evaluate the exposure of the risk. The exposure of the risk is calculated by the probability of occurrence of risk and the impact of risk.

Since 2004 to 2013, the researchers proposed several approaches for the assessment of risk like, cluster analysis [7], which performs K-mean cluster analysis, where the major parameter was the formation of clusters. Then, evidential reasoning approach [8], where the major parameters were degree of belief and disbelief for the risk to be critical. After that, fuzzy logic based risk analysis [9], where the probability and impact of the risk was of major concern. Then, Analytical Hierarchical Process (AHP) [10] for risk assessment, where the intense aspects were probability and impact of the risk and AHP is used to analyze the priorities of those risks.

Table 1 Risk Analysis Methods

Risk Analysis Methods	Technique	Criteria
Cluster Analysis [7]	K-means cluster Analysis	Forms the cluster of risk factors
Evidential Reasoning Approach [8]	Dempster-Shafer Theory of belief functions	Focus on the assessment of evidence strength for a particular assertion .
Fuzzy Logic based Risk Analysis [9]	Analysis based on fuzzy logic.	Focus on the probability and impact of identified risk.
Analytical Hierarchy Process(AHP) [11]	Pairwise Comparison of risk factors.	Evaluates the priority of risk factors.

Cluster analysis [7], gives the empirical evidence that the most well known risk factors are associated with high risk projects using K-means cluster analysis and proved that the requirement risk, planning and control risk and organizational risks are the most eminent risk factors, while complexity risk is most eminent to low risk projects. Evidential Reasoning Approach [8], provides the estimation of risk and identify the sources of risk using the fluctuations in the strength of evidences obtained from the process of software project development. This procedure also evaluates the interrelationship among the risk factors while estimating the software project risk.

Fuzzy Logic Based Risk Analysis [9], describe each risk factor by the probability of risk and impact of risk. Both the measures are described by linguistic terms and modelled by fuzzy sets. In this paper, overall risk of a system is evaluated by the fraction of sum of the probability and impact of each subcomponent of the system and the summation of impacts of each subcomponent.

Further, Analytic Hierarchy Approach [11], provides the pair sampling of risk factors by forming a matrix of n x n risk factors. The resultant normalized matrix gives the precedence of risk factors which interprets the most eminent risk of the software project.

But none of the above approaches considers the scalability issue. If the requirements of the software project changes, risk factors corresponding to it also changes, further the priorities will also be affected but none of the approach overcome this issue.

3 Proposed Approach

This paper proposes a method of classifying risk factors using machine learning approach. The paper uses SVM [12] for classifying risk factors and considers relative values and ranking of risk factors obtained from AHP [13]. The proposed work basically focuses on the binary SVM as the dataset is linearly separable and is classified into two labelled classes.

The ultimate goal of this approach is to provide an ability to reduce the probability of failure of the software project. Classification of risk factors clearly visualizes the criticality of the impact of the risk and helps the developer to identify hard bound risk factors in order to apply the mitigation plans to reduce the probability of impact.

3.1 Software Project Risk Classification Using SVM

The proposed SVM approach provides classification of risk factors of the software project. An iterative process of an overall work helps to achieve the scalability factor and allow the software developer to handle the risk before it degrades the quality of the software project.

The procedure of software project risk classification using support vector machines approach is shown in Fig. 1 which shows the step by step process processes of the proposed method. Number of iterations in this process is based on the change in requirements. As the requirements of the software project changes, the risk factors corresponding to it also changes. There are four steps on which the proposed approach is based are as follows:

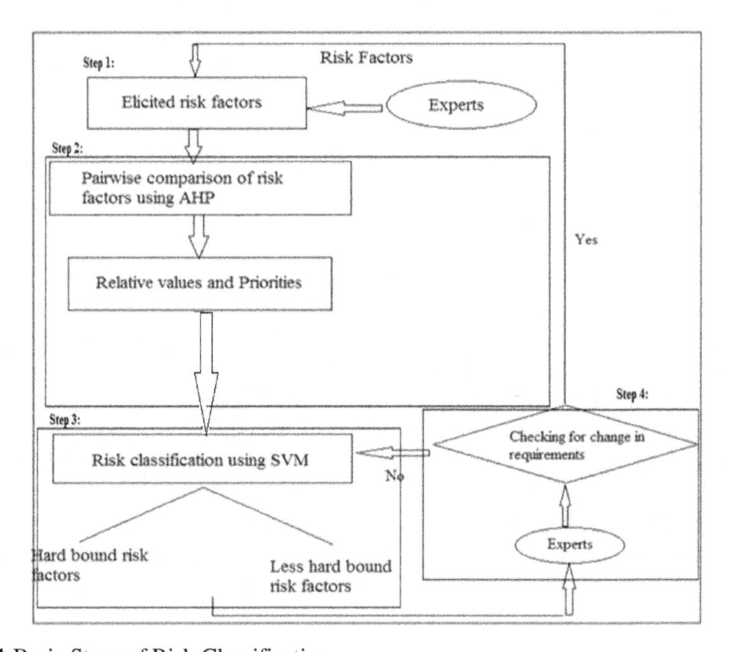

Fig. 1 Basic Steps of Risk Classification

3.1.1 Risk Factors Elicitation

The first step involves in interviewing the experts directly involved in the project risk management process, that is, the risk owners and the project managers. These participants were given a questionnaire on the basis of their experience in the organization. Through the expert judgement, risk factors of the software project are identified [10].

3.1.2 Pair Wise Comparison

The strength of risk interactions is assessed with the help of analytical hierarchy process (AHP) [14]. One of the most useful methods for selecting the risk factor that is becoming more and more important is AHP. This approach was developed by Thomas Saaty in 1980 to help with solving technical and managerial problems [15].

Table 2 Primary Scale for Pairwise Comparison [6]

RELATIVE INTENSITY	DEFINITION	EXPLANATION
1	Of equal value	Two risk factors are of similar value
3	Slightly more value	Experience slightly favoured one risk factor over other
5	Essential or strong value	Experience strongly favoured one risk factor over other
7	Very strong value	A risk factor is strongly favoured and its dominance is demonstrated in practice
9	Extreme value	The evidence favouring one over another is of the highest possible order of affirmation
2,4,6,8	Intermediate values between two adjacent judgements	When compromise is needed
Reciprocals	If 'ij' has any above number assigned to it then 'ji' will have the reciprocal value. Where, 'ij' – ith risk factor compared to jth risk factor. 'ji' – jth risk factor compared to ith risk factor.	

Place n risk factors in the rows and columns of an n x n matrix. Carry out pair wise comparisons among all the risk factors. For this purpose, a primary scale used is shown in Table 2. Now calculate the sum of each column of a matrix. Subsequently divide each element with the sum of their respective columns and calculate the sum of each row. Afterwards, divide each row sum with the number of risk factors to normalize the sum of each row. This calculation gives the eigen values of matrix and is referred to as priority matrix.

This paper uses analytical hierarchy process online system (AHP-OS). AHP-OS is a web based tool [13] to support decision making based on AHP. It helps to calculate the relative values and priorities of the risk factors.

Excel sheet of resultant relative values and priorities of the risk factors is made. Labelling to the risk factors is a given on the basis of assumption, Out of n risk factors,

If n € Even Number,

$$\begin{pmatrix} n-\frac{n}{2}=HB \\ n-\frac{n}{2}=LHB \end{pmatrix} \tag{1}$$

If n € Odd Number,

$$\begin{pmatrix} \left(n-\frac{n}{2}\right)-.5=HB \\ rest=LHB \end{pmatrix} \tag{2}$$

where, HB is hard bound risk factors and LHB is less hard bound risk factors. Finally, excel sheet is converted to attribute-relation file format (.arff) file manually.

3.1.3 Risk Classification

The resultant .arff file is considered as an input to the Waikato environment for knowledge analysis (Weka) tool for classification using LIBSVM. Weka is a machine learning toolkit given by Waikato University, New Zealand. It is open source software written in java, basically used for research and education [16]. SVM is a machine learning approach firstly introduced in 1992 and is a popular approach for classification. Since the dataset of risk factors is linearly separable in this paper, here we consider the linearly separable binary class SVM problem.

Risk factors have to classify among two classes based on the training set of examples whose classification is assumed as stated in step 3.1.2. SVM constructs a decision boundary [17][5] between the class of hard bound risk factors and the class of less hard bound risk factors and maximizes the margin of separation of two classes as shown in Fig. 2.

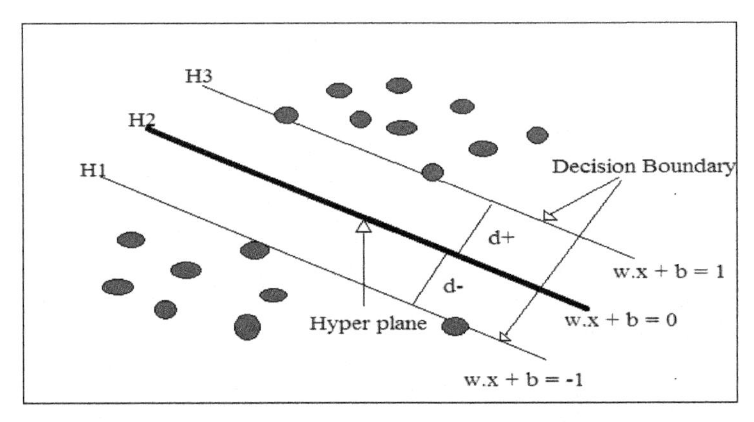

Fig. 2 Best Hyperplane Separator

Suppose there are N training risk factors, $\{(y_1,z_1), (y_2,z_2),....(y_N,z_N)\}$, where y_i € R^d and

$$z_i = \begin{cases} HB & if\ y_i\ in\ class\ 1 \\ LHB & if\ y_i\ in\ class\ 2 \end{cases} \quad (3)$$

where, class 1 is a set of higher bound risk factors and class 2 is a set of less higher bound risk factors.

Prediction Rule: $$w^T y + b = 0 \quad (4)$$

Aim: Learn w and b to achieve the maximum margin.

Now, the major goal is to construct a separating hyper plane $w^T y + b=0$ that separates all the risk factors. From the Figure 2,

$$H1 = (w^T y_i) + b > 0 \quad \text{for} \quad z_i = 1 \quad (5)$$

$$H3 = (w^T y_i) + b < 0 \quad \text{for} \quad z_i = -1 \quad (6)$$

The points that lie on the planes H1 and H3 are known as Support Vectors. Plane H2 is a median in between H1 and H3.

d+ is the shortest distance from the nearest positive point.

d- is the shortest distance from the nearest negative point.

The optimization algorithm to generate the weights proceeds in such a way that only the support vectors determine the weights and thus the boundary. The major goal is to contruct a hyperplane whose margin of separation β is maximized. Let the distance from point (y_0,z_0) to a line:

$Bx + Cy + d = 0$ is: $|By_0 + Cz_0 + d|/\text{sqrt}(B^2 + C^2)$, so, the distance between H1 and H2 is:

$$|w.y + b|/||w|| = 1/||w|| \tag{7}$$

The total distance between H1 and H3 is $2/||w||$.

Hence, in order to maximize the margin, there is a need to minimize $||w||$. With the condition that there are no data points between H1 and H3.

$$\text{Min f: } ½||w||^2 \text{ s.t.}$$

$$g: z_i(w.y_i) - b = 1 \tag{8}$$

As this paper uses weka tool, the result of classification is a confusion matrix which gives the interpretation of correctly and incorrectly classified risk factors.

Checking for the Change in Requirements. After classifying the risk factors and visualizing the results, check for the change in the requirements. If the requirements of the software project have changed, then elicit the risk factors again and repeat step 1 to step 3 to achieve the scalability factor.

3.2 Proposed Algorithm and its Illustration

We proposed an algorithm which is shown below. For a real time software project, namely, Online Shopping System, elicited risk factors by the experts are:

$$R' = \{r_1,r_2,....,r_6\}$$

Now, pairwise comparison of elicited risk factors is carried out using AHP in order to obtain the output in the form of relative values and precedence of risk factors. Further, the dataset of relative values and precedence will be considered as input to SVM. Train SVM by learning the values of w and b and construct a hyperplane $w.y_i + b = 0$ s.t.,

$$w.y_i + b >= +1 \text{ when } z_i = +1$$

$$w.y_i + b <= -1 \text{ when } z_i = -1$$

Where w is a weight vector and b is a bias.

Algorithm 1. A Sketch of the Classification algorithm

Input:

R' = {r_1, r_2, r_3,......, r_4}

The set of elicited risk factors.

R'' = {(y_1,z_1),(y_2,z_2),...,(y_i,z_i)}

i training dataset for input to SVM obtained from AHP.

Output:

$$z_i = \begin{cases} HB & if\ y_i\ in\ class\ 1 \\ LHB & if\ y_i\ in\ class\ 2 \end{cases}$$

Where, class 1 & class 2 is a set of higher bound risk factors (+1) and less higher bound risk factors (-1) respectively.

Begin

1. X = Initialize(R')
   Input risk factors to AHP for pairwise comparison
2. Obtain R'' = {(y_1,z_1),(y_2,z_2),...,(y_i,z_i)}
   Where, a vector y_i € R^n and z_i € Label and is an output of AHP
3. R'' is input to SVM.
4. Construct the hyperplane w.y_i + b = 0, s.t.
   The decision boundary is maximized.
5. Select w and b such that:
   w.y_i + b >= +1 for z_i = +1
   w.y_i + b <= -1 for z_i = -1

where, w is a weight vector and b is a bias.

6. Maximize the margin:
   For any point, (y_0,z_0) which lies on the boundary, its distance from hyperplane is:
   |w.y + b|/||w|| = 1/||w||
   Hence, total distance between decision boundaries is:
   2/||w||
7. Maximizing boundary implies minimizing ||w||
   Min ½||w||^2 s.t.
   z_i(w.y_i + b) -1 >= 0 $\forall i$
8. To minimize w we use Lagrangian Multiplier α
   $L = \frac{1}{2}||w||^2 - \sum_{i=1}^{n} \alpha\, [z_i(w.y_i + b) - 1]$
   Differentiate w.r.t. w and b will give:
   $w = \sum_{i=1}^{n} \alpha_i z_i\, y_i$
   $b = \sum_{i=1}^{n} \alpha_i z_i = 0$
9. Check for the change in requirements and repeat step 1 to 9.

Fig. 3 Classifier Algorithm

After the construction of the separating hyperplane, decision boundaries of the hyperplane have to be maximized in order to clearly classify the risk factors of both the class. Any point which lies on the decision boundary is called support vectors. Hence, for any point, (y_0,z_0) which lies on the boundary, its distance from hyperplane is:

$$|w.y + b|/||w|| = 1/||w||$$

Hence, total distance between decision boundaries is:

$$2/||w||$$

Maximizing boundary implies minimizing ||w||

$$\text{Min} \quad ½||w||^2 \text{ s.t.}$$

$$z_i(w.y_i + b) - 1 >= 0$$

Hence, the risk factors will be classified into two classes i.e., HB: Hardbound Risk Factors and LHB: Less Hard Bound Risk Factors.

In order to improve the scalability, firstly check for the change in requirements. If the requirements are changing, add more risk factors of the online shopping system, and then give its relative value and precedence to SVM. It will automatically take the position of either hard bound or less hard bound risk factor without affecting the other risk factors.

4 Results

The paper presents the result analysis of the proposed approach of classification using support vector machines (SVM) on three real time projects namely project 1, library management system [20], project 2, high speed railway system [18] and project 3, online shopping system [19]. The Fig. 4, Fig. 5 and Fig. 6 below visualize the correctly classified and incorrectly classified risk factors of three projects using weka tool. The highlighted circles depict the incorrectly classified risk factors.

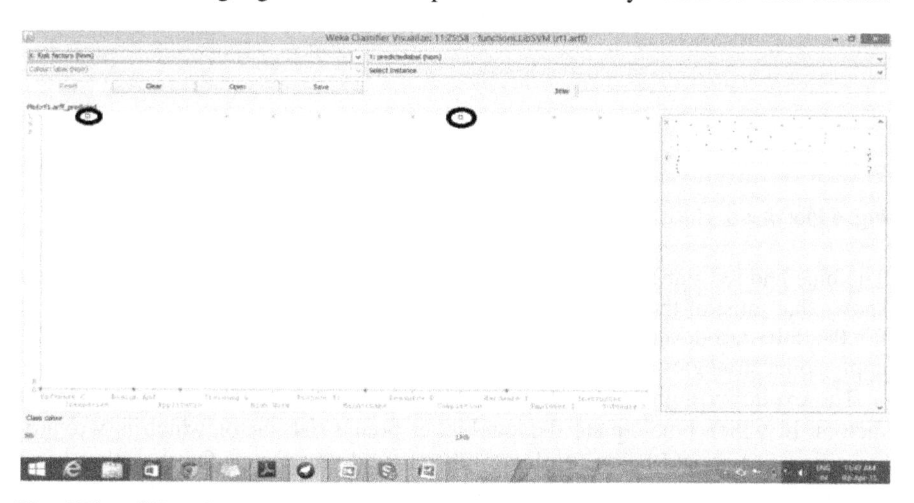

Fig. 4 Plot of No. of Risk Factors and Label for Project 1.

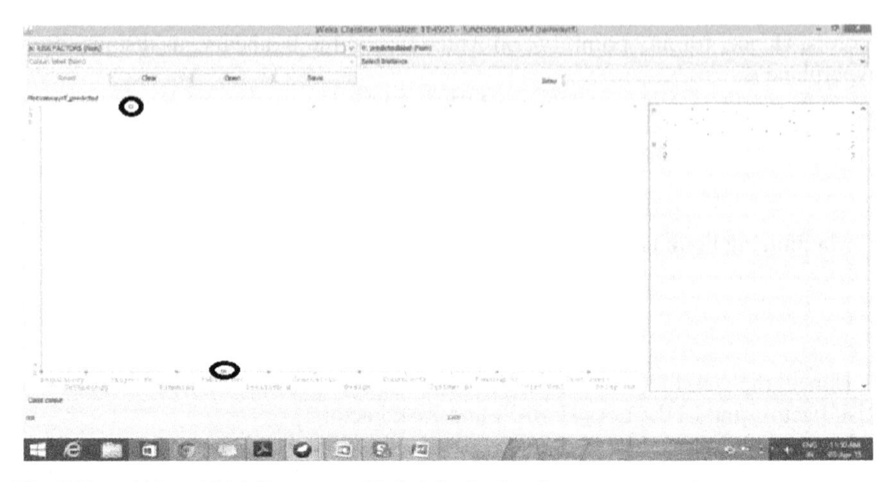

Fig. 5 Plot of No. of Risk Factors and Label for Project 2.

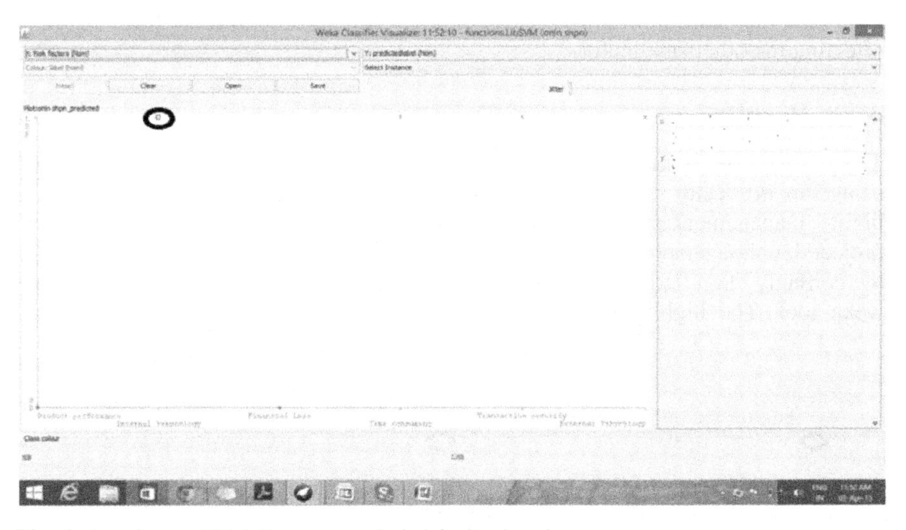

Fig. 6 Plot of No. of Risk Factors and Label for Project 3.

The blue and red star shows the correctly classified risk factors. The overall figure shows that most of the risk factors of each project are correctly classified. Hence, for the software developer it is easy to identify the uncertain risk so that he can apply some strategies to mitigate those risks.

The explicitly highlighted circles in the graph interpret the misclassified risk factors. In which blue square depicts higher bound risk factor which is wrongly classified and should belong to less higher bound risk factor. On the other hand red square depicts less high bound risk factor which is wrongly classified and should belong to higher bound risk factor. Red and blue colour stars shows the correctly classified risk factors.

In order to validate the results, error rate of both the approaches are estimated using equation 8 and shown using a cone graph in Fig. 7, Fig. 8 and Fig. 9. The plotted graph between number of risk factors and error rate describes the gap between the results of AHP and results of SVM. Through this we can visualize the accuracy in the results.

$$\text{Error Rate} = \frac{Current\ value - Expected\ value}{Current\ value} * 100 \quad (8)$$

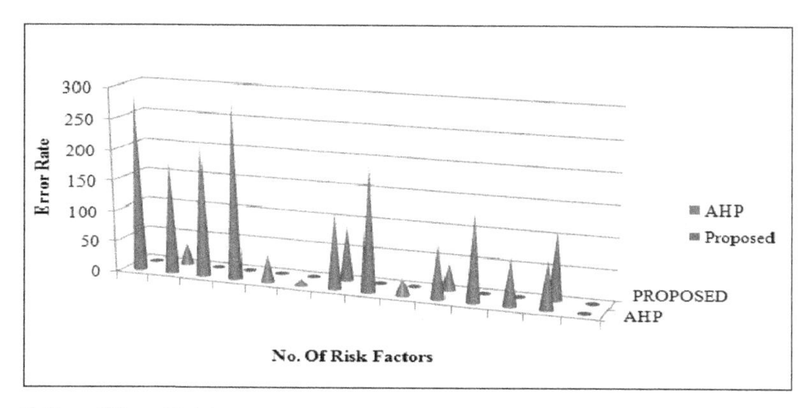

Fig. 7 Plot of No. of Risk Factors and Error Rate for Project 1.

Fig. 7 shows the difference in the error rates of the risk factors using AHP as well as SVM approach for project 1 i.e., library management system. This shows that the classification that is provided by SVM is more accurate than the ranking provided by the AHP.

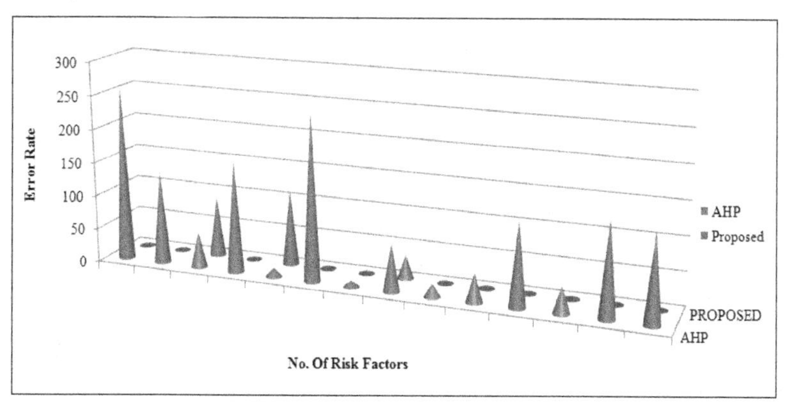

Fig. 8 Plot of No. of Risk Factors and Error Rate for Project 2.

Fig. 8 depicts the relation between number of risk factors of project 2 i.e., high speed railway system and error rates from both the approaches. The graph clearly interprets that the classification using SVM is more accurate.

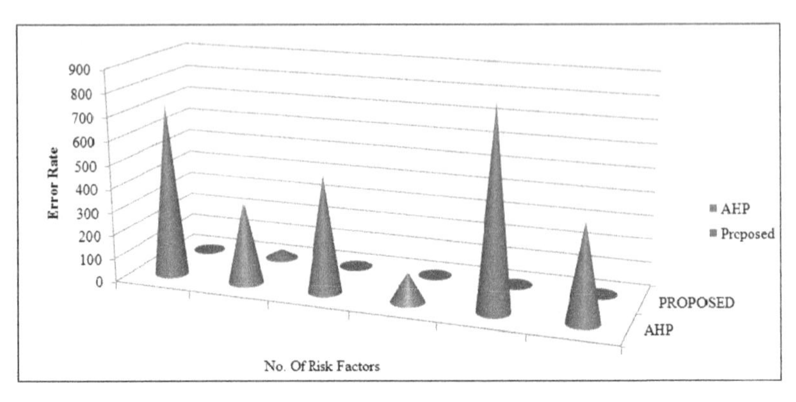

Fig. 9 Plot of No. of Risk Factors and Error Rate for Project 3.

Fig. 9 also depicts the relation between the risk factors and the error rates obtained from AHP and SVM.

In order to show the scalability factor, additional risk factors corresponding to changing requirements follow every step agai

5 Conclusion

In this paper, classification of risk factors using SVM is proposed which is one of the machines learning approach. The machine learning approach (SVM), is used to classify the risk factors into two classes to get the optimal error rates. Firstly, AHP provides the priorities of risk factors which depicts the harmful risk more frequently but while analysing the risk further after prioritization there is a possibility that the ranking of risk factor was not totally correct, hence degrades the performance of the software project. In order to get a strong evidence for a risk factor to be harmful or not secondly, we chooses the way of classifying the risk factors. Hence, initially after the elicitation of risk factors, pairwise comparison of risk factors is done using AHP which gives the priorities. These priorities will further be an input dataset to the SVM approach which classifies the risk factors and gives the result in the form of confusion matrix. The accuracy of the classified result is depicted using TP rate and FP rate. Finally, the error rates of SVM have been positioned with respect to the error rate of AHP through a cone graph. Multiple iterations will be followed if the requirements of the software project get changed and it continues until the result gets converged. The potential advantage of the SVM approach is that it fulfils the scalability factor, as with the changing requirement, the additional risk factors can automatically get its class without affecting the previous classes of risk factors. As in this, we are using machine learning approach, the learning provides accurate results. But on the other hand, as the input dataset to SVM is made manually, hence if risk factors increase, then the effort required by human evaluators grows rapidly.

In the future directions, first, the manual efforts of the developer can be reduced if the risk factors increases rapidly. Second, the risk factors can be classified in multiple classes to make the effort of the developer easier and faster. In spite of AHP for pairwise comparison, we can use any other approach to improve the results.

References

1. Chittister, C., Haimes, Y.Y.: Risk associated with software development: a bolistic framework for assessment and management. IEEE Transaction on Systems, Man and Cybernetics **23**, 710–723 (1993)
2. Boehm, B.: Software risk management. Springer (1989)
3. Vapnik, V.N.: The Nature of Statistical Learning Theory. Springer, New York (1995)
4. Cortes, C., Vapnik, V.: Support-vector networks. Machine Learning **20**, 273–297 (1995)
5. Burges, J.C.: A Tutorial on Support Vector Machines for Pattern Recognition. Data Mining and Knowledge Discovery **2**, 121–167 (1998)
6. Chang, C-C., Lin, C-J.: LIBSVM: A Library for Support Vector Machines, March 2013
7. Wallace, L., Keil, M., Rai, A.: Understanding Software Project Risk: A Cluster Analysis. Information Management **42**, 115–125 (2004)
8. Feng, N., Li, M., Gao, H.: A software project risk analysis model based on evidential reasoning approach. In: World Congress on Software Engineering (2009)
9. Wulan, M., Petrovic, D.: A Fuzzy Logic Based System for Risk Analysis and Evaluation within Enterprise Collaborations. Computers in Industry **63**, 739–748 (2012)
10. Fang, C., Marle, F.: A Simulation-Based Risk Network Model for Decision Support in Project Risk Management. Decision Support Systems **52**, 635–644 (2012)
11. Singh, D., Sharma, A.: Software requirement prioritization using machine learning. In: Proceedings of 26th International Conference on Software Engineering & Knowledge Engineering, SEKE-2014, Vancouver, Canada, pp. 701–704, July 2014
12. Gunn, S.: Support Vector Machines for Classification and Regression, ISIS Technical Report, November 1997
13. Goepel, K.D.: BPMSG's AHP Online System, BPMSG Online System, May 2014
14. Palcic, I., Lalic, B.: Analytical Hierarchy Process as a Tool for Selecting and Evaluating Projects. Int. J. Simul. Model **8**, 16–26 (2009)
15. Saaty, T.L.: Fundamentals of Analytical Hierarchy Process. RWS Publications (1994)
16. Sharma, T.C., Jain, M.: WEKA Approach for Comparative Study of Classification Algorithm. International Journal of Advanced Research in Computer and Communication Engineering **2**, April 2013
17. Jayenthi, S.N.: Efficient Classification Algorithms using SVMs for Large Datasets. Supercomputer Education and Research Center, June 2007
18. Suh, S.D.: Risk Management in a Large-Scale New Railway Transport System Project, June 2000
19. Chu, K.-K., Li, C.-H.: A Study of The Effect of Risk- Reduction Strategies on Purchase Intentions in Online Shopping. International Journal of Electronic Business Management **6**, 213–226 (2008)
20. http://www.slideshare.net/cutestpanchi/library-management-system-33260945

Research and Development of Knowledge Based Intelligent Design System for Bearings Library Construction Using SolidWorks API

Esanakula Jayakiran Reddy, C.N.V. Sridhar and V. Pandu Rangadu

Abstract The traditional design method of bearing is mainly based on the manual design process which invites numerous calculations. The small change in shape or size of assembly component will cause massive chain reaction like revision of blueprint because of many interrelated design issues. Hence, the bearing design needs to be changed in order to match the altered component. Advanced design methods such as CAD/CAM provide solutions for these issues by using of parametric modeling technique. This paper presents a typical knowledge based engineering system for rapid design and modeling of bearings based on operating conditions by integrating commercially available CAD package SolidWorks with Microsoft Access. An inference engine and proper user interface was developed for bearing design for assisting the engineering designers. The developed system proved itself as better application of engineering by utilizing the reuse of the design knowledge.

Keywords Knowledge Based System · Intelligent Design System · Parametric Modeling · SolidWorks API · Computer Aided Systems · Bearing · Macro · Access

---

E. Jayakiran Reddy(✉) · C.N.V. Sridhar
Department of Mechanical Engineering, Annamacharya Institute of Technology and Sciences, Rajampet, Kadapa 516126, Andhra Pradesh, India
e-mail: ejkiran@gmail.com, sridharcnv@yahoo.co.in

V. Pandu Rangadu
Department of Mechanical Engineering, JNTUA College of Engineering, Ananthapuramu 515002, Andhra Pradesh, India
e-mail: pandurangaduv@yahoo.com

S. Berretti et al. (eds.), *Intelligent Systems Technologies and Applications*,
Advances in Intelligent Systems and Computing 385,
DOI: 10.1007/978-3-319-23258-4_27

1 Introduction

Bearing is a widely used general machinery part in almost every industry. Developing the part model for bearing includes the modeling of inner ring, outer ring and rollers by considering the parameters like width, inner diameter, outer diameter, number of rollers. Owing to similar structure attribute of those of the same kind and diverse specifications, suitable bearing can be modeled by parametric modeling technique. Even though CAD became as the inevitable design practice in the recent decades, modeling is a time consuming process because of lack of skilled CAD modeling professionals. Parametric modeling technique rise above this issue as it regenerates the task in very less time in comparison with human. As the parametric modeling technique is dimension driven it facilitates automatic re-use of existing design process based on the results on engineering analysis. Moreover, parametric modeling technique is quick, efficient and interactive in comparison with conventional 3D modeling techniques. So, it is required to research and develop a knowledge based intelligent design system for the modeling of bearings by developing a library.

1.1 Secondary Development Tools for SolidWorks

SolidWorks is a three dimensional parametric solid modeling software developed based on windows. Any programming language supporting OLE (Object Linking and Embedding) and COM (Component Object Model) can be used as development tools of SolidWorks, such as Visual Basic (VB), Visual Basic Application (VBA), VC + +, C# etc. VBA is the simple tool for secondary development as it manages the secondary development during recording and editing macro.

1.1.1 SolidWorks API

API (Application Programming Interface) is a software development tool which enables integration between different applications by providing a code in a programming language within another application [1]. SolidWorks provides lot of API functions based on OLE automation which are the interfaces of OLE and COM to make the secondary development easier to the designer. API facilitates to develop windows based custom stand-alone executable files. Similarly, with the help of specific codes, SolidWorks API offers the automation of modeling, assembling of components. SolidWorks API covers all the functions of the software. Uday Hameed Farhan et al. [2] introduced an automated approach for assemblies of modular fixtures with SolidWorks API with the help of VB. Abhishek et al. [3] developed software application for product design and its CAD model updating with the help of SolidWorks API. Similarly, Tian et al. [4] established, the methods and steps for developing standard parts CAD system based on software re-use with the help of SolidWorks API.

1.1.2 Visual Basic Fundamentals

The secondary development of SolidWorks is based on a reasonably simple, extensively used and prominent programming language called Visual Basic (VB). VB is a primary development programming language with which API can access GUI [5]. VB is an effective engine for generating the macros in all Microsoft software. Thus, VB turned as a handy tool for developing different software packages. Moreover, VB is the engine of Microsoft Access for building the database which leads the programmer to have the better database control [6]. The integration of VB with different windows applications can be achieved with the help of ActiveX DLL (AvtiveX dynamic link libraries) programmers which helps in developing new menus, tools and toolbars into the application environment [7]. The most helpful feature of integration of VB and SolidWorks for this research is that it enables the user to develop their own GUI. By means of VB to transfer API function of SolidWorks, it is not only possible to construct, modify, delete the parts and part features but also to extract entities like dimension, geometry, topology etc.

2 Literature Review

Fast and easy solid modeling can be attained by using macro programming and parametric modeling technique. Construction of the standard parts library can be achieved with the help of these tools as they can simplify the complexity of the modeling by automating the commonly performed tasks using a predefined algorithm [8]. Wei et al. [9] recommended a method to generate the general parts based on the practice and assemble by developing and executing a part library on SolidWorks platform. This tool drastically reduced the modeling time by providing the model library of the desired standard parts. Bo et al. [10] attempted for developing the standard parts library of SolidWorks with VB along with plug-ins to fulfill the needs of the users. Avitus et al. [11] developed secondary development of SolidWorks for standard components based on database and demonstrated the development by automating the molding of the standard bolt. Tian et al. [12] established a 3D standard parts library with the help of functions of SolidWorks API as secondary development. Wang et al. [13] was able to change the existing design of the bottle by interfacing the CAD software with parametric sketching by controlling the database software remotely.

3 Methodology for Modeling Automation and Implementation

SolidWorks has the capability to reproduce all the geometric part data in the form of the text. This text can be used to modify the part model according to user input with the help of API. In this research work, Microsoft Access and SolidWorks

API are integrated for retrieving the data from database to SolidWorks to facilitate automatic and/ parametric modeling.

3.1 Selecting the Design Plan

The design plan of designing and modeling of Bearing is given as follows: firstly, a defined standard part is selected and is modeled in CAD software according to its dimension size and saved in database. Then its variables are parameterized according to the user needs. Secondly, the objects of API are called to revise the associated variable parameters restricted in the program through VB, which exchanges information of the model's geometry or topology and concludes the progress to build up the model of new one. It is necessary to create a database of variable parameters based on Microsoft Access as the variable parameters are specific. The calling of data from the database triggers itself in terms of Data Control or ActiveX Data Object Data Control (ADODC) in Visual Basic.

3.2 Preparing the Microsoft Access Database

In order to eliminate the data transfer errors, Microsoft Access database management system has been chosen as database as both VB and Access are from Microsoft Corporation. The database has been prepared with all relevant data like outer ring diameter, inner ring diameter, width, average life, basic static load, basic dynamic load, maximum speed, corresponding SKF and/ ISI number etc. as shown in the Fig. 1 for designing and modeling the bearings. The database is built in such a way that, the user needs to input only minimum data like loads, working speed, expected average life, so that all the relevant feature dimensions should be retrieved from the database.

SKF_ID	Dia Max	Dia Min	Width	ISI_ID	Max_Speed
6200	30	10	9	10BC02	20000
6201	32	12	10	12BC02	20000
6202	35	15	11	15BC02	16000
6203	40	17	12	17BC02	16000

Fig. 1 Microsoft Access Database of Deep Groove Ball Bearing

3.3 Developing the Standard Models and Their Macro Files

Macro programming is used for the development of standard models. Macro file can store any operation occurs on the SolidWorks screen like drawing a curve, trimming a line, revolving a surface etc. in the form of code. Macro has the ability of recording and replaying the operation. Macro uses the 3D coordinate system to store the dimensions and the position of any entity in the component. Additionally, the generated code in macro is in the form of a module of a VBA which helps in linking the database to macro.

Before building the standard part, the characteristics of the model must be analyzed and the basic features must be established. Then, manually develop the standard CAD part with the standard size parameters. If the relations between parameters of the model are found then equation should be established among them. The procedure for standard part modeling is given as follows: At first, the standard 3D CAD model of the component with all the features, dimensions and constrains has been generated while recording its macro file. Soon after recording, the macro file need to be modified to parameterize itself to fit into VB code of SolidWorks API for generating the appropriate model of the component whenever it is called by the user. For the generating the altered or new CAD model, the dimensions can be retrieved from the database based on the inputs given by the user in GUI and design calculation. Likewise, separate macros are developed for different standard bearings. The developed bearing library in this research work consists of Deep Groove Ball Bearing, Single Row Angular Contact Ball Bearing, Double Row Angular Contact Ball Bearing, Single Row Thrust Ball Bearing, Double Row Thrust Ball Bearing, Self Aligning Ball Bearing, Roller Bearing, and Needle Bearing.

3.4 Creating Human-Machine Interface

SolidWorks allows the user to develop their own GUI with features and functions of VB language. Every bearing variety consists of their own GUI that can be appeared by clicking the appropriate bearing button in the GUI of the SolidWorks that are developed with the help of *New Macro Button* function of SolidWorks. Fig. 2 shows the developed bearing buttons. In total, 8 varieties of bearings are considered in the construction of bearing library as said in the section 3.3. Each bearing variety consists of a separate button. As a sample, the GUI of the deep groove ball bearing is shown in the Fig. 3. For user ease, 3D part image and tooltip of the selected bearing are added in the GUI. While developing the GUI, it is significant to consider Microsoft ActiveX Data Objects 2.5 Library as reference.

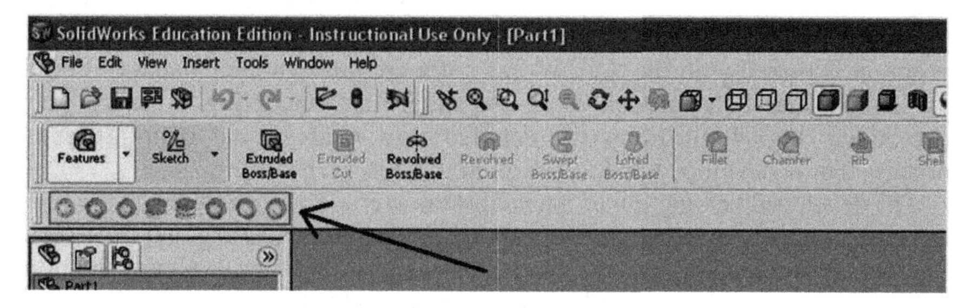

Fig. 2 Developed bearing buttons in the GUI of SolidWorks.

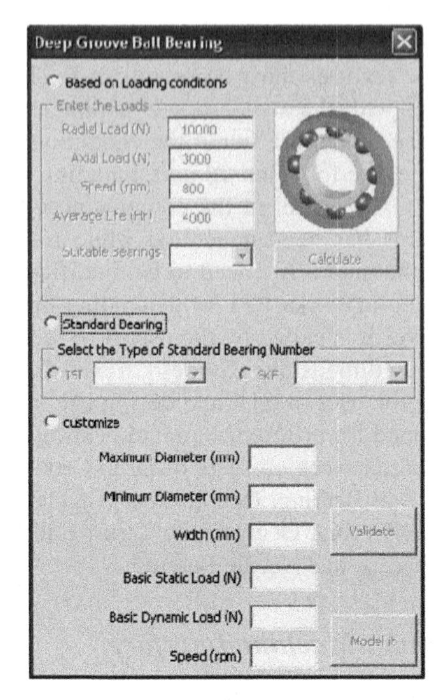

Fig. 3 Developed GUI of the deep groove ball bearing

3.5 Program Development for Design Calculation

The developed intelligent design system in this research work can support the logical algorithms to do the respective design calculations. It is well-known fact that, in the process of designing, the dimensions of bearings will change as per the working or loading condition whenever there is a change in assembly component, but their geometry never changes. Therefore, it is tedious and time consuming to recalculate the suitable dimensions whenever there is a change in component. In order to reduce the time consumption, a computer program is developed for design calculation purpose. The output from this program is linked with the SolidWorks software through the database in order to generate the CAD model. Thus, the design and modeling work has been automated. Fig. 4 illustrates the detailed flow chart of the Bearing design process that is developed and followed in this research work.

The developed bearing library facilitates the user to design and model the bearing based on three options: *loading conditions*, *Standard Bearing* and *Customize*. The option *loading condition* facilitates the user to input the loading conditions like radial load, axial load, working speed (rpm) and average life as shown in the Fig. 5. After entering the input data the user need to click the *Calculate* button to calculate the suitable bearing based on loading conditions. At the backend with the help of standard imperial formulas, developed computer program and intelligent design system, the suitable bearing will be selected. The result will be displayed in the GUI in combo box field as shown in Fig. 6.

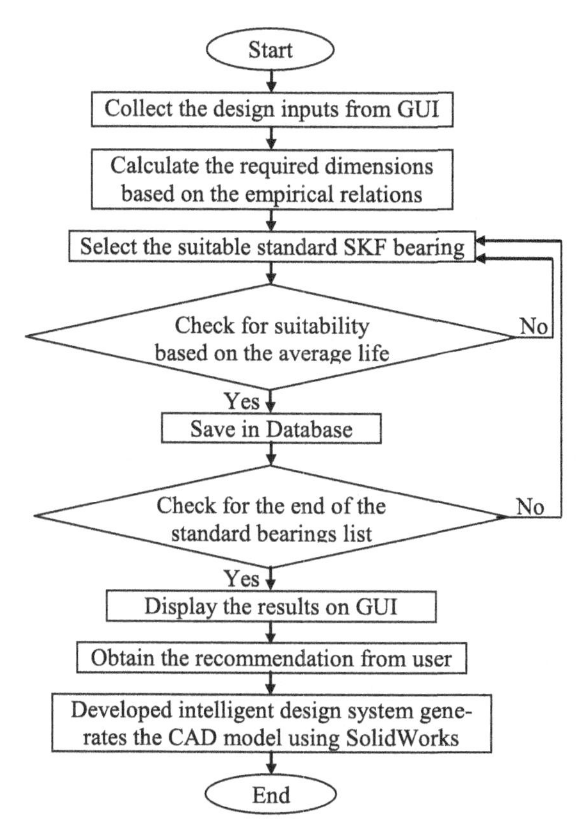

Fig. 4 Flow chart of the Bearing design process

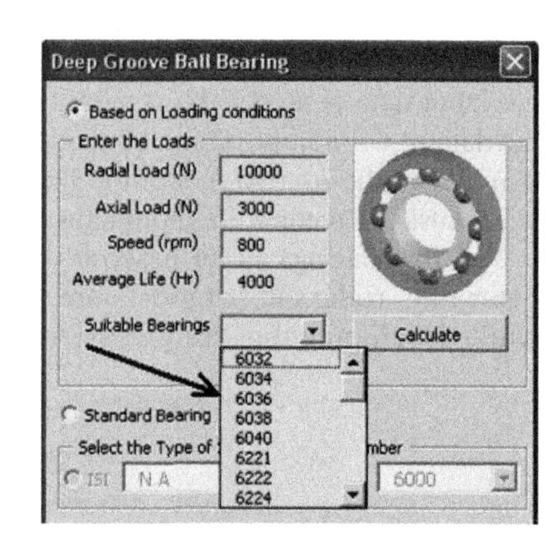

Fig. 5 GUI to collect the input data

Fig. 6 Display of suitable bearing

The number in the combo box is the SKF bearing standard number and they are sorted from best suitable to least suitable from top to bottom. By default, the best suitable bearing will be modeled. If user wishes to model some other from the list, an option is provided to selecting the alternative. Upon selection, the corresponding dimensions and ISI number can be seen in the same GUI at the bottom portion as shown in the Fig. 7. Soon after clicking the *Model it* button, CAD model will be generated as shown in the Fig. 8.

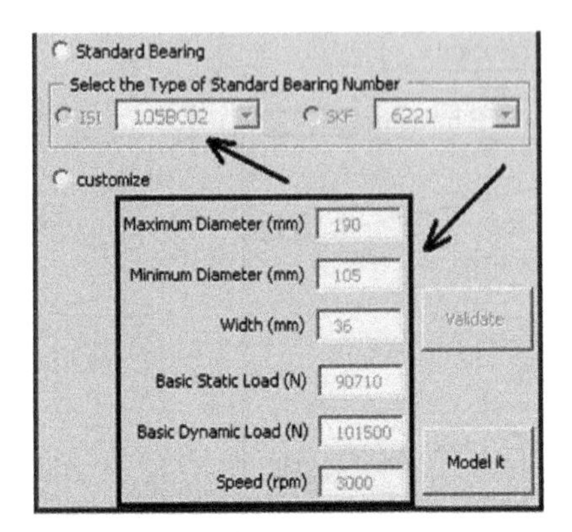

Fig. 7 ISI number and corresponding dimensions

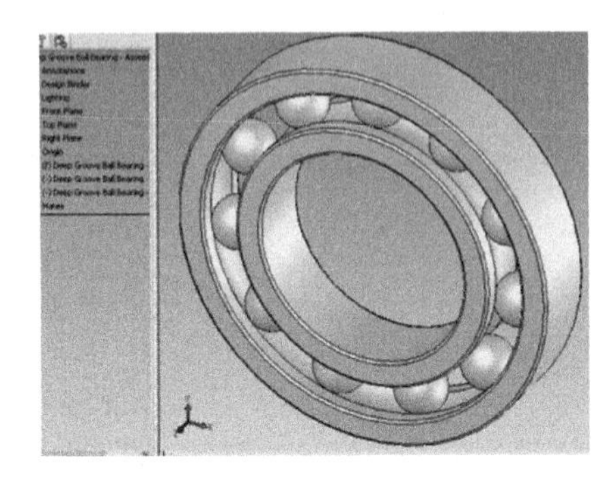

Fig. 8 Generated CAD model

The next modeling option for the user is *Standard Bearing*, which enables the user to select and model the standard bearing based on their SKF or ISI number. Upon selecting the standard bearing from the corresponding combo boxes the respective dimensions can be available as shown in the Fig. 7. Soon after selection, by clicking the *Model it* button the selected CAD model can be generated.

Another modeling option is *Customize*, which enables the user to input the dimensions of the bearing like outer ring diameter, inner ring diameter and width. Soon after entering the input data, user should validate the input data by clicking the *Validate* button. The computer program behind this button will check whether the input data is as per the standards or not. Fig. 9 shows the facility of entering the data in *Customize* option of modeling. If the entered data is valid then the system allows the user to model the bearing by enabling the *Model it* button, otherwise it asks to reenter the input data.

As the bearings is an assembly consists of different parts like inner ring, outer ring, roller etc., at most care has been taken to handle the assemble parts while developing this design system.

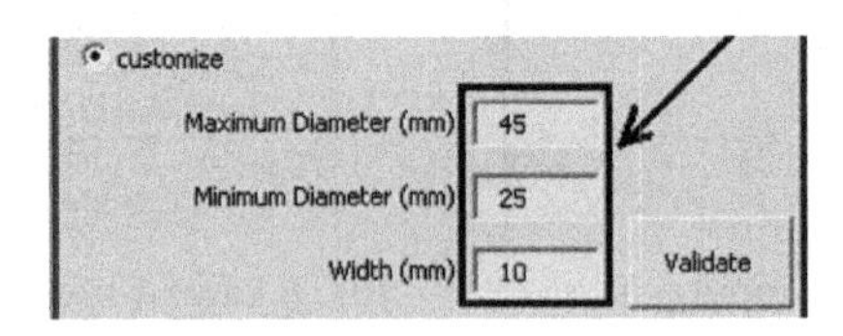

Fig. 9 Customize option of modeling

4 Conclusion

This research work is a preliminary attempt to develop intelligent design system for bearing library construction. The presented method decreases the modeling time. The compilation of presented computer code is easy and meets the design requirements and efficiency. The proposed method is significant in engineering applications. Although the presented example is reasonably moderate, future work can be extended to carry out other complex bearings to fulfill needs of designers.

References

1. Prince, S.P., Ryan, R.G., Mincer, T.: Common API: using visual basic to communicate between engineering design and analytical software tools. In: ASEE Annual Conference (2005)
2. Farhan, U.H., Tolouei-Rad, M., O'Brien, S.: An Automated Approach for Assembling Modular Fixtures Using SolidWorks. International Scholarly and Scientific Research & Innovation **6**(12), 365–368 (2012)
3. Lad, A.C., Rao, A.S.: Design and Drawing Automation Using Solid Works Application Programming Interface **2**(7), 157–167 (2014)
4. Jian-li, T., Shen-xiao, L., Hui, F.: CAD system design on standard part based on software reuse. In: Fourth International Symposium on Knowledge Acquisition and Modeling (KAM), pp. 229–232 (2011)
5. Deitel, H.M., Deitel, P.J., Nieto, T.R.: Visual Basic 6 how to program. Prentice Hall, Upper Saddle River (1999)
6. Kerman, M.C., Brown, R.L.: Computer programming fundamentals with applications in visual basic 6.0. Addison-Wesley, Reading (2000)
7. Farhan, U., O'Brien, S., Tolouei Rad, M.: SolidWorks Secondary Development with Visual Basic 6 for an Automated Modular Fixture Assembly Approach. International Journal of Engineering **6**(6), 290–304 (2012)
8. Jayakiran Reddy, E., Sridhar, C.N.V., Pandu Rangadu, V.: Knowledge Based Engineering: Notion, Approaches and Future Trends. American Journal of Intelligent Systems **5**(1), 1–17 (2015)
9. Liu, W., Zhou, X., Niu, Q., Ni, Y.: A Convenient Part Library Based On SolidWorks Platform. International Journal of Mechanical, Aerospace, Industrial and Mechatronics Engineering **8**(12), 1851–1854 (2014)
10. Sun, B., Qin, G., Fang, Y.: Research of standard parts library construction for solidworks by visual basic. In: International Conference on Electronic & Mechanical Engineering and Information Technology, pp. 2651–2654 (2011)
11. Titus, A., Xue Bin, L.: Secondary Development of Solid works for Standard Components Based on Database. International Journal of Science and Research **2**(10), 162–164 (2013)
12. Chen, T., Yan, X., Zhonghai, Yu.: The Research and Development of VB and Solidworks-Based 3D Fixture Component Library. Applied Mechanics and Materials **300–301**, 301–305 (2013)
13. Wang, S.-H., Melendez, S., Tsai, C.-S.: Application of parametric sketching and associability in 3D CAD. Computer-Aided Design and Applications **5**(6), 822–830 (2008)

Discovering Context Using Contextual Positional Regions Based on Chains of Frequent Terms in Text Documents

Anagha Kulkarni, Vrinda Tokekar and Parag Kulkarni

Abstract While assigning importance to terms in Vector Space Model (VSM), most of the times, weights are assigned to terms straightaway. This way of assigning importance to terms fails to capture positional influence of terms in the document. To capture positional influence of terms, this paper proposes an algorithm to create Contextual Positional Regions (CPRs) called Dynamic Partitioning of Text Documents with Chains of Frequent Terms (DynaPart-CFT). Based on CPRs, Contextual Positional Influence (CPI) is calculated which helps in improving F-measure during text categorization. This novel way of assigning importance to terms is evaluated using three standard text datasets. The performance improvement is at the expense of small additional storage cost.

1 Introduction

In past few years, analysis of unstructured data, where context determination is vital, has gained importance due to tremendous rise in volume of documents. Most of the unstructured data contains text. The technique of assigning appropriate categories to text documents is called text categorization or classification (TC). *Content based* TC does not consider complex relationships between the contents of the documents.

A. Kulkarni(✉)
Cummins COE for Women, Pune, India
e-mail: anagha.kulkarni@cumminscollege.in

V. Tokekar
Institute of Engineering and Technology, DAVV, Indore, India
e-mail: vrindatokekar@yahoo.com

P. Kulkarni
EkLAT, Pune, India
e-mail: paragindia@gmail.com

S. Berretti et al. (eds.), *Intelligent Systems Technologies and Applications*,
Advances in Intelligent Systems and Computing 385,
DOI: 10.1007/978-3-319-23258-4_28

Context based TC attempts to use *unsaid* or *implicit* information, thereby improving relevance of documents to the category.

k Nearest Neighbour (kNN) [2] and Support Vector Machine (SVM) [20], popular classification algorithms with text data, use Vector Space Model (VSM) [32]. In VSM, every document is transformed into a vector in high-dimensional term space. Importance of a term is represented by assigning a weight to the term. Most common way of assigning a weight to a term is by counting its frequency of occurrence in the document, called Term Frequency (TF) measure. TF suffers from drawback of poor discrimination power.

Terms used in a document convey meaning or context of a document. Context has been studied by many researchers. It is a very powerful concept. Depending on the application, it is a location, weather, blood pressure, harmony, time, environment or even history of events [3, 4, 8, 11, 18, 33]. From text documents' perspective, it could be author, title, readability index, digital trace of user's actions and so on [13, 31]. It may also be the information about category of the text document [24] or the terms surrounding an important term [22]. In short, context is a set of preferences which are infinite and are partially known. In this research work, clue is taken from positions of terms and their contextual positional influence (CPI) in the document is calculated.

By intuition, everybody knows that all the terms having equal TF may not contribute equally in deciding context of the document. *Role of a term towards building context of the document is called contextual positional influence (CPI).* For example, consider two sentences - 'Steve runs faster than Tom' and 'Tom runs faster than Steve'. In the first sentence, influence or importance of 'steve' is regarded to be more due to its occurrence before 'tom'. Thus, CPI of 'steve' is more than that of 'tom'. Hence it becomes more important to understand involvement and importance of a term in the context. Therefore, this paper attempts to design a weighting scheme based on the CPI of terms in a document.

To calculate CPI of terms, each text document needs to be divided in contextual positional regions (CPR). *CPR is the segment or region in which term is present.* It is called *CPR*, as it is not simply a region, but an indicative of contextual significance of terms. This paper proposes CPR creating algorithm based on discourse segmentation - Dynamic Partitioning of text documents with Chains of Frequent Terms (DynaPart-CFT). DynaPart-CFT creates segments (or regions) in a dynamic manner and CPI is assigned to terms.

In view of the above discussion, the present research paper is divided in five sections. Section 3 presents details of proposed algorithms and their storage complexities. Section 4 presents results and discussions. Conclusions are stated in section 5. Next section reviews literature applicable to above mentioned work.

2 Literature Review

Popular techniques for term weighting are TF, TF-Inverse Document Frequency (TF-IDF) [29], Relevance Frequency (RF) [25], Term Rank Identifier (TRI) [6] and so on. Many researchers have commonly used these techniques [21]. However, most important limitation of all these techniques is that they do not consider positional significance of terms. This important aspect on which importance of a term depends is ignored by all the above mentioned techniques.

Some progress in weighting terms using positional significance has been done. Kulkarni et al. have shown that terms more relevant to the document occur in the earlier partitions of the documents [23]. Enhancing weights of these terms improves classification results. The authors divided each document in 2, 3 and 4 partitions irrespective of length of the document and found that by enhancing weights of the terms occurring in earlier partitions, average F-measure got improved.

Gawrysiak et al. include word position or even its formatting in document representation along with words and their TF [14]. They state that it may be useful in analysis of longer documents. Murata et al. calculated term weight using location and detailed information [30]. Location information included the information about the location of the term in the document. More weight is given to the term if it appears in title or at the beginning of the document. Detailed information includes whether it is a proper noun and more weight is given to proper nouns.

For partitioning a document, tiles based on topic are created in [16]. Chain of repeated terms is considered for tiling a document. Chain of repeated terms indicates a subtopic flow. When such chain breaks, a subtopic is assumed to have ended and a tile is created.

3 DynaPart-CFT and Storage Complexity

CPRs are created in text documents before CPI of every term is calculated. CPR creation and CPI calculation are discussed below.

3.1 DynaPart-CFT

In literature, it is found that most of the approaches for TC straightaway assign weights to the terms. This paper proposes a new method of assigning weights to terms. To improve F-measure, CPI of terms is calculated using CPRs.

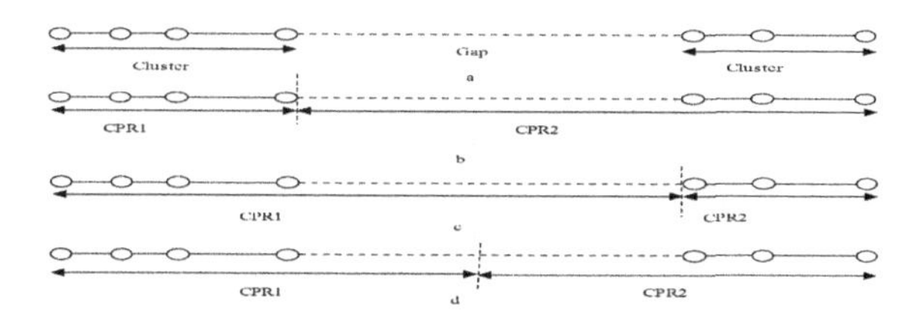

Fig. 1 Clusters of frequent terms and a gap between two clusters

To create CPRs, DynaPart-CFT uses judgment similar to the one used by human beings. When set of words read earlier and the set of words being read presently are different, human beings recognize change in subtopic or context. Even though a document contains three logical sections, *introduction*, *middle section* and *conclusion* [10, 34], terms used in these sections may be different. In such case, the subtopics that are discussed could be different. Therefore, a CPR is created when a subtopic changes. DynaPart-CFT identifies frequent terms in every document. Then, it locates chains of frequent terms. One chain is one CPR. Figure 1(a) shows two clusters of frequent terms with a gap. A circle represents a term in the document. A gap is the set of terms which are not frequent. It shows three ways of creating CPRs:

1. A gap can be included in next CPR (shown in figure 1(b))
2. A gap can be included in previous CPR (shown in figure 1c)
3. First half of gap can be included in previous CPR and second half in next CPR (shown in figure 1d)

DynaPart-CFT creates CPRs as shown in figure 1(b).

Mathematically, set of frequent terms (FT), chain of frequent terms (CFT) and a gap between two CFTs (Gap) are represented by (1), (2) and (3) respectively.

$$FT = \{t_j : w_j | w_j > \tau\} \tag{1}$$

where FT is set of frequent terms t_j in document d having weight w_j and τ is minimum number of times t_j must occur in d.

$$CFT^p = \{FT | dist(t^p_{j+1} - t^p_j) \leq \Delta\} \tag{2}$$

where dist() is a function that finds distance (number of infrequent terms) between two consecutive frequent terms t_j and $t_{(j+1)}$. Therefore, CFT is a set of FT such that distance between two consecutive frequent terms is less than or equal to pre-specified maximum distance, Δ.

$$Gap^p_{p-1} = \{t_l : w_l | w_l \leq \tau\} \tag{3}$$

where gap between two CFTs is the set of consecutive terms t_l whose weight w_l is less than or equal to τ. Finally, number of CPRs $f_{CPR}(d)$ and size of each CPR $size_p$ is given by (4) and (5).

$$f_{CPR}(d) = max(1, 2, 3, 4) \tag{4}$$

where max(1,2,3,4) indicates maximum number of CPRs that can be possibly created.

$$size_p = \begin{cases} n(CFT^p) & , if\ p = 1 \\ n(Gap^p_{p-1}) + n(CFT^p) & , otherwise \end{cases} \tag{5}$$

where n() indicates number of terms in that set.

For example, consider a document with Newid=8510 in Reuters-21578 dataset (hereafter called SampleDoc). SampleDoc is shown in Appendix along with CPRs obtained by DynaPart-CFT. Newid is the identification number (ID) the story has in the Reuters-21578. These IDs are assigned to the stories in chronological order. For this document, total number of terms are 153 (after pre-processing), $f_{CPR}(d) = 4$ and $size_p$ is variable. It depends on chains of frequent terms.

CPI calculating algorithms are discussed next.

3.2 CPI Calculating Algorithms

TF alone does not characterize the significance of a term in the document. Terms that occur in the 'introduction' may reappear in the 'middle' and 'conclusion' section. These terms may be assisted by some new terms to define the context. In 'conclusion' section, the whole discussion is summarized. Frequently appearing terms may appear again in this section. In other words, relevant terms appear more frequently in the document. They start appearing from the beginning of the document. Thus, while calculating CPI of the terms, their CPR should be considered.

3.2.1 CC1

The algorithm assigns maximum CPI (equal to number of CPRs) to terms appearing in the first CPR. CPI is reduced by one in every subsequent CPR such that CPI assigned to the terms in last CPR is one. CPI of terms in SampleDoc is 4 for first 5 terms, 3 for next 84 terms, 2 for next 33 terms and 1 for last 31 terms. This can be put mathematically as follows for DynaPart-CFT:

$$CPI_t^p = f_{CPR}(d) - p + 1 \tag{6}$$

Thus, in d, total CPI of t is given by

$$f_{CPI}(d,t) = \sum_{p=1}^{f_{CPR}(d)} f_t^p \times CPI_t^p \tag{7}$$

where f_t^p is frequency of occurrence of t in p.

3.2.2 CC2

Compactness measures how well a term is distributed [35]. If a term is compact, it is less relevant. On the other hand, Callan felt that partitions reflecting inverse paragraph frequency (similar to inverse document frequency) may improve the classification results [5]. As per inverse paragraph frequency, weight assigned to a term reduces if the term is present in all the partitions whereas as per CC2 the CPI of a term increases if it is present in more than one CPR (that is, if it is distributed).

The proposed algorithm modifies Xue and Zhou's approach [35]. First, this concept is applied to CPRs created as discussed in section 2.1. Then, if a term is present in more than one CPR, it is considered as distributed. CPI of such term is calculated using (6). If a term is compact, its CPI is calculated using TF.

For example, in SampleDoc term 'florjancic' appears at positions 105 and 136 out of total 153 terms. It is a distributed term as initially it appears in third CPR and then in fourth CPR. Therefore, CPI of 'florjancic' is 2+1=3. CPI is calculated according to (6) and (7).

3.3 Storage Complexity

In addition to storage requirements of TF, DynaPart-CFT requires additional storage. It is required to find frequently occurring terms in every document. If m most frequent terms are found in the document, an array of size m is required per document to store those m terms. If the document has g gaps of large size, then an additional array of size g is also required per document. Therefore for DynaPart-CFT, additional storage complexity is O(m+g) per document. For z documents, total complexity would be $O(z \times (m+g))$. This is an extra storage cost in addition to required by TF calculation.

Experiments are conducted to evaluate DynaPart-CFT and CC1 and CC2 using standard text datasets. Results are presented in the next section.

4 Results and Discussions

In this paper, three standard text datasets: Reuters-21578 [27], 20-Newsgroups [26] and RCV1 [28] are used for testing. The data in these datasets is classified using most popular classifiers: Weka's lazy classifier IBk (implements kNN) [15] and LIBSVM (implements SVM) [7]. Many researchers have mentioned these two classifiers as widely used classifiers with text datasets [1, 17, 19, 25, 35]. kNN is a simple, instance based, lazy classification technique. It makes prediction based on k training instances which are closest to the unlabeled (test) instance. This is done at prediction time. Thus in case of high-dimensional and large datasets, it is not efficient [9]. Yet, it produces accurate results. SVM is a supervised classification algorithm. The most important property of SVM is that it selects support vectors required for classification even from high-dimensionality of feature space [12, 17, 20]. This makes it suitable for TC. Some researchers have experimented on choosing kernels for natural language processing related applications. They have found that polynomial kernels of degree 2 or so are popular [7].

kNN is run for k=5 and k=10 neighbors (for Reuters-21578 and 20-Newsgroups) and for k=5 neighbors (for RCV1). SVM is run with two kernel functions namely, linear and polynomial. Rest of the parameters are set as default. Polynomial kernel is run with degrees 1, 2 and 3. TF is used as baseline for comparison of results obtained using DynaPart-CFT and CPI calculating algorithms.

Using DynaPart-CFT, experiments are carried out to find ten most frequently occurring terms per document. Large gaps between chains of frequent terms are also found. It is observed that, even a very large document does not have more than four or five gaps of size more than ten terms. The algorithm identifies at most three largest gaps to have maximum 4 CPRs in the document. Size of ten terms or more is considered as a large gap. Generally if stop words are included, then this large gap will be equivalent to one or two sentences.

Results are evaluated using weighted average F-measure [29]. They are discussed below:

1. **Effect of CPR and CPI on Reuters-21578 Dataset:**
   Reuters-21578 dataset contains 21578 articles. Reuters-21578 contains TOPICS as one of the categories of documents. This paper has focused on five sub-categories of TOPICS namely 'acq', 'crude', 'earn', 'grain' and 'trade'. The distribution of TOPICS in the dataset is skewed. The 'ModApte' split is used to split training and testing documents. As per this split, the number of training and test documents are 4070 and 1638 respectively. Some documents do not have any text. They are ignored. The largest file size before pre-processing is 14KB. After pre-processing (stop word removal and stemming) training documents, 11697 distinct terms are extracted [29]. For dimensionality reduction, terms having collection frequency greater than 2.5% across the dataset are selected. 459 terms are extracted to form the vocabulary. It is observed that some terms are used very frequently to represent a TOPIC throughout the dataset. It is also observed that after pre-processing, all the documents have less than 1000 terms. As shown in figure 2(a), approximately 86% documents have less than 100 terms. It indicates that documents in this dataset have small size.

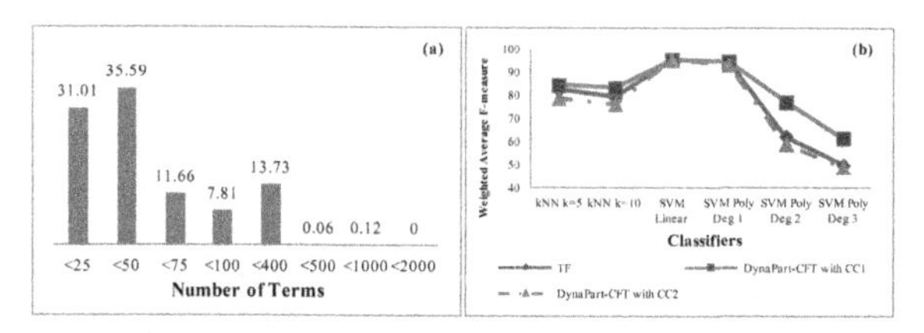

Fig. 2 Distribution of documents according to sizes and comparison of weighted average F-measure obtained by TF and DynaPart-CFT with CC1 and CC2 in Reuters-21578

   When CPI of terms is calculated using CC1, terms occurring in earlier CPRs have more weight than terms in subsequent CPRs. The number of CPRs in a document and their sizes are decided by chains of frequent terms in that document. It is observed that F-measure improves with DynaPart-CFT with CC1 as shown in figure 2(b).

   When CPI of terms is calculated with CC2, more weight is given to distributed terms than compact terms. CPI of compact terms is calculated using TF. F-measure is poor. This is so, because the CPRs are formed on the basis of chains of frequent terms. The frequent terms occurring in two CPRs may or may not be similar. Since F-measure has decreased, we can say that frequent terms are not distributed in this dataset.

2. **Effect of CPR and CPI on talk.* Newsgroups Dataset:**
   20-Newsgroups dataset contains 19997 articles collected from the Usenet newsgroup collections. It contains news from twenty categories. This paper has focused on all subcategories in 'talk' (talk.* Newsgroups). Unlike Reuters-21578, distribution of categories in talk.* Newgroups is uniform. 600 documents out of 1000 from every subcategory are used as training documents and remaining 400 as test documents. After pre-processing, 24647 distinct terms are extracted. For dimensionality reduction, terms having frequency of occurrence greater than 150 across the dataset are selected. 690 terms are extracted to form the vocabulary. The largest file size without pre-processing is 71KB. It is observed that after pre-processing, 1.5% documents have more than 1000 terms. As shown in figure 3(a), around 86% documents have more than 100 but less than 500 terms. It indicates that documents in this dataset have large size and different terms are used in the documents.

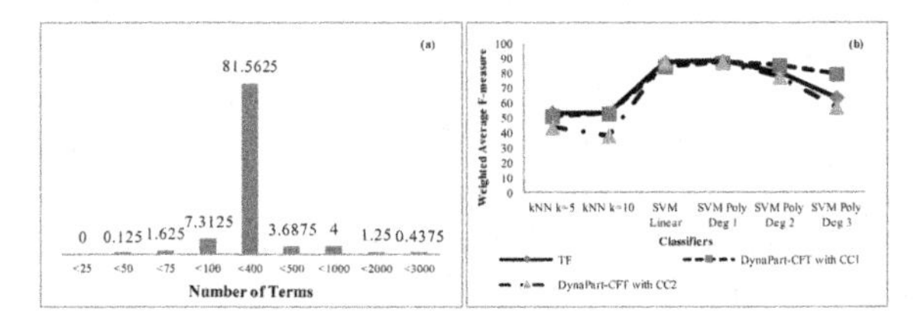

Fig. 3 Distribution of documents according to sizes and comparison of weighted average F-measure obtained by TF and DynaPart-CFT with CC1 and CC2 in talk.* Newsgroups

   When CPI of terms is calculated using CC1, terms occurring in earlier CPRs have more weight than terms in subsequent CPRs. The number of CPRs in a document and their sizes are decided by chains of frequent terms in that document. It is observed that F-measure improves for DynaPart-CFT with CC1 as shown in figure 3(b).

   When CPI of terms is calculated with CC2, more weight is given to distributed terms than compact terms. CPI of compact terms is calculated using TF. F-measure does not improve for any classifier as every chain of frequent terms contains different terms. In this dataset, documents contain a very small number of repeating terms.

3. **Effect of CPR and CPI on RCV1 Dataset:**
   Reuters Corpus, Volume I (RCV1) is an archive of 806791 manually categorized newswire stories made available by Reuters, Ltd. for research purposes. The stories are manually coded using three categories: TOPICS, INDUSTRIES and REGIONS. This paper has focused on TOPICS category. Totally, there are 15 TOPICS: C, CE, CEG, CEGM, CEM, CG, CGM, CM, E, EG, EGM, EM,

G, GM and M where C is Corporate/Industrial category, E is Economics category, G is Government/Social category and M is Markets category. Around 36700 documents do not contain any TOPIC. They are ignored in this paper. Remaining documents are split into training and test data sets using training/test split suggested by [28]. All documents from August 20, 1996 to August 31, 1996 (total 22272) are included in training data set. As the data set contains very large number of documents, DynaPart-CFT with CC1 and CC2 is tested for 98308 documents from test dataset. The evaluation also proves that the proposed technique is highly scalable.

The largest file size before pre-processing is 36KB. After pre-processing of training documents, 53173 distinct terms are extracted. For dimensionality reduction, terms having collection frequency of more than 20% are selected. 187 terms are extracted to form the vocabulary. It is observed that some terms are used very frequently to represent a TOPIC throughout the dataset. Thus, only a few terms are enough to classify the documents correctly. It is also observed that after pre-processing, 99.95% documents have less than 1000 terms. As shown in figure 4(a), approximately 55% documents have 75 to 400 terms. 44% documents have less than 75 terms.

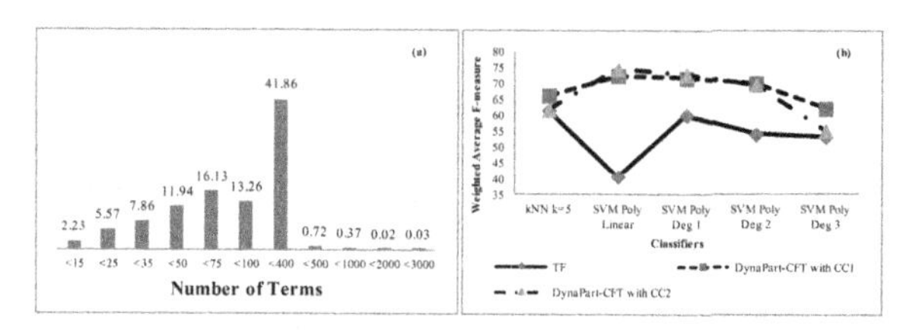

Fig. 4 Distribution of documents according to sizes and comparison of weighted average F-measure obtained by TF and DynaPart-CFT with CC1 and CC2 in RCV1

It is observed that F-measure improves for all classifiers with CC1 and CC2 both as shown in figure 4(b). This indicates that most of the terms in the documents are distributed.

5 Conclusion

This research paper proposes a novel idea of calculating CPI of terms using CPRs. The research reported in this paper suggests that this helps in improving weighted average F-measure (based on macro-weighted averaging). Experiments show that proposed algorithms improve categorization results.

DynaPart-CFT has shown good performance. For Reuters-21578 dataset and talk.* Newsgroups, CC1 performs well. With CC2, F-measure does not improve. It can be concluded that, documents in these datasets contain a very small number of repeating terms across CPRs. For RCV1, F-measure has improved with both CPI calculating algorithms. This dataset contains documents having repeating terms across CPRs. CC1 performs well with all datasets. This indicates that terms that appear from the beginning of the document are more relevant and if their CPI is calculated using CC1, F-measure improves.

It is difficult to recognize correct boundaries in text documents. In future, it is very important to identify logical boundaries automatically. It would be interesting if this can be achieved using human way of thinking. Experiments could be carried using inverse paragraph frequency to see if it improves performance further.

References

1. Aggarwal, C.C., Zhai, C.: A survey of text classification algorithms. In: Mining text data, pp. 163–222. Springer (2012)
2. Aha, D.W., Kibler, D., Albert, M.K.: Instance-based learning algorithms. Machine learning **6**(1), 37–66 (1991)
3. Brézillon, P.: Context in problem solving: a survey. The Knowledge Engineering Review **14**(01), 47–80 (1999)
4. Brown, P.J., Bovey, J.D., Chen, X.: Context-aware applications: from the laboratory to the marketplace. IEEE Personal Communications **4**(5), 58–64 (1997)
5. Callan, J.P.: Passage-level evidence in document retrieval. In: Proceedings of the 17th Annual International ACM SIGIR Conference on Research and Development in Information Retrieval, pp. 302–310. Springer, New York (1994)
6. Chand, K.P., Narsimha, G.: An integrated approach to improve the text categorization using semantic measures. In: Jain, L.C., Behera, H.S., Mandal, J.K., Mohapatra, D.P. (eds.) Computational Intelligence in Data Mining-Volume 2. SIST, vol. 32, pp. 39–47. Springer, Heidelberg (2015)
7. Chang, C.C., Lin, C.J.: Libsvm: a library for support vector machines. ACM Transactions on Intelligent Systems and Technology (TIST) **2**(3), 27 (2011)
8. Chen, G., Kotz, D., et al.: A survey of context-aware mobile computing research. Tech. rep., Technical Report TR2000-381, Dept. of Computer Science, Dartmouth College (2000)
9. Cover, T., Hart, P.: Nearest neighbor pattern classification. IEEE Transactions on Information Theory **13**(1), 21–27 (1967)
10. Dao, N.: A new class of functions for describing logical structures in text. PhD thesis, Massachusetts Institute of Technology (2004)
11. Dey, A.K.: Understanding and using context. Personal and ubiquitous computing **5**(1), 4–7 (2001)
12. Dumais, S., Platt, J., Heckerman, D., Sahami, M.: Inductive learning algorithms and representations for text categorization. In: Proceedings of the Seventh International Conference on Information and Knowledge Management, pp. 148–155. ACM (1998)
13. Dumitrescu, A., Santini, S.: Think locally, search globally; context based information retrieval. In: IEEE International Conference on Semantic Computing, ICSC 2009, pp. 396–401. IEEE (2009)
14. Gawrysiak, P., Gancarz, L., Okoniewski, M.: Recording word position information for improved document categorization. In: Proceedings of the PAKDD Text Mining Workshop (2002)
15. Hall, M., Frank, E., Holmes, G., Pfahringer, B., Reutemann, P., Witten, I.H.: The weka data mining software: An update. SIGKDD Explorations **11**(1) (2009)

16. Hearst, M.A.: Multi-paragraph segmentation of expository text. In: Proceedings of the 32nd annual meeting on Association for Computational Linguistics. Association for Computational Linguistics (1994)
17. Hotho, A., Nürnberger, A., Paaß, G.: A brief survey of text mining. Ldv Forum **20**, 19–62 (2005)
18. Hull, R., Neaves, P., Bedford-Roberts, J.: Towards situated computing. In: First International Symposium on Wearable Computers, Digest of Papers, pp. 146–153. IEEE (1997)
19. Ikonomakis, M., Kotsiantis, S., Tampakas, V.: Text classification using machine learning techniques. WSEAS Transactions on Computers **4**(8), 966–974 (2005)
20. Joachims, T.: Text categorization with support vector machines: Learning with many relevant features. In: Nédellec, C., Rouveirol, C. (eds.) Machine Learning: ECML-98. LNCS, vol. 1398, pp. 137–142. Springer, Heidelberg (1998)
21. Kou, G., Peng, Y.: An application of latent semantic analysis for text categorization. International Journal of Computers Communications & Control **10**(3), 357–369 (2015)
22. Kulkarni, A., Tokekar, V., Kulkarni, P.: Identifying context of text documents using naïve bayes classification and apriori association rule mining. In: 2012 CSI Sixth International Conference on Software Engineering (CONSEG), pp. 1–4. IEEE (2012)
23. Kulkarni, A., Tokekar, V., Kulkarni, P.: Text classification by enhancing weights of terms based on their positional appearances. International Journal of Computer Applications **78**(9), 23–26 (2013)
24. Kulkarni, A., Tokekar, V., Kulkarni, P.: Discovering context of labelled text documents using context similarity coefficient. Procedia Computer Science **49C**(9), 118–127 (2015)
25. Lan, M., Tan, C.L., Su, J., Lu, Y.: Supervised and traditional term weighting methods for automatic text categorization. IEEE Transactions on Pattern Analysis and Machine Intelligence **31**(4), 721–735 (2009)
26. Lang, K.: Newsweeder: learning to filter netnews. In: Proc of 12th Intl Conference on Machine Learning, pp. 331–339 (1995)
27. Lewis, D.: Reuetrs-21578 text categorization test collection, dist 1.0 (1997)
28. Lewis, D., Yang, Y., Rose, T.G., Li, F.: Rcv1: a new benchmark collection for text categorization research. The Journal of Machine Learning Research **5**, 361–397 (2004)
29. Manning, C.D., Raghavan, P., Schütze, H.: Introduction to information retrieval, vol. 1. Cambridge University Press, Cambridge (2008)
30. Murata, M., Ma, Q., Uchimoto, K., Ozaku, H., Utiyama, M., Isahara, H.: Japanese probabilistic information retrieval using location and category information. In: Proceedings of the Fifth International Workshop on Information Retrieval with Asian languages, pp. 81–88. ACM (2000)
31. Navrat, P., Taraba, T.: Context search. In: 2007 IEEE/WIC/ACM International Conferences on Web Intelligence and Intelligent Agent Technology Workshops, pp. 99–102. IEEE (2007)
32. Salton, G., Wong, A., Yang, C.S.: A vector space model for automatic indexing. Communication of the ACM **18** (1975)
33. Schilit, B.N., Theimer, M.M.: Disseminating active map information to mobile hosts. IEEE Network **8**(5), 22–32 (1994)
34. Stovall, J.G.: Writing for the Mass Media. 6th edn. Pearson Education (2006)
35. Xue, X.B., Zhou, Z.H.: Distributional features for text categorization. TKDE **21**(3), 428–442 (2009)

Appendix

SampleDoc: Newid=8510 in Reuters-21578

Brunswick Corp expects 1987 first quarter **{CFT1}** sales to be up "dramatically" and profits to "do well," chairman and president Jack Reichert said after a securities analysts meeting. He declined to be more specific. In the 1986 first quarter, Brunswick reported earnings of 23.8 mln dlrs or 57 cts a share on sales of 396.7 mln dlrs. Reichart noted that results of its two newly-acquired boat manufacturing companies will be included in the company's first quarter report. Brunswick expects its recreation centers to benefit from increased attention to the sport of bowling resulting from acceptance in the 1988 Summer Olympics of bowling as an exhibition sport and as a medal sport in the 1991 Pan American Games, he said. He said field testing of a new bowling concept involving electronic features is being readied for test marketing this summer, and if successful, could "materially benefit" operations. Brunswick is currently test marketing in California a health club facility adjoining a bowling center, he said. Turning to its defense operations, **{CFT2}** Reichert said he expects the division to receive significant contracts in the near future. At 1986 year end, Brunswick's defense contract backlog stood at 425 mln dlrs. Frederick Florjancic, vice president-finance, told analysts Brunswick was disappointed two credit rating services recently downgraded the company's debt which stood at about 665.4 mln dlrs **{CFT3}** at 1986 year end. "We are confident we can service our debt and bring it down in the very near term," based on strong cash flow from Brunswick's expanded boat operations, Florjancic said. Shareholders at the company's April 27 annual shareholders meeting will be asked to approve an increase in the authorized common shares outstanding to 200 mln from 100 mln shares, a company spokesman said. **{CFT4}**

Most frequent terms found by DynaPart-CFT with their TF are shown below. {brunswick,7}, {mln, 6}, {company, 5}, {dlr, 4}, {bowl, 4}, {test, 3}, {sport, 3}, {share, 3}, {quarter, 3}, {operation, 3}.

Data Integration of Heterogeneous Data Sources Using QR Decomposition

Harikumar Sandhya and Mekha Meriam Roy

Abstract Integration of data residing at different sites and providing users with a unified view of these data is being extensively studied for commercial and scientific purposes. Amongst various concerns of integration, semantic integration is the most challenging problem that addresses the resolution of the semantic conflicts between heterogeneous data sources. Even if the data sources may belong to similar domain, due to the lack of commonality in the schema of databases and the instances of databases, the unified result of the integration may be inaccurate and difficult to validate. So, identification of the most significant or independent attributes of each data source and then providing a unified view of these is a challenge in the realm of heterogeneity. This demands for proper analysis of each data source in order to have a comprehensive meaning and structure of the same. The contribution of this paper is in the realization of semantic integration of heterogeneous sources of similar domain using QR decomposition, together with a bridging knowledge base. The independent attributes of each data source are found that are integrated based on the similarity or correlation amongst them, for forming a global view of all the data sources, with the aid of a knowledge base. In case of an incomplete knowledge base, we also formulate a recommendation strategy for the integration of the possible set of attributes. Experimental results show the feasibility of this approach with the data sources of same domain.

1 Introduction

Every organization designs its databases to suit one or more autonomous applications with varied and multiple representations. The schema of the tables designed,

H. Sandhya(✉) · M.M. Roy
Department of Computer Science and Engineering, Amrita Vishwa Vidyapeetham, Amritapuri, Kollam 690525, Kerala, India
e-mail: sandhyaharikumar@am.amrita.edu, mekharoy90@gmail.com

S. Berretti et al. (eds.), *Intelligent Systems Technologies and Applications*, Advances in Intelligent Systems and Computing 385,
DOI: 10.1007/978-3-319-23258-4_29

the type of instances stored and the applications developed vary from one database to another which amass hundreds of separate, disconnected data sources in varying formats. This exponential growth of heterogeneous data sources distributed at various sites have led to the requirement of semantic integration of data for information sharing and collaboration. This process becomes inevitable not only for business prospects(when two similar companies need to merge their databases) but also for scientific(combining patient details of different hospitals for knowledge discovery) reasons. The semantics of each data source can be inferred from only a few information sources such as creators of the database, documentation and some sort of annotation of the application and associated schema and data [1]. Extracting semantic information manually is a very cumbersome task due to the non-accessibility of people and confidentiality. Hence two data sources are matched for schema based on data stored and some existing prior and partial knowledge about the data. [2] describes about a significant research to describe information sources and the trade offs of query processing explored in various fields. It also discusses about the Local-as-View (LAV) and Global-as-View(GAV) approach. However, Global view once evolved may undergo changes due to the dynamic and frequent changes in the autonomous data sources. So, a periodic check on the mappings found should be done in order to leverage the global schema for an efficient querying system. The data sources being very high dimensional, a strategy to bring only the significant attributes at the global level, that form the basis of the data source, should be devised.

The key to automated integration is to be rigorous about capturing semantic metadata. The most important challenge in semantic integration is understanding the structure and meaning of the data. A large scale integration will inevitably involve thousands or hundreds of thousands of semantic mappings. At this scale, an effective mechanism to retrieve the basis of each data source need to be devised. Though some sort of knowledge base may be available, the basis can be definitely considered as significant attributes to be taken at the global level. There are different types of semantic heterogeneities like name heterogeneity, structural heterogeneity etc. Name heterogeneity regards name conflicts in different source schemas (i.e., synonyms, homonyms), while structural heterogeneity concerns the representation of the information in different schemas. If we can compare the structure of the two data sources, we can throw light onto the semantic matching of the two sources. There are different types of structural conflicts like type conflicts, key conflicts and domain conflicts. Type conflicts represents the same piece of information using different constructs in different schemas. Our focus here is to deal with the type conflicts where in the aim is to understand the structure of the data from the same domain represented in different forms in different schemas.

For instance, consider the two schemas given in Fig. 1 taken from [3] represented in different constructs where the data is about house listing information of North Kentucky and Texas from their real estate website. They are from the same domain and the data in these databases might be having the same meaning but represented under different attribute names and types.
There should be a method to find the independent attributes in each local schema. Bringing all the attributes from the data sources to form the global schema is absurd

House_location	Price($)	Rooms	Bedrooms	Baths	Levels	...	Directions
301 BRACKEN	159000	6	3	2	2	...	AA HWY S, L HWY 1159, R HWY 8, L MAIN, R 4TH, L BRACKEN
...	...	...	...	...	...	...	...
32 RILEY RD	129900	5	3	1	1	...	AA TO ALEX-SILVER GROVE EX TO ALEXANDRIA HOUSE ON R ACROSS FROM FAIR GROUNDS

A. NORTH KENTUCKY REAL ESTATE DATASET

Firm_name	...	House_location	County	Price($)	Bedrooms	Full_baths	Living rooms	...	Highlights
Westfall Real Estate	...	County Road #226 Giddings, TEXAS 78942	Lee	115000	3	2	1	...	Tree Covered Acreage
...	...	...	...	...	...	...	...	...	...
CENTURY 21 Lakeland Realtors	...	830 Miller Circle Meridian, TEXAS 76665	Bosque	94900	3	2	1	...	Quiet Country Subdivision

B. TEXAS REAL ESTATE DATASET

Fig. 1 Real Estate database

and causes superfluous complications in all means. Here, data integration involves combining of independent attributes from different sources. A much more generalized way of representing the data in its substructure form without losing the semantics is devised here.

The major contributions of this paper are:

1. QR decomposition of each data source after normalization.
2. Finding the independent attributes of each data source.
3. Compute the best possible integration of the independent attributes using knowledge base and similarity measure.

2 Related Work

Semantic integration has been extensively studied and various techniques based on machine learning aspects have been applied to form a fruitful merging of the data sources. [4] comes up with a solution on how to classify the attributes from different sources to target attributes, when there is a need to transform one representation to another to bring all to a common representation in data integration. Here a Clio tool is used for the semi-automatic schema mapping. In [5], for identifying attribute correspondences, a semi-automated semantic integration method is found. It needs the proficiency to automatically ascertain the likelihood of attributes pointing to the real-world data that are the same. In [6], a GAV approach is followed by iteratively applying multiple L1 linear regression and Singular Value Decomposition for finding the most prominent attributes in each data source. A global schema is designed consisting of the integration of most significant attributes of different relational schemas of a particular domain. In [7] the uncertainty in data-integration systems is to be handled at three levels as demanded by the authors. Instead of the structured form, the queries to the system would probably be keywords. In such cases, inaccurate data might be extracted and given. Hence the concept of probabilistic schema mappings was introduced: by-table semantics and by-tuple semantics. In [8], the authors have put forward a method using data semantics for data integration and pattern discovery for environmental research. The method revolves around a metadata approach and takes benefit of the data interrelationships which is depicted as a semantic network. An algorithm is given which expands naturally, supplement and filter the semantic network with the help of usage patterns. [9] presents an architecture of integration between information systems that, allows to extract the ontology that represents the information. The architecture has a component that allows to merge ontologies both in each information system-creating the shared ontology-and between several information systems-creating the global ontology.

In [10], the authors identified that many challenges are not addressed such that the clustering algorithms rely on a single instance property, a consistent score for an attribute match is not produced and hierarchical relationships between the data are not considered. To solve these, they introduced a tool called GeoSim which comprises of GeoSimG and GeoSimH used for finding the semantic similarity between geospatial schemas. In [11], the authors have found a general logic-based framework for semantically registering resources with ontologies and discussed about certain properties of the resulting semantic registrations. In [12], authors show how to integrate the semantic parsing techniques for simple foundation units into a connected semantic graph that helps in understanding the meaning of the text and also put forward an approach to integrate new documents with an available semantic knowledge base. In [13], authors present a classification that covers most of the existing approaches. The authors distinguish between schema-level and instance-level, element-level and structure-level, and language-based and constraint-based matchers. They use it to set apart and be on a level with a variety of previous match implementations keeping in mind that it will be helpful to programmers and to researchers. Very few research [14] has attempted to fully explore the use of attribute values to perform attribute identification instead of concentrating on the schema for integration. However identification of significant attributes from each data source is quite mandatory in a high dimensional data source. This identification can not only lead to reduced attributes at global level but will also give a comprehensive understanding of the data source.

3 Semantic Integration Using QR Decomposition

Let $D_1, ..., D_k$ be k different data sources and $L_1, ..., L_k$ be the respective local schemas, then the formation of global schema using QR decomposition aims to find the optimal mapping amongst the local schemas. Let D_i be a relational data source with n attributes $a_1, a_2, ..., a_n$ and m rows with either $m \geq n$ or $m < n$. Any relational schema can be considered as a vector space model by preprocessing the attributes and tuples. Hence the relational data source D is converted into a matrix form by applying a preprocessing step as given in 3.4.

3.1 QR Decomposition

An efficient method for finding the independent attributes is required and that can represent the data in its proper substructure form without losing its semantics. One such method is QR decomposition. This is an efficient method for decomposing a matrix A into a product A = QR of an orthogonal matrix Q and an upper triangular matrix R [15]. The result of QR decomposition helps get rid of the redundancy in the data sources. It is essential to remove the redundant data and bring out only the significant data to the forefront. QR decomposition with column pivoting is the algorithm used here as it is useful when A is rank deficient, or is suspected of being so. The method provides a numerical stability [15]. Two attributes which are mathematically called as vectors are orthogonal if the angle between them is a right angle [16]. In the schema given in Fig. 1, the house location attribute may or may not determine living rooms. This can be established by the fact that one vector(house location) is orthogonal to other vector(living rooms). QR decomposition is a technique to establish this fact. A minimum threshold value showing the variation of the attributes or vectors is chosen in QR decomposition with column pivoting to distinguish between the independent and dependent attributes.

The next objective is to provide an integrated view of the heterogeneous data sources with the help of a knowledge base.

3.2 Knowledge Base and Similarity Measure

Knowledge Base: Knowledge base is a centralized repository which has information about the schema matching of different data sources. Here knowledge base is used as a guide to form the global schema after picking the independent attributes from every schema. A knowledge base requires inputs from experts. The construction of knowledge base is out of scope and is assumed to be available in prior. For instance, consider the knowledge base given in Fig. 2 for the Real Estate schema given in Fig. 1. This knowledge base contains information for not all the attribute relations meaning that it is incomplete. For the attributes which does not have any information from the knowledge base, a recommendation strategy is devised with the help of a similarity measure.

Similarity Measure: Those independent attributes about which no information is available in the knowledge base, a suggestion in the form of a vector similarity measure is recommended for verification by the decision makers. It is up to the decision maker whether to include the recommendation in the global schema or not. Otherwise it is just a suggestion given based on the similarity measure. A vector based

Real Estate-North Kentucky	Real Estate-Texas
House location	House location
Price	Price
Rooms	?
Bedrooms	Bedrooms
Baths	Full baths
Levels	?
mls	mls
House type	?
Firm phone	?
Agent phone	?
Hoa fee	?
Semi annual fee	?
?	Firm name
?	Office
?	Fax
?	id
?	Living
?	Dining
?	Garage

Fig. 2 Knowledge base for Real Estate

similarity measure is adopted here to quantify the similarity between two attribute valued vectors.

$$\cos\theta = \frac{A \cdot B}{\|A\|\,\|B\|} \tag{1}$$

For instance, consider the following table which shows how the recommendation strategy works for an attribute(independent) when the knowledge base is incomplete in case of a Real Estate schema.

Table 1 Recommendation strategy

Data source 1 attribute	Similarity measure of data source 1 with data source 2	Recommendation from data source 2
Baths	0.76-Price, 0.33-House location, 0.19-id	Price

3.3 Architecture and Query Processing

The Fig. 3 represents the architecture for the semantic integration. All the heterogeneous data sources are modelled as a matrix with normalized data. Here the attributes are assumed to be of numerical datatype. Then QR decomposition with column pivoting is performed on each of the matrix. The interest is actually in the Q of the matrices which gives the basis that span the vectors or the attributes in the matrix. This Q is then used to find the similarity between the attributes in the local schemas. For all those entries in the knowledge base which is a repository of information about relation

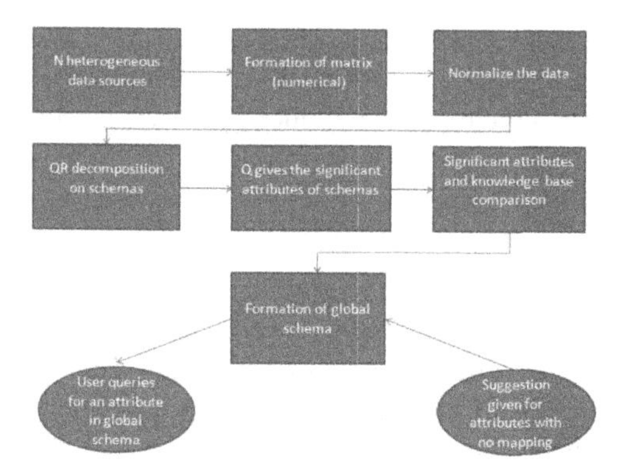

Fig. 3 Architecture for semantic integration of heterogeneous data sources

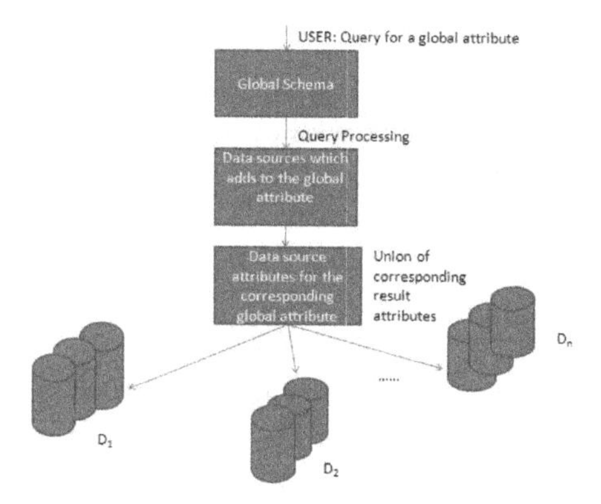

Fig. 4 Architecture of query processing at global schema

between attributes of schemas, forms an entry in the global schema of independent attributes. Recommendation or suggestion for independent attributes missing in the knowledge base given based on the similarity measure of attributes. Recommendation is given with an implication to provide a suggestion for a possible entry in the global schema with the help of a similarity measure when there is no information available in the knowledge base about the corresponding independent attribute.

The Fig. 4 is the pictorial representation of what happens and how the result is obtained when a user gives a query to view a global schema attribute. The user gets an integrated view of the heterogeneous data sources and the user can query for any global attribute. The global attribute will be a union of the different data sources. The global schema is formed by identifying the independent attributes of each of the data sources using QR decomposition with column pivoting and with the help of a knowledge base and the recommendation strategy. When the user queries for a global attribute, the global schema is referred and the corresponding attributes from the respective data sources that forms the global attribute are identified. These attributes are fetched and their union of values are given as a result of the querying of the global attribute.

3.4 Algorithm

Preprocessing the Data Source: To begin with, each data source modelled as a matrix is a vector space and the attributes of each data source quantified as the vectors. Attributes with numerical values are only considered for integration after applying the normalization as follows:

$$x = \frac{x - \mu}{\sigma} \tag{2}$$

After the normalization of the numerical attributes, a preprocessed data source is obtained which further goes through the following algorithms. QR decomposition is applied on a normalized data.

Algorithm 1. QR decomposition with column pivoting

{n is the number of columns in the matrix}
{A_k is the columns in the matrix A which is a preprocessed data source}
Begin
$l = 1$
for k:1 to n **do**
  W_k=A_k
  for j:1 to k **do**
    R_{jk}=$Q^T_{j-1}A_k$
    W_k=W_k-($R_{jk}Q_{j-1}$)
  end for
  if $W_k \geq$ threshold value **then**
    $R_{kk} = \| W_k \|_2$
    $Q_l = W_k / R_{kk}$
    $l = l + 1$
  end if
end for
End

Algorithm 2. Formation of Global Schema

{DS is the data source}
{nds is the number of heterogeneous data sources}
{PDS is the preprocessed data sources}
{SA is the matrix with significant attributes}
{g is the number of global schema attributes}
{a is a data source say DS_1}
{b is a data source say DS_2}
{c is the number of attributes in DS_1}
{d is the number of attributes in DS_2}
Input: DS_1, DS_2, ..., DS_n, Knowledge Base
Output: Global Schema, Recommendation table
Begin
for i from 1 to nds **do**
  for each DS_i **do**
    PDS_i contains the preprocessed data source
    Perform QR decomposition with column pivoting on PDS_i
    $PDS_i = Q_i R_i$
    $SA_i = Q_i$
  end for
end for
g=0
for each SA_{a_c} **do**
  for each SA_{b_d} **do**
    if SA_{a_c} maps SA_{b_d} in knowledge base **then**
      Insert G_g, SA_{a_c}, SA_{b_d} into the global schema
    else if unmapped **then**
      a suggestion for the attribute of the data source given by calculating the similarity
      $\cos\theta = \frac{A \cdot B}{\|A\|\|B\|}$
      Recommendation: Select the attribute with the maximum similarity value
    end if
  end for
end for
user query any attribute in the global schema
End

4 Experimental Analysis

For experimental purpose, the real datasets like Bank, German, Hepatitis and Real Estate datasets have been taken. The experiments are mainly done on the datasets available from UCI repository and Illinois Semantic Integration Archive.

Experimentation have been done by incorporating the following changes in the dataset:

Table 2 Execution time for QR decomposition with column pivoting in different cases of a Hepatitis dataset

	QR with column pivoting
columns of dataset interchanged	0.1450s
making the dataset rank deficit	0.1700s
rank deficit and columns of dataset interchanged	0.2910s

The following are the results when semantic integration is done on different datasets with different cases:

Table 3 Execution times and similarity in different datasets

	Bank Dataset	German Dataset	Hepatitis Dataset
complete dataset	15.13s	2.19s	2.72s
columns interchanged of complete dataset	13.87s; mismatch=2/17	2.79s; mismatch=2/21	2.35s; mismatch=3/20
subset of the complete dataset	12.84s	1.87s	2.24s
columns interchanged of subset of the complete dataset	12.26s; mismatch=2/17	2.83s; mismatch=3/21	2.51s; mismatch=4/20

Mismatch: Mismatch is a measure of how many dissimilarities are there from the knowledge base when only the recommendation strategy(similarity measure) is used.

Experimental results of Real Estate dataset and Bank dataset:
North Kentucky attributes: House location, Price, Rooms, Bedrooms, Baths, Levels, mls, House type, Firm phone, Agent phone, hoa fee, Semi annual fee
Texas attributes: Firm name, Office, Fax, id, mls, House location, Price, Bedrooms, Full baths, Living, Dining, Garage
North Kentucky independent attributes: House location, Price, Rooms, Bedrooms, Baths, Levels, mls, House type, Firm phone
Texas independent attributes: Firm name, Office, Fax, id, mls, House location, Price, Bedrooms, Dining

In Bank dataset with 17 attributes, we get 15 attributes as independent from the data sources after the QR decomposition.

Mapping and Recommendation: Mapping is based on the independent attributes and their relation given in the knowledge base. If the knowledge base contains relation between the independent attributes then they are mapped and if the knowledge base does not have relation for the significant attributes then it gives a recommendation.

Global attribute	North Kentucky Real Estate	Texas Real Estate	Mapped or not(Yes/No)	Recommendation
G1	House location	House location	Yes	
G2	Price	Price	Yes	
G3	Bedrooms	Bedrooms	Yes	
G4	mls	mls	Yes	
G5	Baths		No	Price

Fig. 5 A snapshot of the Real Estate global schema

Global attribute	Data source 1	Data source 2	Mapped or not(Yes/No)	Recommendation
G1	A1		No	B10
G2	A2		No	B7
G3		B3	No	A15
G4	A4	B4	Yes	
G5	A5	B5	Yes	
G6	A6	B6	Yes	
G7	A7	B7	Yes	

Fig. 6 A snapshot of the Bank dataset global schema

5 Conclusion

A novel approach towards semantic integration of heterogeneous data sources have been proposed by leveraging the strategy of QR decomposition. QR decomposition aids in extracting the most significant or independent attributes from each data source. Formation of global schema is based on these attributes together with a knowledge base consisting of the mapping amongst the attributes in the local schema. A recommendation strategy for more possible mappings amongst the attributes is also suggested for the decision makers to update the knowledge base.

6 Future Work

The future work aims at expanding the strategy in distributed environment for practical relevance. Functional dependency existing amongst the attributes of each data source can be explored in order to improve the query processing at Local-as-View and Global-as-View.

Acknowledgments We would like to thank Dr. M.R. Kaimal for his valuable suggestions and support through out the duration of this work.

References

1. Bergamaschi, S., Castano, S., Vincini, M., Beneventano, D.: Semantic Integration of Heterogeneous Information Sources. Data and Knowledge Engineering **36**, 215–249 (2001)
2. Alon, H., Rajaraman, A., Joann, O.: Data integration-the teenage years. In: VLDB 2006, pp. 9–16

3. Illinois Semantic Integration Archive: http://pages.cs.wisc.edu/anhai/wisc-si-archive/summary.name.html
4. Naumann, F., Ho, C.-T., Tian, X., Haas, L., Megiddo, N.: Attribute classification using feature analysis. In: International Conference on Data Engineering(ICDE) (2002)
5. Li, W.-S., Clifton, C.: SEMINT-A Tool for Identifying Attribute Correspondences in Heterogeneous Databases Using Neural Networks. Data and Knowledge Engineering **33**, 49–84 (2000)
6. Harikumar, S., Reethima, R., Kaimal, M.R.: Semantic integration of heterogeneous relational schemas using multiple L1 linear regression and SVD. In: International Conference on Data Science and Engineering(ICDSE), pp. 105–111, August 2014
7. Dong, X., Halevy, A.Y., Yu, C.: Data integration with uncertainty. In: VLDB 2007, pp. 687–698 (2007)
8. Chen, Z., Gangopadhyay, A., Karabatis, G., McGuire, M., Welty, C.: Semantic Integration and Knowledge Discovery for Environmental Research. Journal of Database Management, 43–67, January–March 2007
9. Guido, A.L., Paiano, R.: Semantic Integration of Information Systems. International Journal of Computer Networks and Communications(IJCNC) **2**, January 2010
10. Partyka, J., Khan, L.: Content-based geospatial schema matching using semi-supervised geosemantic clustering and hierarchy. In: Fifth IEEE International Conference on Semantic Computing, pp. 247–254, September 2011
11. Bowers, S., Lin, K., Ludascher, B.: On integrating scientific resources through semantic registration. In: 16th International Conference on Scientific and Statistical Database Management, pp. 349–352, June 2004
12. Bethard, S., Nielsen, R., Martin, J.H., Ward, W.: Semantic integration in learning from text. In: 22nd International Conference on Artificial Intelligence, AAAI 2007 (2007)
13. Rahm, E., Bernstein, P.A.: A Survey of Approaches to Automatic Schema Matching. The VLDB Journal **10**, 334–350 (2001)
14. Chua, C.E.H., Chiang, R.H.L., Lim, E.-P.: Instance-based Attribute Identification in Database Integration. The VLDB Journal **12**, 228–243 (2003)
15. QR decomposition: http://en.wikipedia.org/wiki/QR_decomposition
16. Meyer, C.D.: Matrix Analysis and Applied Linear Algebra. SIAM (2000)

An Intelligent Model for Privacy Preserving Data Mining: Study on Health Care Data

Jisha Jose Panackal and Anitha S. Pillai

Abstract Critical challenge in developing a privacy protection mechanism is to preserve maximum information because protection mechanisms normally impact on the quality of data and which are served not accordingly with the data utility. Practical solutions to address various socio-economic needs with special emphasize on the utility of data have not been devised yet. To publish maximum information while protecting the privacy, we propose an intelligent mechanism and this paper includes a comprehensive study and explores how effectively the privacy of individuals can be protected with minimum information loss. Empirical evaluations on original health care data related to Indian Population show the effectiveness of the new approach, namely Adaptive Utility-based Anonymization (AUA).

1 Introduction

Information can be shared or published for the process of knowledge discovery, if there is a one to fit all mechanism to protect the privacy of individuals. Anonymization is one of the promising Privacy Preserving Data Mining (PPDM) techniques in order to deal with the problem of disclosure risk and it is extensively used in the data publishing scenario. Even though, some remarkable developments are noticed in this field, there exist only few academic studies based on Anonymization to effectively handle the problem of disclosure risk and the problem of information loss evenly using original data set. We explored the need for development of such mechanism according to Indian context and a detailed discussion is available in [1].

J.J. Panackal(✉)
Vidya Academy of Science and Technology, Thrissur 680 501, Kerala, India
e-mail: jishapanackal@gmail.com

A.S. Pillai
Hindustan University, Chennai 603 103, Tamil Nadu, India
e-mail: anithasp@hindustaniniv.ac.in

S. Berretti et al. (eds.), *Intelligent Systems Technologies and Applications*,
Advances in Intelligent Systems and Computing 385,
DOI: 10.1007/978-3-319-23258-4_30

To address the existing privacy issues in National Family Health Survey data NFHS-3, we proposed a model namely, Adaptive Utility-based Anonymization (AUA) as mentioned in [2]. The study [3] proposes an implementation framework which is based on AUA model. The main objective of this paper is to present how effectively the privacy of individuals is protected with minimum information loss based on the utility of data NFHS-3. The rest of the paper is organized as follows: Section 2 is used to discuss some of the related work and the research objectives. Section 3 presents the proposed approach namely, AUA. In section 4 results are included and finally by section 5 we conclude the paper with some future research directions.

2 Related Works

According to [4] and [5], the privacy is addressed in two distinct scenarios distributed and centralized. In distributed environment, the data may be distributed and two or more parties can perform data mining operations on the union of these databases. Cryptographic techniques are widely used to achieve privacy in this scenario. In centralized scenario, the data is published and it can be used for variety of purposes. Studies like [6], [7] and up to [12] show that the privacy is addressed by various techniques including Anonymization, Perturbation, Condensation, Randomisation and Fuzzy-based methods in the data publishing scenario.

Even though the data is not encrypted, some sort of precaution should be taken regarding the privacy of individuals, before publishing the data. The techniques such as data hiding, compression, generalization, swapping and permutation are commonly applied mechanisms to protect privacy in this scenario. K-anonymity as mentioned in [13] is the classical approach under this category and is widely accepted mechanism to protect privacy to some extent. Later the studies, [14], [15] and [16] strengthen the process of anonymization by extending the capability of k-anonymity model.

The very recent studies like [17] and [18] also mention the need for protecting the privacy of individuals and suggests some useful mechanisms to PPDM. The one to fit all mechanisms to achieve privacy as per different socio-economic needs has to be developed. Thus PPDM is becoming a hot area of research today to address various privacy issues by preserving maximum information. A complete review of literatures can also be seen in [19].

3 Proposed Model: AUA

The proposed method namely, Adaptive Utility-based Anonymization (AUA) model presented in [2], is a solution to the problem of disclosure risk. The problem of disclosure risk on NFHS-3 data set is explained in [1]. This paper presents a study of AUA on health care data, NFHS-3 which explores various experimental results and shows the possibility of treating it as an adaptive anonymization approach focusing on utility of data.

3.1 Methodology

Over generalization and sensitive values suppression will definitely affect the data accuracy. Therefore, we are focusing our study based on quasi-sensitive associations as well as on quasi-quasi associations. Several objective measures of pattern interestingness exist related to association mining which is based on the structure of discovered patterns and the statistics underlying in a given data set. Some of the measures namely, support and confidence [20] are defined as follows and these form the basis of AUA model.

3.1.1 Support

An objective measure for association rules of the form $X => Y$ is rule support, representing the percentage of instances from a database that the given rule satisfies. This is taken to be the probability $P(X \cup Y)$, where $X \cup Y$ indicates that an instance contains both X and Y, that is, the union of itemsets X and Y. That is, a support gives the frequency of the occurrence of a value combination in a dataset. Here the micro-data set NFHS-3 is denoted as Oiginal Table (OT).

Definition 3.1. *The support of a value combination q_i in OT, denoted by $supOT(q_i)$, is the percentage of the records in OT which contains all the attribute values in q_j.*

3.1.2 Confidence

Another objective measure for association rules is confidence, which assesses the degree of certainty of the detected association. This is taken to be the conditional probability $P(Y|X)$, that is, the probability that an instance containing X also contains Y.

Definition 3.2. *The confidence of a value combination q_i in OT, denoted by $confOT$*
(q_i), is the probability that the records in OT which contain q_i also contain q_j.

A confidence, or certainty, of 50% means that if a person belongs to the state Karnataka, there is a 50% chance that he will be of age 23 as well. A 1% support means that 1% of all of the transactions under analysis showed that Karnataka and 23 occurred together.

3.1.3 Quasi-Sensitive Associations

The first step in the AUA process is based on quasi-sensitive association and is described in this section. The support and confidence are calculated by also considering the sensitive value of each instances. Thus, every instances with probability Disease-"yes" is compared with Disease-"no" for finding the association betwen attribute values.

Definition 3.3. *Let QI be the set of quasi-identifying attributes in the data set OT and SA be the sensitive attributes, and attribute set* $QI' \subseteq QI$ *and* $SA' \subseteq SA$*. For any values* $q' \in QI'$ *and* $s' \in SA'$*, the associations* $q' \Rightarrow s'$ *is termed as quasi-sensitive associations in OT.*

3.1.4 Quasi-Quasi Associations

The second step that is Anonymization process in AUA model is based on quasi-quasi associations and is defined as follows.

Definition 3.4. *Let QI be the set of quasi-identifying attributes in the data set OT any subset* $QI' \subseteq QI$ *may be associated with rest of the attributes in QI, denoted as* $(QI\text{-}QI') \subseteq QI$*. For any subset* $q_1' \in QI$ *and* $q_2' \in (QI - QI')$*, the associations* $q_1' \Rightarrow q_2'$ *is termed as quasi-quasi associations.*

3.1.5 Apriori Like Approach

Apriori [20] employs an iterative approach known as a level-wise search, where k-itemsets are used to explore $(k + 1)$-itemsets. First, the set of frequent 1-itemsets is found by scanning the database to accumulate the count for each item, and collecting those items that satisfy minimum support. The resulting set is denoted as L_1. Then L_1 is used to find L_2, the set of frequent 2-itemsets which is used to find L_3, and so on until no more frequent k-itemsets can be found. Finding each L_k requires one full scan of the database.

The proposed method is based on Association Mining and is similar to Apriori, but it is used only to de-identify the risky records which are under disclosure risk. ie, our approach splits the entire data set into two sets, namely, frequent set and non-frequent set based on the quasi-sensitive associations. Then we are using a new anonymization approach based on the following two criterions:

1. Based on the utility of data, for example, the study may be based on the value of quasi-identifier attributes such as State-wise, Age-wise, Religion-wise or Education-wise.
2. Based on the quasi-quasi associations between non-frequent and frequent sets of data.

In short, an iterative approach used in proposed AUA model includes two rounds of association mining, one is based on quasi-sensitive associations in one set and the other is based on quasi-quasi associations between two sets.

3.2 *Adaptive Utility-based Anonymization (AUA) model*

This section reiterates our approach namely AUA which is introduced in [2] and explains the process of anonymization. The approach can broadly be interpreted as a two step process as follows:

Step 1: Filtering of uniquely identifiable records with less than min_sup.

As the initial step, the procedure separates the QI data set into two subsets namely, frequent-QI set and non-frequent-QI set based on quasi-sensitive associations. Since the non-frequently occurring subset is likely being identified, and the records with these QI values are vulnerable to disclosure risk. The association between each attribute in QI vector is analyzed based on the basic principle used in Apriori algorithm. Apart from the usual procedure of Apriori, our approach filtered out the non-frequent-QI set from the entire set.

Step 2: Anonymization process based on the Utility of data.

Anonymize the non-frequently associated attributes based on the utility of data and also based on quasi-quasi associations. Here different versions of anonymization models are generated by retaining as much as information possible. Suppression of certain attribute values is done rather than generalization, in order to address all kinds of possible attacks including background knowledge attack. The algorithm splits the QI vector which consists of the attributes State, Age, Religion, Education into two sets: frequent-attribute-set that satisfies k-anonymity property and the non-frequent-attribute-set that are under disclosure risk.

Each tuple in the second set is given to the anonymization process and the anonymization process also checks the user preferences on any one of the attributes among the QI set. For instance, if the data set is used for a State-wise study on particular disease, the State attribute values have to be retained and that user can prefer state attribute for making the study fruitful. Thus, he/she will get an anonymized version without affecting the State attribute values.

4 Experiments and Results

4.1 Data

The sample of the study, NFHS-3 initially contained 124,385 instances. In that 472 instances are having sensitive attribute value "yes" for Disease-TB and are highly sensitive. The QI values associated with these 472 instances and similar combination of QI values shared with instances having non sensitive attribute value "no" for Disease-TB are identified and those instances are filtered out in order to get the resultant 15,563 instances. And these instances are treated as original data set, termed as OT for the evaluation purposes.

4.2 Software and Tools

Using C#.Net a user interface is created and OT is given as input to the AUA process and four sets of anonymized versions such as AT_Version1, AT_Version2,

AT_Version3 and AT_Version4, based on user preferred QI attribute are obtained. Uniquely identifiable tuples in NFHS-3 are filtered out by checking the support ≤ 3 and performed the suitable anonymization. Both original and anonymized datasets are analyzed using Weka and are tested in terms of classification.

4.3 Results

Although, privacy and information are two contradictory goals, AUA method effectively maintains a balance between these two. This section provides results based on certain measures: Privacy and Information, in order to prove the effectiveness of our approach. The table 1 shows the sample of anonymized health care data with user preferred attribute 'State'.

Table 1 Sample of anonymized Health Care data with user preferred attribute 'State'

No	State	Age	Education	Religion	TB-"yes"	TB-"no"	Conf
1	Kerala	40	Secondary	Muslim	"yes"(1)	"no"(4)	(5)
2	Tamil Nadu	34	Higher	Christian	"yes"(2)	"no"(23)	(25)
3	Karnataka	*****	*****	*****	"yes"(1)	"no"(1)	(2)
4	Manipur	37	Primary	Christian	"yes"(3)	"no"(10)	(13)
5	Arunachal	43	No education	*****	"yes"(1)	"no"(0)	(1)

4.3.1 Privacy

Privacy can be measured as the confidence of an attacker to identify an individual from a set of instances. This can be tested by checking whether the model generates anonymous results, which overcomes possible kinds of known attacks. Our results were compared with the results of standard k-anonymity algorithms and identified that same number of instances are under disclosure risk. Anonymization of these instances eliminates the linking attack problem. Since we have adopted suppression instead of generalization it overcomes the problem of background knowledge attack. The model also eliminates homogeneity attack since we maintain both sensitive attribute values "yes" and "no" in each group. Thus, our model is checked with all the three known attacks and it is found that any individual can be linked only with maximum of 33% confidence. This again can be improved if min-sup is set accordingly.

4.3.2 Information

Information is a measure of purity, which represents the amount of information that would be needed to specify whether a new instance should be classified correctly. The information gain is calculated on each attribute values and is shown in fig. 1.

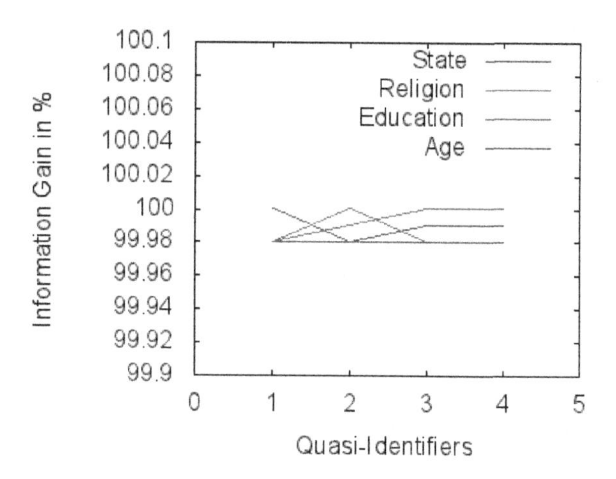

Fig. 1 Information Gain

Table 2 Classification Accuracy on Original and Anonymized Datasets

Dataset	(NaiveBayes)	(RandomForest)
OT	96.73	96.87 ∘
AT-version1-State-wise	96.81	96.87
AT-version2-Age-wise	96.82	96.88
AT-version3-Education-wise	96.81	96.88
AT-version4-Religion-wise	96.80	96.88

∘, • statistically significant improvement or degradation

4.3.3 Classification Accuracy

The sample set is tested with two paradoxical measures known as privacy and information loss. The privacy is ensured by maintaining the k-anonymity principle. Here the min_sup is set as 3 therefore each individual record can be identified with maximum of 33% confidence. The table 2 shows that the classification accuracy is almost similar for original data set and its anonymized versions as well. The results were obtained using two well known classifiers namely, Naive Bayes and Random Forest.

5 Conclusions and Future Prospects

Empirical evaluations show that the proposed method preserves maximum information while protecting privacy of individuals. Apart from classical approaches, the AUA approach is a two step iterative process based on association mining. Using support and confidence, we have performed association mining on given data set either for filtering risky instances and for retaining maximum information based on utility of data. From this iterative process, multiple versions of anonymized data sets can be served positively according to the user's need. The study addressed some of the known attacks namely linking attack, homogeneity attack and background knowledge attack. Along with technological advancements, vulnerability of data may also increase day-by-day and this is to provide certain indicators for future directions to researchers as well as practitioners.

Acknowledgments We express our sincere gratitude towards International Institute for Population Sciences (IIPS), Mumbai for making the data available for the study.

References

1. Panackal, J.J., Pillai, A.S., Krishnachandran, V. N.: Disclosure risk of individuals: a *k*-anonymity study on health care data related to indian population. In: IEEE International Conference on Data Science and Engineering, pp. 221–226 (2014)
2. Panackal, J.J., Pillai, A.S.: Adaptive utility-based anonymization model for privacy preserving data mining. In: International Conference on Data Mining and Warehousing(ICDMW), Data Mining Algorithms, pp. 220–228. Elsevier (2014)
3. Panackal, J.J., Pillai, A.S.: An intelligent framework for protecting privacy of individuals: empirical evaluations on data mining classification. In: International Conference on Hybrid Intelligent Systems (HIS). IEEE (2014)
4. Malik, M.B., Ghazi, M.A., Ali, R.: Privacy preserving data mining techniques: current scenario and future prospects. In: Third International Conference on Computer and Communication Technology. IEEE (2012)
5. Sharma, M., Chaudhary, A., et al.: A review study on the Privacy Preserving Data Mining Techniques and Approaches. International Journal of Computer Science and Telecommunications **4**(9) (2013)
6. Liu, L., Kantarcioglu, M., Thuraisingham, B.: The applicability of the perturbation based privacy preserving data mining for real-world data. Data and Knowledge Engineering (2008)
7. Samarati, P.: Protecting respondents privacy in Microdata release. IEEE Transactions on Knowledge and Data Engineering, pp 1010–1027 (2002)
8. Li, T., Li, N.: Towards Optimal k-anonymization. Data and Knowledge Engineering (2008)
9. Mukkamala, R., Ashok, V.G.: Fuzzy-based methods for privacy-preserving data mining. In: Eighth International Conference on Information Technology: New Generations (2011)
10. Zhu, Y., Peng, L.: Study on k-anonymity models of sharing medical information. IEEE (2007)
11. Liu, J., Luo, J., Huang, J.Z.: Rating: privacy preservation for multiple attributes with different sensitivity requirements. In: 11th IEEE International Conference on Data Mining Workshops (2011)
12. Fernández-Alemn, J.L., Señor, I.C., Lozoya, P.Á.O., Toval, A.: Methodological Review-Security and Privacy in electronic health records: A systematic literature review. Journal of Biomedical Informatics (2013)
13. Sweeney, L.: Achieving k-anonymity privacy protection using generalization and suppression. International Journal on Uncertainty, Fuzziness and Knowledge-based Systems, 571–588 (2002)

14. Zhu, D., Li, X.-B., Wu, S.: Identity disclosure protection: A data reconstruction approach for privacy preserving data mining. Decision Support Systems, 133–140 (2009) (Elsevier)
15. Yang, W., Qiao, S.: A novel anonymization algorithm: Privacy Protection and Knowledge Preservation. Expert Systems with Applications, 756–766 (2010) (Elsevier)
16. Sun, X., Sun, L., Wang, H.: Extended k-anonymity models against sensitive attribute disclosure. Computer Communications, 526–535 (2011) (Elsevier)
17. Taneja, S., Khanna, S., et al.: A Review on Privacy Preserving Data Mining: Techniques and Research Challenges. International Journal of Computer Science and Information Technologies **5**(2), 2310–2315 (2014)
18. Borhade, S.S., Shinde, B.B.: Privacy Preserving Data Mining Using Association Rule With Condensation Approach. International Journal of Emerging Technology and Advanced Engineering **4**(3), 292–296 (2014)
19. Panackal, J.J., Pillai, A.S.: Privacy preserving data mining: an extensive survey. In: ACEEE International Conference on Multimedia Processing, Communication and Information Technology (2013)
20. Han, J., Kamber, M.: Data Mining Concepts and Techniques, 2nd (edn). Elsevier

Categorisation of Supreme Court Cases Using Multiple Horizontal Thesauri

Sameerchand Pudaruth, K.M. Sunjiv Soydaudah and Rajendra Parsad Gunputh

Abstract Text classification is a branch of Artificial Intelligence which deals with the assignment of textual documents to a controlled group of classes. The aim of this paper is to assess the use of a controlled vocabulary in the categorisation of legal texts. Controlled vocabularies such as the Medical Subject Headings, Compendex and AGROVOC have been proved to be very useful in the fields of biomedical research, engineering and agriculture, respectively. In this work, a number of lexicons are created for some pre-defined areas of law through an automated approach. The lexicons are then used to categorise cases from the Supreme Court into eight distinct areas of law. We then compared the performance of these lexicons with each other. We found that lexicons which have a mixture of single words and short phrases performs slightly better than those consisting simply of single words. Weights were also assigned to the terms and this had a significant positive impact on the classification accuracy. The number of words in each thesaurus was kept constant. A hierarchical classification was also attempted whereby cases were first classified into either a civil case or a

S. Pudaruth(✉)
Department of Ocean Engineering & ICT, Faculty of Ocean Studies,
University of Mauritius, Moka, Mauritius
e-mail: s.pudaruth@uom.ac.mu

K.M. Sunjiv Soydaudah
Electrical and Electronic Engineering Department, Faculty of Engineering,
University of Mauritius, Moka, Mauritius
e-mail: ssoyjaudah@uom.ac.mu

R.P. Gunputh
Department of Law, Faculty of Law and Management, University of Mauritius,
Moka, Mauritius
e-mail: rpgunput@uom.ac.mu

S. Berretti et al. (eds.), *Intelligent Systems Technologies and Applications*,
Advances in Intelligent Systems and Computing 385,
DOI: 10.1007/978-3-319-23258-4_31

criminal case. Civil cases were then further classified into company, labour, contract and land cases while criminal cases were classified into drugs, homicide, road traffic offences and other criminal offences. Our best model achieves a global accuracy of 78.9 %. Thus, we have demonstrated that it is possible to get good classification accuracies with legal cases through the use of automatically generated thesauri. This outcome of this research can become an integral part of the eJudiciary project that has already been initiated by the government. In line with the vision of the Judiciary, we are hereby in the process of creating an intelligent legal information system which will benefit all legal actors and will have a definite positive impact on the legal landscape of the Republic of Mauritius. Lawyers, attorneys and their assistants would spend less time on legal research and hence they would have more time to prepare their arguments for their case. We are optimistic in believing that this will make the whole business of providing justice more effective and more efficient through the reduction in postponement of cases and a reduction in the average disposal time of cases.

Keywords Supreme Court · Cases · Categorisation · Horizontal Thesaurus

1 Introduction

The Supreme Court of Mauritius is the highest court of Mauritius. It is said to be a court of unlimited jurisdiction which means that it can hear any civil or criminal case. Every year, more than 9000 new cases are lodged at the Supreme Court but only about 8500 cases are disposed of in each year. Over the years, this has resulted in a massive backlog of outstanding cases. As at December 2013, there were 10,023 such cases [1]. One of the main reasons which cause such delays are the unprepared barristers who repeatedly request for more time from the Judges in order to put their case in order. Barristers and attorneys are often behind schedule because of the lack of proper tools to assist them in their daily legal preparations. The aim of this research is to categorise Supreme Court cases using multiple horizontal legal thesauri. Criminal cases are categorised into homicide, drug offences, road traffic offences and other criminal offences. Civil cases, as well, are classified into business law (company), law of immovable property (land and tenancy issues), labour law (employment issues) and contract law. The law of negligence (tort law) was adjunct to contract law as they share much similarities. We believe that organising judgements in this way will allow legal professionals to make more effective and efficient legal research. This, in turn, will make them better prepared and consequently this will reduce to some extent the delays in the clearance of cases.

In its simplest form, a thesaurus can be a list of word and/or terms that are specific to a domain. Generally, a thesaurus also contains links and relationships between the stored words/terms. Medical Subject Headings (MeSH) [2] and Compendex [3] are among the two most used thesauri in the world. The former is intended for life sciences while the latter is designed for engineering materials.

Many other areas have their own specific thesaurus: AGROVOC [4] exists for agricultural and food studies, AAT Thesaurus [5] for Art and Architectural disciplines, etc. As to date, EuroVoc [6] is the only legal thesaurus that is available in the English language and which is used for documenting the activities of the European Union and the European Court of Justice. Since the legal system and its legal terminologies of different countries are likely to differ, it is very difficult to create a single international taxonomy for law unlike the sciences. Thus, in this research, we have created our own authority lists consisting of multiple lexicons for each area of law through an automated approach. The number of terms in each lexicon was kept fixed. These lists were then used to classify law cases into one of the existing categories.

The structure of this research paper is as follows. The problem statement is described in Section 2. In Section 3, we describe some of the related works on controlled vocabularies in different fields of knowledge. In Section 4, we describe how the corpus and the lexicons were constructed and how these were used for classification. Section 5 describes all the different experiments that were performed and we also evaluate and justify all our findings. Section 6 concludes the paper, summarises our achievements and suggest some directions for future works in this emerging field of law.

2 Problem Statement

The Judiciary, the Executive and the Legislature are the three pillars of a sovereign democratic state. In order to ensure our fundamental rights that are entrenched in the Constitution of the Republic of Mauritius, such as the right to life, right to liberty, right to freedom of expression, right to a fair trial, right to defend oneself in person, etc., it is incumbent on any judicial system to be able to provide easy and fast access to justice to everyone. One way in which this can be achieved is through a more efficient system to dispose of cases and to deliver judgments. In line with this vision the Judiciary has been implementing several measures such as putting time limits for the clearance of commercial cases, family cases, drug cases and cases at the Assizes. A fast track process has been initiated for child victims. The Bail and Remand Court (BRC) is now operational 7 days a week. The Judiciary is providing continuous formal training for all its personnel. Legal research officers have also been recruited to assist Judges in writing their judgment. In the last few years, the Judiciary has been undertaking major reforms in the matter of e-filing and management of cases. The first phase (I) of the eJudiciary project was completed and launched in 2013. Barristers and attorneys are now able to submit a case and the relevant documents at any time from any location via the eJudiciary portal [22]. Payments can also be made online. Major enhancements have also been made recently. Nevertheless, despite the introduction of so many features with a view to administer to justice swiftly, the number of undisposed cases is steadily on the rise. The aim of this research work is the creation of a legal information system which will help Judges, attorneys, barristers, legal research assistants and other legal personnel discover relevant information in less time so that justice can be delivered more expeditiously. In this paper, we assess the reliability of using legal thesauri for the classification of cases into their respective categories.

3 Literature Review

Controlled vocabularies such as subject headings, controlled lists, synonym rings, authority lists, taxonomies, dictionaries, thesauri, ontologies, folksonomies and library classification schemes have been received significant interest from the research community for the cataloguing and retrieval of information [14][17]. The design of controlled vocabularies has also received much attention [11][17]. The main role of a controlled vocabulary is to produce a set of authoritative terms that can be used to link different concepts. However, we should always keep in mind that most of the end-users are usually not experts and hence they need to be provided with some training so that they know what the right terms to use are.

Subject headings such as the Medical Subject Headings (MeSH) have been used extensively for the annotation of scientific data [10]. In [16], the authors described how they used MeSH and the K-Nearest Neighbour classifier for the retrieval of biomedical information. They concluded that such a system can produced results that are good as human annotators. Romeo *et al.* [9] conducted a series of experiments by using a combination of three biomedical dictionaries and three machine learning classifiers. They also came to the conclusion the use of dictionaries in text mining applications can produce superior results. Bleik *et al.* [7] used concept graphs and ontologies to capture the semantic relationships between biomedical documents. Their graph-based classifier performed better than using machine classifiers alone.

Folksonomy refers to the act of social tagging whereby end-users (usually laypersons) describes resources using their own words and terms without showing concerned for accuracy, ordering and brevity. This is a common practice on social networks where users tag resources uploaded by themselves or by other people or organisations. This is also commonly referred to as social indexing or folk categorisation. Tagging is not an authoritative way of annotating resources. It is also often not collaborative in nature. However, in the last few years, it has started to attract the interest of some researchers as a potential means of classifying resources. Zubiaga *et al.* [18][19] first demonstrated that social tagging can be used a powerful means of collecting precious information which can then be used for classification. They also observed that it is possible to further increase the accuracy of free tagging system by categorising the users into two groups, namely, categorisers and describers. Categorisers are users who produce quality tags while describers are those whose tags are less accurate and therefore carry less weight. Magableh *et al.* [20] identified some of the flaws with folksonomies and proposed an approach to address these.

Golub has described a framework for the automatic classification of web pages using subject headings and flat lists [12]. However, the framework was not implemented and thus the effectiveness of this approach in this area could not be assessed. The global astrophysics community has also started to realise the potential benefits of controlled vocabularies. Astrophysics is a very broad field of knowledge which has many sub-disciplines. Tags have not proven to be useful in this scientific discipline [13]. Thus, Gray *et al.* [13] have used a more formal type of controlled vocabulary to link user queries to vocabulary concepts. Each concept has a definition, a list of close synonyms and its relationship to other concepts.

The use of controlled vocabularies in the legal field is quite new. Indeed, literature on this field is very sparse. Piotrowski and Senn are building a multi-lingual controlled vocabulary (Italian, French, and German) for old legal documents from Switzerland [15]. The aim of this research is to facilitate the task of human annotators as well as to assist users in document retrieval irrespective of the language used. The system is also expected to be robust to minor variations in spelling. The EuroVoc is a legal thesaurus which has been used for the manual annotation of legal texts from the European Parliament and from Parliaments for EU states. It is available in twenty-three languages of the European Union. Recently, Saric *et al.* [8] have used the EuroVoc thesaurus and support vector machines for the classification of Croatian legal documents. They report reasonably high accuracy values. They also addressed the problem of class sparsity and proposed some solutions.

4 Methodology

Our dataset for this study consists of 294 Supreme Court cases from the year 2013. This dataset contains 1322 pages of legal information with a total of 643761

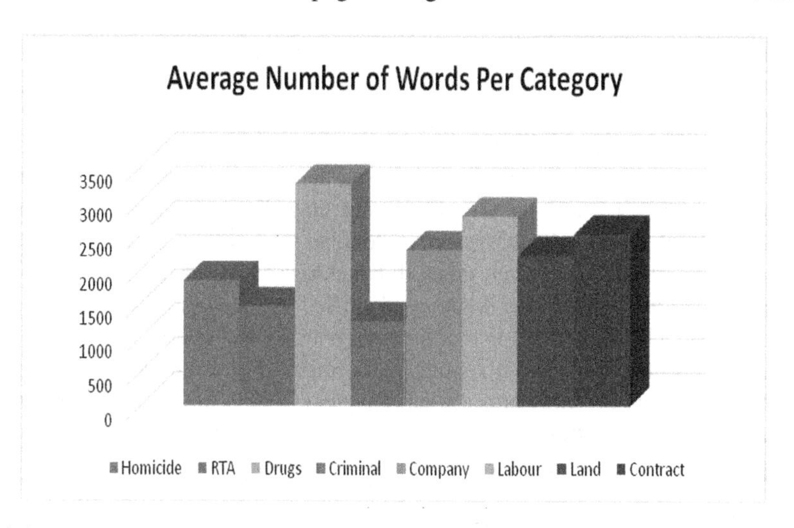

Fig. 1 Average Number of Words per Category (Source: Authors)

words. All the cases can be downloaded freely from the Supreme Court website [1]. Initially, there were 506 cases for the year 2013, however, all categories with less than 10 cases were removed for this study.

Fig. 1 shows the average number of words in each category. We note that drugs cases have the highest average (3000+) while criminal cases have the lowest average. Company, Labour, Land and Contract cases also have high averages compared with homicide and RTA cases. This information can be used as a feature in a machine learning classifier in our future works.

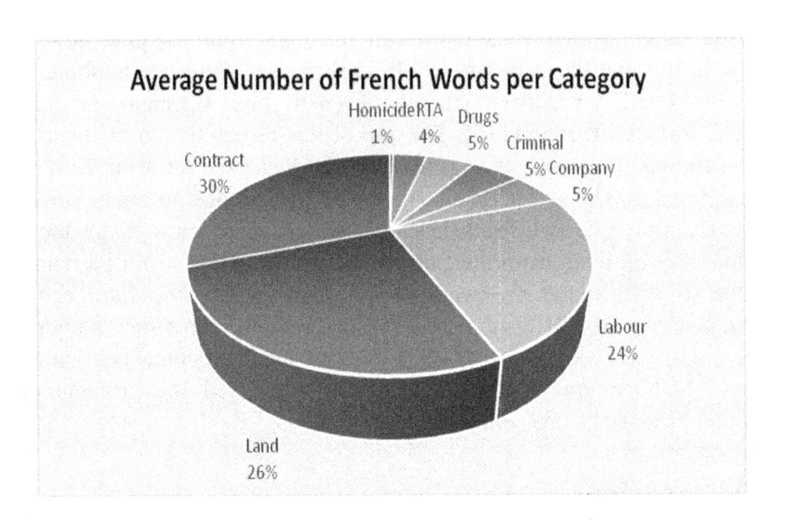

Fig. 2 Average Number of French Words per Category (Source: Authors)

Fig. 2 shows the average percentage of French words for each category. For example, homicide cases have the least number of French words (with an average 1 French word per case) while contract cases have the highest number of French words (with an average of 75 words per case). In general, civil cases tend to have higher number of French words than criminal cases. The presence of French words in Mauritian Supreme Court judgments make the task of classification even harder by converting it into a multi-lingual classification rather than a mono-lingual one. Mauritian Creole words are also present in some cases but in insignificant amounts. However, this will be addressed in our future works.

Another major problem that we have identified when dealing with legal cases is class sparsity. The legal field is incomparably broad and even cases which fall in the same area of law can be disparately dissimilar. For example, the Companies Act 2001 contains 30 chapters, 39 sub-chapters, 365 sections and thousands of sub-sections. The vocabulary used in the different sections are quite different. For example, the terms used for the section on the duties and powers of directors are very different from sections on amalgamation, foreign companies and others. Consequently, many areas of law have only a handful of representative cases while a vast majority of sub-areas have no cases at all.

Table 1 Details of Cases Dataset

Level 1 Categories	Level 2 Categories	No. of Cases	No. of Cases
Criminal Law	Homicide	14	144
	Road Traffic Offences	38	
	Drugs	44	
	Other Criminal Offences	48	
Civil Law	Company Law	22	150
	Labour Law	17	
	Land Law	55	
	Contract Law	56	

(Source: Authors)

Table 1 shows that among the 294 cases, 144 are criminal law cases and 150 are civil law cases. Land law has the highest number of cases while homicide cases are the least. Our first aim was to predict the level 2 category of each of these 294 cases. In order to achieve, we created eight different lexicons (thesaurus), one for each area of law. Relevant *gazetted* acts and sections were used for each area of law. For example, the microthesaurus for company law was created by automatically extracting the first one hundred most frequent words from the Companies Act 2001 and the Insolvency Act 2009. The thesaurus for land law and contract law also contains French words. This is because Civil Code is written entirely in French. Exceptionally for contract law, we selected only the first fifty most frequent terms (which were all in French) and translated them to English using the WordReference online French-English dictionary [21]. Thus, the total number of words in the contract thesaurus was also brought up to one hundred.

Table 2 shows the legal sources that were used for each legal category. Ten words which are highly representative of this category have also been listed. We note that the word '*penal*' appear in both homicide and other criminal offences categories. The work '*property*' also appears in land law and other criminal offences categories. This is a common difficulty in document categorisation problems.

Table 2 Legal sources for each level 2 thesaurus

#	Level 2 Categories	Source	List of 10 Frequent Words
1	Homicide	Sections 215-223 and Sections 239-247 of Criminal Code Act 1838	manslaughter, murder, blows, homicide, wounds, guilty, penal, servitude, verdict, death
2	Road Traffic Offences	Road Traffic Act (RTA) 1962	vehicle, licence, motor, road, driving, trailer, vehicles, traffic, registration, driver
3	Drugs	The Dangerous Drugs Act 2000	dangerous, drug, drugs, preparations, premises, detained, imprisonment, conviction, fine, video
4	Other Criminal Offences	Criminal Code Act 1838 except sections used for Homicide	assault, punished, penal, punishment, property, criminal, intention, consent, larceny, prisoner
5	Company Law	The Companies Act 2001 The Insolvency Act 2009	company, shares, registrar, director, shareholders, register share, meeting, board, constitution
6	Labour Law	The Employment Rights Act 2008 Employment Relations Act 2008	worker, employer, remuneration, agreement, employment, years, workers, leave, employed, termination
7	Land Law	Landlord and Tenant Act 1999 Sections 516-889 of the Mauritius Civil Code	succession, rent, propriétaire, owner, biens, tenant, usufruct, heir, property, landlord
8	Contract Law	Sections 1101-1233 and Sections 1382-1386 of the Mauritius Civil Code	obligation, créancier creditor, intérêts, convention, contrat, contract, dommages, damage, débiteurs, debt

(Source: Authors)

5 Research and Findings

In this section, we describe the different classification experiments that were performed. The results are presented in the form of confusion matrices (CM). The experiments were classified into four scenarios (scenario 1 - scenario 4).

	company	contract	criminal	drugs	homicide	labour	land	rta
company	<16>	4	.	.	.	.	2	.
contract	12	<35>	.	2	.	3	1	3
criminal	.	1	<14>	21	10	.	.	2
drugs	.	.	.	<43>	.	.	.	1
homicide	.	.	2	1	<10>	.	1	.
labour	3	4	.	.	.	<9>	.	1
land	1	14	.	.	.	.	<40>	.
rta	.	1	.	2	.	.	1	<34>

Fig. 3 Confusion Matrix for Words (Source: Authors)

	company	contract	criminal	drugs	homicide	labour	land	rta
company	<15>	4	.	.	.	.	3	.
contract	10	<37>	.	1	1	4	2	1
criminal	.	.	<22>	9	11	.	1	5
drugs	.	.	.	<42>	.	.	.	2
homicide	.	.	.	2	<12>	.	.	.
labour	2	.	.	.	.	<15>	.	.
land	2	8	.	.	.	2	<43>	.
rta	.	2	.	1	.	1	.	<34>

Fig. 4 Confusion Matrix for Weighted Words (Source: Authors)

In order to generate the confusion matrix (CM) displayed in Fig. 3 (**scenario 1**), we used a thesaurus of 100 words for each of the eight areas of law. The number of common words between each thesaurus and a case is determined. The predicted category is the one with the highest number of matches. Words are counted only once even if they appear multiple times in the case. Each word in the list was also given a weight of 1. A row in the confusion matrix represents the actual values while a column represents predicted values. For example, the value <35> in row 2 and column 2 indicates that out of 56 contract cases, 35 have been correctly classified. The number <12> in row 2 and column 1 indicates that 12 cases that are actually contract cases have been misclassified as company law cases. Also, the number <14> in the row named '*land*' and column named '*contract*' tells us that 14 land cases have been misclassified as contract cases. The overall accuracy for scenario 1 was 68.37% (201/294).

The algorithm was then modified to give more importance to the words at the top of the list and lesser weight as we go down the list. An exponential decay function was used for this purpose. Thus the first word was assigned a top weight of 2.718 (e^1), the 10th element a weight of 1.584, the 50th element a weight of 0.1437 and the 100th one was assigned a weight of 0.0076 (very close to zero). From Fig. 4 (**scenario 2**), we can see that there has been a slight improvement in the classification accuracies of contract, criminal, homicide, labour and land cases. Road traffic offences (rta) have stayed at 34 while company and drugs have decreased by a count of 1. In general, the accuracy have gone up from 68.37% to 74.83% (220/294), an increase of 6.46%.

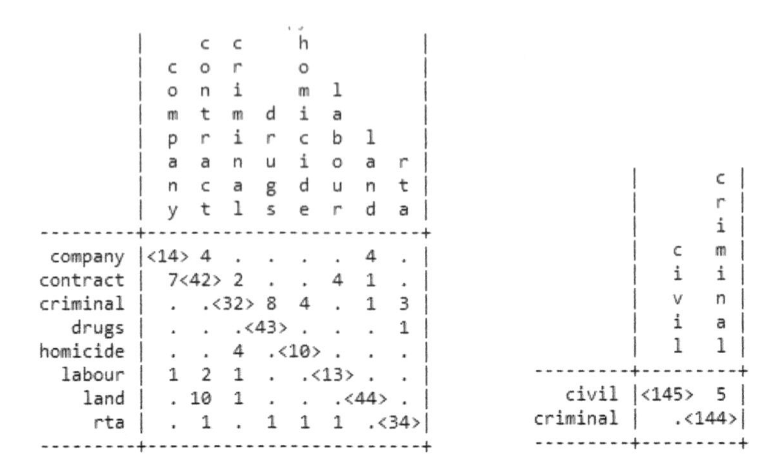

	company	contract	criminal	drugs	homicide	labour	land	rta
company	<14>	4	.	.	.	.	4	.
contract	7	<42>	2	.	.	4	1	.
criminal	.	.	<32>	8	4	.	1	3
drugs	.	.	.	<43>	.	.	.	1
homicide	.	.	4	.	<10>	.	.	.
labour	1	2	1	.	.	<13>	.	.
land	.	10	1	.	.	.	<44>	.
rta	.	1	.	1	1	1	.	<34>

Fig. 5 CM for Weighted Words & Bigrams (Source: Authors)

	civil	criminal
civil	<145>	5
criminal	.	<144>

Fig. 6 Confusion Matrix for Civil & Criminal Law (Source: Authors)

In the **third scenario**, shown in Fig. 5, bigrams (frequent combinations of two words) and trigrams (frequent combinations of three words) for each area of law were appended to each authority list. The bigrams and trigrams were extracted from the sources mentioned in Table 2 using an automated approach. All bigrams and trigrams were given a weight of 1.49, independent of their location in the list. This corresponds to the weight of the 11th element in the original list. The classification accuracy has now jumped to 78.91% (232/294), showing the discriminative power of short phrases over single words. Some frequent bigrams and trigrams for each area of law are shown below in Table 3.

We also attempted a hierarchical classification of the cases (**scenario 4**). The cases were firstly divided into only two categories, i.e. either a criminal case or a civil case. To create the thesaurus for criminal cases, we combined the microthesaurus for each of the four areas of criminal law. The same procedure was adopted for the civil law category. Fig. 6 shows that 289 cases out of 294 have been correctly classified, giving a classification accuracy of 98.3% for the first level classification. All criminal cases have been correctly classified, however, 5 civil cases have been classified as criminal cases. Three of these are contract cases while the two others are labour cases.

We compared the performance of scenario 3 and scenario 4 using recall, precision and the F1-score. Recall is the number of correct documents retrieved over the total number of documents actually belonging to a specific category. Precision is the number of correct documents retrieved over the total number of documents retrieved. F1-score is usually termed as the harmonic mean of recall and precision. It can be considered as an average of these two measures and it provides a general view of the accuracy of that category.

Table 3 Frequent Bigrams and Trigrams

Level 2 Categories	List of Frequent Bigrams and Trigrams
Homicide	penal servitude, lesser sentence, compelling circumstances self defense, involuntary homicide, wilful act, actual necessity lawful authority, human being, unlawful killing
Road Traffic Offences	motor vehicle, driving licence, public place, traffic sign, registered owner, breathtest, provisional driving licence, speed limit, authorised vehicle, dangerous driving
Drugs	dangerous drug, dangerous drugs, video recording, aggravating circumstances, narcotic drugs, per dosage, smuggling unit, controlled delivery, drugs act, pharmacy act
Other Criminal Offences	penal servitude, criminal procedure, offensive weapon without lawful authority, person charged, police officer offenders act, criminal code, class contravention
Company Law	companies act, share register, private company, registered office, relevant interest, best interests, official receiver, fair value, voting rights, special meeting, global business licence
Labour Law	severance allowance, job contractor, sick leave, annual leave allowance payable, public holiday, pensions act, trade union retirement age, hourly basis, labour welfare, gross earnings, basic wage, relations act, remuneration order
Land Law	doit de retention, lien de connexité, legal title, exclusive possession, a titre de propriétaire, division in kind, affidavit of prescription, unlawfully occupying, deed transcribed, action en revendication, apports en societé,
Contract Law	parties contractantes, obligation principale, condition suspensive, bonne foi, action solidaire, contractée solidairement, obligation contractée, sans prejudice, toute obligation, créancier solidaire, mauvaise foi, bonnes moeurs

(Source: Authors)

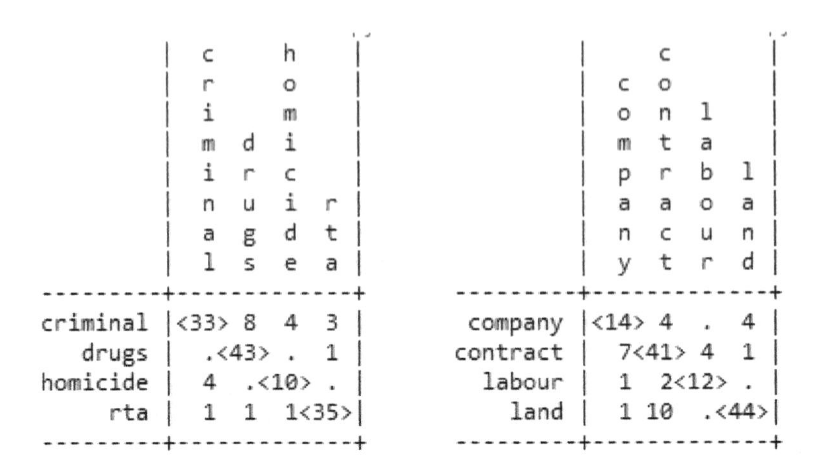

Fig. 7 Confusion Matrix for Criminal (Source: Authors)

Fig. 8 Confusion Matrix for Civil Law (Source: Authors)

Table 4 Performance Measures for Scenario 3 and 4

	Level 2 Categories	Scenario 3			Scenario 4		
		Recall	Precision	F1-Score	Recall	Precision	F1-Score
1	Criminal	0.67	0.80	0.73	0.69	0.87	0.77
2	Drugs	**0.98**	0.83	**0.90**	**0.98**	0.83	0.90
3	Homicide	0.71	0.67	0.69	0.71	0.67	0.69
4	RTA	0.89	**0.89**	0.89	0.92	**0.90**	**0.91**
5	Company	**0.64**	**0.64**	**0.64**	**0.64**	**0.61**	**0.62**
6	Contract	0.75	0.80	0.73	0.77	0.72	0.75
7	Labour	0.76	0.72	0.74	0.80	0.75	0.77
8	Land	0.80	0.88	0.84	0.80	0.90	0.85
	Average	**0.78**	**0.78**	**0.77**	**0.79**	**0.78**	**0.78**

(Source: Authors)

Table 4 shows that drugs cases and RTA cases have the highest recall and precision values while company cases have the lowest recall and precision values. This means that drugs cases and RTA cases are on average more different from other cases while company cases share a lot of similarities with other types of cases. The average recall, precision and F1-score all turns around 78% for both scenarios. The overall accuracy for scenario 4 (hierarchical classification) is 78.9%, which is the same as scenario 3. This means that hierarchical classification does not bring any benefit. This is because of the two-level procedure where errors in the first level are propagated to the second level and this tends to reduce the success rate. Hierarchical classification can be effective only if we are able to get a 100% accuracy in the first stage.

6 Conclusion and Recommendations

Controlled vocabularies, thesauri and dictionaries have been used aggressively in domains such as bioinformatics, astrophysics, agriculture and medicine. Because of the regional nature of law and because of its high language dependencies, it is very difficult to have a common international thesaurus for law. Hence, we had to create our own thesauri. We then showed how we can use these horizontal thesauri for the classification of Supreme Court cases from the Republic of Mauritius. In particular, we compared the performance of several legal lexicons for eight different areas of law. Our best experiment produced an accuracy of 78.9%, a precision of 78%, a recall of 79% and an F1-score of 78%. We have to be careful when comparing our results with those obtained from the literature as our dataset is different from theirs. However, in general, our results fair less well compared to machine learning approaches where reported accuracies are commonly found to be above 80%. Legal documents are more challenging to classify. Unlike news articles, the size of legal cases is not fixed. They vary from one page up to 60 pages. A unique characteristic of legal cases from Mauritian courts is that they are often bilingual, i.e. they are written using a combination of English and French language. Furthermore, cases often does not neatly fall into one category only. Thus, in our future work, we intend to use machine learning classifiers to tackle some of the challenges that have been outlined. Both approaches will then be compared. We shall also investigate on the possibility of creating a hybrid approach whereby part of the work is done through a thesaurus and the other part by a machine learning classifier.

References

1. Supreme Court. Annual Report of the Judiciary 2013. Republic of Mauritius (2013)
2. National Library Of Medicine. Medical Subject Headings (online). United States of America (2014). http://www.nlm.nih.gov/mesh/ (accessed 01 March 2015)
3. Elsevier. Engineering Village [online]. The Netherlands (2014). www.engineeringvillage.com/ (accessed 01 March 2015)
4. Food And Agriculture Organisation. Agrovoc (online). FAO, United Nations (2014). http://aims.fao.org/standards/agrovoc/ (accessed 01 March 2015)
5. The Getty Research Institute. Art and Architecture Thesaurus (online). Los Angeles, California, US (2014). http://www.getty.edu/research/tools/vocabularies/aat/index.html/ (accessed 01 March 2015)
6. Europa. EuroVoc, the EU's Multilingual Thesaurus [online]. Bruxelles, Belgium (2014). http://eurovoc.europa.eu/drupal/ (accessed 01 March 2015)
7. Bleik, S., Mishra, M., Huan, J., Song, M.: Text Categorization of Biomedical Data Sets using Graph Kernels and a Controlled Vocabulary. IEEE/ACM Transactions on Computational Biology and Bioinformatics **10**(5), 1211–1217 (2013)

8. Saric, F., Basic, B.D., Moens, M.F., Snajder, J.: Multi-label classification of croatian legal documents using EuroVoc thesaurus. In: Proceedings of the SPLeT - Semantic Processing of Legal Texts: Legal Resources and Access to Law Workshop Location, Reykjavik, Iceland, May 27, 2014
9. Romero, R., Iglesias, E.L., Borrajo, L., Marey, C.M.R.: Using dictionaries for biomedical text classification. In: Rocha, M.P., Rodríguez, J.M.C., Fdez-Riverola, F., Valencia, A. (eds.) 5th International Conference on Practical Applications of Computational Biology & Bioinformatics (PACBB 2011). AISC, vol. 93, pp. 365–372. Springer, Heidelberg (2011)
10. Schijvenaars, B.J.A., Schuemie, M.J., van Mulligen, E.M., Weeber, M., Jelier R., Mons, B., Kors, J.A.: A concept-based approach to text categorization. In: Proceedings of Text REtrieval Conference (TREC) 2005 Genomics Track (2005)
11. Svenonius, E.: Design of Controlled Vocabularies. Encyclopedia of Library and Information Science (2003). doi:10.1081/E-ELIS120009038
12. Golub, K.: Using Controlled Vocabularies in Automated Subject Classification of Textual Web Pages, in the Context of Browsing. Theory and Practice of Digital Libraries (TCDL) **2**(2) (2006)
13. Gray, A.J.G., Gray, N., Ounis, I.: Searching and exploring controlled vocabularies. In: Proceedings of the ACM Workshop on Exploiting Semantic Annotations in Information Retrieval (ESAIR), pp. 1–5 (2009)
14. Lee-Smeltzer, K.H.J.: Finding the Needle: Controlled Vocabularies, Resource Discovery and Dublin Core. Library Collections, Acquisitions & Technical Services **24**, 205–215 (2000)
15. Piotrowski, M., Senn, C.: Harvesting indices to grow a controlled vocabulary: towards improved access to historical legal texts. In: Proceedings of the 6th EACL Workshop in Language Technology for Cultural Heritage, Social Sciences and Humanities, Avignon, France, pp. 24–29, April 24, 2012
16. Trieschnigg, D., Pezik, P., Lee, V., de Jong, F., Kraaji, W., Wandrebholz-Schuhmannd, D.: MeSH-Up: Effective MeSH Text Classification for Improved Document Retrieval. Bioinformatics **25**(11), 1412–1418 (2009)
17. Harping, P.: Introduction to Controlled Vocabularies: Terminology for Art, Architecture and Other Cultural Works. Getty Research Institute, Los Angeles (2010)
18. Zubiaga, A.M.: Harnessing Folksonomies for Resource Classification. Thesis (PhD). National University of Distance Education, Spain (2012)
19. Zubiaga, A., Korner, C., Strohmaier, M.: Tags vs Shelves: from social tagging to social classification. In: Proceedings of the 22nd ACM International Conference on Hypertext and Hypermedia, Eindhoven, The Netherlands, June 6–9, 2011
20. Magableh, M., Cau, A., Zedan, H., Ward, M.: Towards a multilingual semantic folksonomy. In: Proceedings of the IADIS International Conference on Collaborative Technologies, pp. 178–182, Freiburg, Germany, July 2010
21. Wordreference (2015). http://www.wordreference.com/fren/ (accessed 14 March 2015)
22. Ejudiciary Mauritius (2015). https://www.ejudiciary.mu/ (accessed 17 March 2015)

Gender Profiling from PhD Theses Using k-Nearest Neighbour and Sequential Minimal Optimisation

Hoshiladevi Ramnial, Shireen Panchoo and Sameerchand Pudaruth

Abstract Author profiling is a subfield of text categorisation in which the aim is to predict some characteristics of a writer. In this paper, our objective is to determine the gender of an author based on their writings. Our corpus consists of 10 PhD theses which was split into equal sized segments of 1000, 5000 and 10000 words. From this corpus, a total of 446 features were extracted. Some new features like combined-words, new words endings and new POS tags were used in this study. The features were not separated into categories. Two machine learning classifiers, namely the k-nearest neighbour and a support vector machines classifier were used to assess the practicability and utility of our study. We were able to achieve 100% accuracy using the sequential minimal optimisation (SMO) algorithm with 40 document parts. Surprisingly, the simple and lazy k-nearest neighbour (kNN) classifier which is often discarded in gender profiling studies achieved a 98% accuracy with the same group of documents. Furthermore, 5-NN and 7-NN even outperformed SMO when using 400 document parts of 1000 words each. These values are much higher than those obtained in previous studies. However, we have used a new dataset and the results are therefore not directly comparable. Thus, our experiments provide further evidence that it is possible to infer the gender of an author using a computational linguistic approach.

Keywords Gender profiling · Text classification · Machine learning

---

H. Ramnial(✉) · S. Panchoo
School of Innovative Technologies and Engineering, University of Technology, Port Louis, Mauritius
e-mail: hoshila@education.mu, s.panchoo@umail.utm.ac.mu

S. Pudaruth
Department of Ocean Engineering and ICT, Faculty of Ocean Studies, University of Mauritius, Port Louis, Mauritius
e-mail: s.pudaruth@uom.ac.mu

S. Berretti et al. (eds.), *Intelligent Systems Technologies and Applications*,
Advances in Intelligent Systems and Computing 385,
DOI: 10.1007/978-3-319-23258-4_32

1 Introduction

Authorship attribution is a branch of artificial intelligence which attempts to predict the author of a given piece of text while author profiling endeavours to predict some characteristic of an author [1][2]. This may include the age, gender, nationality and even personality of authors. With the growing amount of texts on the internet and the growing problem of threats such as pedophiles lurking for unwary adolescents in chat rooms or cyber-bullying on social networks, it is becoming more and more important to devise a system which could allow law enforcement bodies to sketch digital profiles of such persons [3][4]. It was reported in Chaski [5] that the computational linguistic approach to stylometry has already found its way to the Federal courts of the United States of America.

The question of analysing differences between the writing styles of male and female writers have been occupying researchers for many years [6][7][8]. However, in recent years, there has been a new surge in the study of such differently due large to the increasing availability of large standard datasets, stylometric tools, machine learning tools and increasing power of computers [9]. Although many studies have already been carried out on gender profiling, most of them use datasets from informal contexts such as tweets [10], posts on social networks [11], chat logs [12][13], emails [14][15][16] and blogs [17][18][19]. Indeed, very little studies have been done using formal writings. Bergsma *et al.* [20] classified 6000 research papers as either written by a male research or a female one but they achieved a very low accuracy of 48.2% using SVM (support vector machines).

In this study, our focus is on gender profiling, whereby we make use machine learning classifiers to predict the gender of authors from equal-sized document segments extracted from 10 PhD theses. Using supervised machine learning classifiers, we have been able to show with very high accuracy that it is possible to differentiate between the writings of male and female writers even for formal non-fiction works.

This paper is organised as follows. In Section 2, we described related works that have been done on gender profiling. The data collection, pre-processing steps and corpus creation is described in Section 3. Section 4 describes the results of the intensive experiments that were conducted on the corpus. The concluding remarks and future research are found in Section 5.

2 Literature Review

Even though men and women speak the same language, experimental evidence suggests that women communicate differently than men [21]. In the same way, Singh [22] found that male speech was lexically richer and often tend to use longer phrases, whereas female speech used more verbs and shorter sentence structures. Thus, a pattern matching problem which will identify male and female is particularly suited for machine learning. Many author profiling research focuses on the prediction of a small number of traits, for example, on gender [6] and gender & age [23][24]. Some of them are described below.

The work of Argamon *et al.*, [25][26] showed that statistical analysis of word usage in documents could be used to determine the author's gender, age, native language and personality type. The authors made use of a blog corpus of 19200 documents with an average of 7250 words per blog. They made use of functions words to increase accuracy in gender prediction and thus achieved an accuracy of 76.1% using the Bayesian Multinomial Regression (BMR). Mukherjee and Liu [17] used 3100 blog posts and achieved an accuracy of 88.6% using SVM. They used a new feature selection method called Ensemble Feature Selection (EFS) which they showed to be superior to traditional feature selection methods such as information gain, mutual information and chi-square.

Another study by Estival *et al.*, [14] was to determine whether emails of various languages (English, Spanish and Arabic) can be used to differentiate between different people. Their corpus contains 9,836 emails from 1033 different authors with an average of 3260 words per author. They were able to identify the age, gender, native language, level of education, country of origin and personality traits of an author using author profiling. They made used of many stylometric features, for example, function words, PoS (parts of speech) tags, set phrases (e.g. greetings, farewells) and structural and character-based features. They were able to achieve an accuracy of 56% for age using SMO, 69% accuracy for gender using SMO, 84% accuracy for language using Random Forest, 80% accuracy for education using Bagging, 81% accuracy for country using SMO and an average of 55% accuracy for different personality traits. Extended versions of this work with 17864 email messages and 2063 authors produced an overall accuracy of 84.2% [15][16].

Corney *et al.*, [27] tried to predict gender in emails using SVM. They made use of a corpus of 4369 emails with a minimum of 50 words and a maximum of 200 words written by 325 different authors in English. They used 5 different stylometric features and were able to achieve 64%, 69.6%, 70%, 68.1% and 70.2% accuracy when using function words, word-based, character-based, structural and gender-preferential features, respectively. A similar work was performed by de Vel *et al.* [28] with the same dataset but with a major improvement of 10.6% in the overall classification accuracy. Koppel *et al.*, [29] apply the parts of speech feature on an average of 34,320 words document written by female and male authors. They made use of 566 documents. They were able to achieve an accuracy of 80% when trying to deduce the gender of an author of formal documents using the Exponential Gradient (EG) algorithm. Koppel *et al.* [36] classified 10000 blogs using a simple nearest neighbour based on cosine similarity and obtained recall of 83% and a precision of 90%.

Kucukyilmaz and Cambazoglu [12] collected 250,000 chat messages from 1500 users and used kNN, Naïve Bayes, covering rules and back propagation to predict the gender of the authors. Their best experiment with Naïve Bayes achieved an accuracy of 81.9%. Moreover, Rangel *et al.*, [31] used a large and realistic collection of blog posts and chat logs. They used a total of 236600 words for the English language training set and 75990 words for the Spanish language. They successfully identified the gender, age and emotion of many of the anonymous sexual predators. They made use of the word length, punctuation, unique words, PoS tags and content words for prediction. They were able to achieve an accuracy of well above 50% using SVM.

Another study by Schler *et al.* [18] used 37,478 blogs obtained from bloggers.com where their minimum number of words was 200. They made use of features such as non-dictionary words, parts-of-speech, function words, content words and hyperlinks to predict age and gender of the authors. They placed more emphasis on content words and they were able to achieve up to 80% accuracy on gender prediction using the Multi-Class Real Window (MCRW) algorithm which they claimed to be more efficient than SVM. Lim *et al.* [19] used 6000 English and Spanish blogs and achieved an accuracy of 69% and 74% respectively using SVM on predicting the gender of the authors.

A work that is very similar to ours was accomplished by Cheng *et al.* [32]. They used a total of 545 features, ordered into 5 feature subsets, and three different machine learning classifiers to predict the gender from standard datasets. They found that SVM produced the best accuracies with both 545 (85.1%) and 157 (82.1%) features. The accuracy increases proportionally with the number of instances in the dataset and the number of words in each sample. Daelemans [33] made an interesting comment on stylometric research in that the purpose of research should not only be '*who has got the best classifier or the best set of features*' but rather researchers should place more emphasis on providing explanation on classification results.

3 Methodology

3.1 Corpus

Ten PhD theses were downloaded from the British Library [34]. The theses belonged to 10 different authors and all of them were from British universities. We selected only theses which had more than 40000 words. Among the 10 theses, 5 of them belonged to female writers and 5 to male writers. Furthermore, all the PhD theses belonged to the field 'ICT in Education'. We believe that using similar materials would make the task of classification more difficult. To our knowledge, this issue has never been raised before but biased datasets can have a placebo effect on the true performance of classifiers.

The documents were processed manually in order to select only the document body, i.e., only content from the introduction to the conclusion. The rest are discarded. The document is then split into equal-sized segments of 1000, 5000 and 10000 words using our own document segmentor software. For example, if the theses contained 45000 words from introduction to conclusion and had to be split into four equal-sized fragments of 10000 words, then the first three set of 10000 words are selected as well as the last set of 10000 words. The same set of 10000 words are then divided into 2 sets of 5000 words and 10 sets of 1000 words. Next, we removed all document noise such as citations, quotations, weird texts and numbers using an automated approach.

3.2 *Stylometric Features*

446 features are extracted from each document segment and stores them in a csv file, which is then converted into an arff format for further processing in Weka [35]. Two popular machine learning classifiers, namely the k-nearest neighbour and support vector machines (Weka's SMO) were used for predicting the gender. The experiments were then repeated with a subset of 153 features. These 153 features were selected using a heuristic approach.

Table 1 Sample of features used for gender profiling

#	**Features**	**Description**
1	Character statistics	total number of characters in each document, etc
2	Word statistics	mean word length, vocabulary richness, etc
3	Sentence statistics	mean and standard deviation of sentence length, etc
4	Punctuation statistics	number of occurrences: . , : ; ! ? ' ' " " -
5	Special symbols statistics	number of occurrences: ( ) [ ] / \ % # & @ < > + - ~ *
6	Parts of Speech tags	verbs, nouns, adverbs, determiners, articles, etc
7	Function words	after, and, but, because, not, shall, that, what, you, etc
8	Word endings	`ll, n`t, ed, ing, ion, ly, s
9	Combined words	challenge/agent, computer-based, etc
10	Specific words per sentence	mean number of and, the, which, what, etc per sentence

Table 1 shows the 10 categories of features that were used in this study. However, the features were not differentiated based on their categories when they were fed to the classifiers. For example, the word statistics feature included several features like the mean, variance and standard deviation of the word length, vocabulary richness, number of unique words, maximum word length, etc.

4 Results and Evaluation

In this section, we describe the results of all the experiments that were performed with different sizes of documents, different number of documents and with the k nearest neighbour (kNN) and support vector machines (SMO) algorithms.

A cross-validation technique with 10 folds was used for all experiments. With 10,000 words and 446 features, 1-NN and 3-NN had the same performance values as SMO (98%). For 5,000 words and 446 features, the differences between 3-NN (98%) and SMO (99%) were insignificant. Surprisingly, 5-NN and 7-NN (94%) did better than SMO (93%) by 1% for 1000 words and 446 features.

Overall, the accuracy values for all the experiments were very high showing that it is indeed possible to distinguish between the gender of authors via their writing style if the appropriate features are used. The best confusion matrix (100%) for gender attribution occurs with SMO using 153 features and 10,000 words. This is shown in figure 1 below. The worst prediction accuracy of 87% for gender attribution was obtained using 1-NN with 153 features and 1,000 words. This is shown in figure 2 below. In this study, we do not use content words, yet we were able to achieve very high performance for gender prediction of 100% when using 10,000 words, 153 features and SMO. Therefore, it can be concluded that it is possible to distinguish between female and male authors using stylometric analysis.

```
=== Confusion Matrix ===
Predicted
  a  b   <-- classified as
 40  0 |  a = Female
  0 40 |  b = Male       Actual
```

Fig. 1 SMO, 153 features &10000 words

```
=== Confusion Matrix ===
  Predicted
   a    b   <-- classified as
 175   25 |   a = Female
  28  172 |   b = Male       Actual
```

Fig. 2 1-NN, 153 features & 1000 words

From Figure 2, we can see that among the 400 instances, 25 document fragments belonging to female authors were wrongly classified as belonging to male authors and 28 of the document segments belonging to male authors were wrongly classified as belonging to female authors.

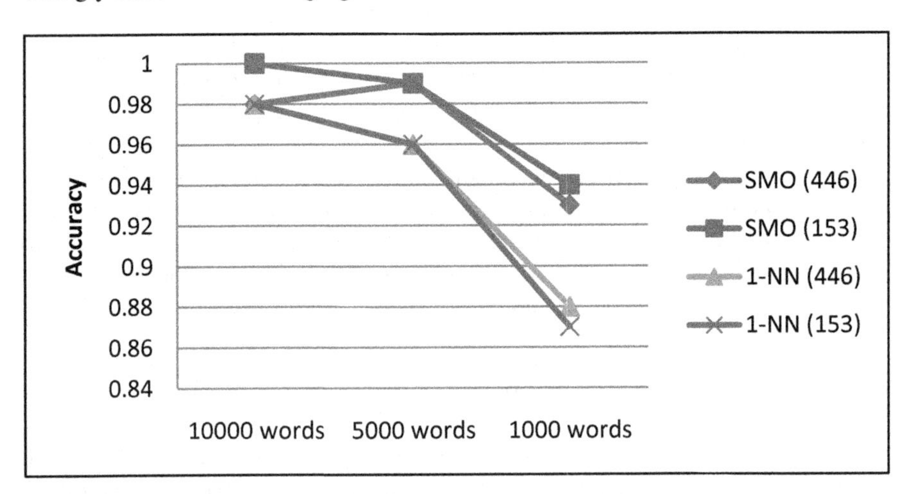

Fig. 3 Effect on accuracy of using different document size

Figure 3 shows the effect of varying document size while using the different number of features and algorithms for gender prediction. It can be observed that the performance of the classifier decreases when the document size is decreased. Using SMO with 153 features gives the best accuracy for every document size.

5 Conclusions

Accurate prediction of the gender of an unknown author has found applications in digital forensics, law enforcement, threat analysis, plagiarism detection, and literary science. In our present study, we performed various experiments in order to determine the best classifier for gender profiling, the best set of features and the right number of words to use to achieve an acceptable level of performance. Previous studies have concentrated mainly on works on fiction and informal writings from online blogs, chat logs and emails. Our corpus is a new one and involves the splitting of ten PhD theses into equal sized segments. From our extensive experiments, we show that with the right combination of machine learning classifier (kNN/SMO), document size (1000/5000/10000) and set of features (156/443), we can achieve 100% accuracy in predicting the gender of authors which demonstrates that there exists significant differences between the writing style of male and female writers. In general, we conclude that the bigger the document size and the larger the feature set, the higher is the accuracy. The strength of our approach is its independence on content. Thus, these techniques can be easily applied to texts from multiple domains. Together with an age predicting module, this system can also be used to discern potential cheating cases where students have used report or essay writing services and submit the received works as their own in order to clear a module or to earn a degree. The main limitation of our study is the requirement for large document sizes which makes it unsuitable for predicting the gender from shorter length messages such as sms and emails. As future research, we will experiment with a larger dataset and smaller document sizes and adapt the techniques, if necessary, to deal with texts of smaller sizes.

References

1. Mikros, G.K.: Authorship Attribution and Gender Identification in Greek Blogs. Methods and Applications of Quantitative Linguistics 21 (2012)
2. Segarra, S., Eisen, M., Ribeiro, A.: Authorship Attribution through Function Word Adjacency Networks. Cornell University Library, Computation and Language (2014)
3. Corney, M.: Analysing E-mail Text authorship for Forensic Purposes. Master of Information Technology Thesis. Queensland University of Technology (2003)
4. Gressel, G., Hrudya, P., Surendran, K., Thara, S., Aravind, A., Poornachandran, P.: In Proceedings of Notebook for PAN at CLEF 2014 (2014)
5. Chaski, C.E.: The Computational-Linguistic Approach to Forensic Authorship Attribution. Law and Language: Theory and Practice. Düsseldorf: Düsseldorf University Press (2006)
6. Koppel, M., Schler, J., Argamon, S., Messeri, E.: Authorship attribution with thousands of candidate authors. In: Proceedings of the SIGIR 2006 Proceedings of the 29th annual international ACM SIGIR conference on Research and development in information retrieval, pp. 659–660. New York, NY, USA (2006)

7. Abbasi, A., Chen, H.: Visualizing authorship for identification. In: Mehrotra, S., Zeng, D.D., Chen, H., Thuraisingham, B., Wang, F.-Y. (eds.) ISI 2006. LNCS, vol. 3975, pp. 60–71. Springer, Heidelberg (2006)
8. Abbasi, A., Chen, H.: Writeprints: A stylometric approach to identity-level identification and similarity detection in cyberspace. ACM Transactions on Information Systems, **26**(2), Article 7 (2008)
9. Koppel, M., Schler, J., Argamon, S.: Computational Methods in Authorship Attribution. Journal of the Americal Society for Information Science and Technology **60**(1), 9–26 (2009). John Wiley & Sons
10. Mechti, S., Jaoua, M., Belguith, L.H., Faiz, R.: Machine Learning for classifying authors of anonymous tweets, blogs, reviews and Social media. In: Proceedings of the PAN@CLEF, Sheffield, England, September 2014
11. Peersman, C., Daelemans, W., Vaerenbergh, L.V.: Predicting age and gender in online social networks. In: Proceedings of the 3rd international workshop on search and mining user-generated contents, pp. 37–44 (2011)
12. Kucukyilmaz, T., Cambazoglu, B.B., Aykanat, C., Can, F.: Chat mining for gender prediction. In: Yakhno, T., Neuhold, E.J. (eds.) ADVIS 2006. LNCS, vol. 4243, pp. 274–283. Springer, Heidelberg (2006)
13. Lin, J.: Automatic author profiling of online chat logs. Naval Postgraduate School, Monterey (2007)
14. Estival, D., Gaustad, T., Hutchinson, B., Pham, S.B., Radford, W.: TAT: an author profiling tool with application to Arabic emails. In: Proceedings of the Australasian Language Technology Workshop 2007, pp. 21–30 (2007)
15. Estival, D., Gaustad, T., Pham, S.B., Radford, W., Hutchinson, B.: Author profiling for English emails. In: Proceedings of the 10th Conference of the Pacific Association for Computational Linguistics, PACLING 2007, pp. 262–272 (2007)
16. Estival, D., Gaustad, T., Hutchinson, B., Pham, S.B., Radford, W.: Author Profiling for English and Arabic Emails. Natural Language Engineering, Cambridge University Press (2008)
17. Mukherjee, A., Liu, B.: Improving gender classification of blog authors. In: Proceedings of the 2010 Conference on Empirical Methods in Natural Language Processing, pp. 207–217. MIT, Massachusetts, October 9–11, 2010
18. Schler, J., Koppel, M., Argamon, S., Pennebaker, J.W.: Effects of age and gender on blogging. AAAI Spring Symposium Computational Approaches to Analyzing Weblogs, pp. 199–205 (2006)
19. Lim, W., Goh, J., Thing, V.L.L.: Content-centric age and gender profiling. In: Proceedings of the Notebook for PAN at CLEF 2013 (2013)
20. Bergsma, S., Post, M., Yarowsky, D.: Stylometric analysis of scientific articles. In: Proceedings of the 2012 Conference of the North American Chapter of the Association for Computational Linguistics: Human Language Technologies (NAACL-HLT), pp. 327–337. Stroudsburg, USA (2012)
21. Corney, M., Vel, O., Anderson, A., Mohay, G.: Gender preferential text mining of e-mail discourse. In: Proceedings of the 18th Annual Computer Security Applications Conference (ACSAC 2002), pp. 282–292. Las Vegas, USA (2002)
22. Singh, S.: A Pilot Study on Gender Differences in Conversational Speech on Lexical Richness Measures. Literary and Linguistic Computing **16**(3), 251–264 (2001)
23. Koppel, M., Argamon, S., Shimoni, A.R.: Automatically Categorizing Written Texts by Author Gender. Literary and Linguistic Computing, **17**(4) (2002)

24. Maharjan, S., Shrestha, P., Solorio, T., Hasan, R.: A straightforward author profiling approach in MapReduce. In: Bazzan, A.L., Pichara, K. (eds.) IBERAMIA 2014. LNCS, vol. 8864, pp. 95–107. Springer, Heidelberg (2014)
25. Argamon, S., Koppel, M., Fine, J., Shimoni, A.R.: Gender, genre, and writing style in formal written texts. Text - Interdisciplinary Journal for the Study of Discourse **23**(3), 321–346 (2003)
26. Argamon, S., Koppel, M., Pennebaker, J.W., Schler, J.: Automatic profiling the author of an anonymous text. Communications of the ACM **52**(2), 119–123 (2009)
27. de Vel, O., Corney, M., Anderson, A., Mohay, G.: Language and gender author cohort analysis of e-mail for computer forensics. In: Proceedings of the digital forensic research workshop (2002)
28. Koppel, M., Schler, J., Argamon, S., Winter, Y.: The Fundamental Problem of Authorship Attribution. English Studies **93**(3), 284–291 (2012). Taylor & Fancis
29. Rangel, F., Rosso, P., Koppel M., Stamatatos, E., Inches, G.: Overview of the author profiling tasks at PAN 2013. In: Notebook for PAN at CLEF 2013 (2013). http://www.clef-initiative.eu/documents/71612/2e4a4d3a-bae2-47f9-ba3c-552ec66b3e04 (accessed March 3, 2015)
30. Cheng, N., Chandramouli, R., Subbalakshmi, K.P.: Author gender identification from text. In: Proceedings of the IEEE Symposium on Computational Intelligence and Data Mining Conference, April 2009, Digital Investigation, vol. 8, no. 1, July 2011, pp. 78–88. Elsevier Ltd (2009)
31. Daelemans, W.: Explanation in computational stylometry. In: Gelbukh, A. (ed.) CICLing 2013, Part II. LNCS, vol. 7817, pp. 451–462. Springer, Heidelberg (2013)
32. The British Library: THE BRITISH LIBRARY - The world's knowledge (2015). http://www.bl.uk/ (accessed April 11, 2015)
33. Weka: WEKA, The university of Waikato (2015). http://www.cs.waikato.ac.nz/ml/weka/ (accessed March 28, 2015)

Bridging the Gap Between Users and Recommender Systems: A Change in Perspective to User Profiling

Monika Singh and Monica Mehrotra

Abstract One of the prevalent research challenges in the field of recommender system is to do better user profiling. There are some advanced user profiling techniques found in the literature to achieve the same. User profiling aims to understand the user well and as a result recommending the most relevant items to the user, where relevant means items returned as a result of intelligent techniques from various fields, mainly from data mining. This work is an attempt to answer the question "who understands a user the most?" The three obvious answers are Recommender System's high end approaches (e.g. data mining and statistical approaches), neighbors of the user or the user herself. The correct answer would be the last one, which is a user knows herself the best. In this direction, we propose to make users empowered and responsible for registering their preferences and sharing the same at their discretion. More personalized solutions can be offered when a user tells what she prefers and can contribute explicitly to the recommendation system's results generation. When a user is given the handle to communicate her preferences to the recommender system, more personalized recommendations can be given which are not only relevant (as tested by sophisticated evaluation matrices for recommender systems) but also plays wonder to users' satisfaction.

1 Introduction

Recommender System as a research field flourished well with many techniques and features borrowed from other disciplines. Some of these disciplines are data mining, statistics, psychology and human computer interaction (HCI). Recommender Systems as an applicative area of data mining are used to recommend users in an online world, a few highly relevant items from a vast pool of available options. A significant research

M. Singh(✉) · M. Mehrotra
Jamia Millia Islamia, New Delhi, India
e-mail: {monikasingh.jmi,drmehrotra2000}@gmail.com

S. Berretti et al. (eds.), *Intelligent Systems Technologies and Applications*,
Advances in Intelligent Systems and Computing 385,
DOI: 10.1007/978-3-319-23258-4_33

and development is seen in recommender system field. This work is an attempt to re-visit the field again in quest of finding some unseen/ underdeveloped opportunities specifically to do better user profiling.

When we talk about Recommender System, we observe three core entities involved in a typical e-commerce interaction as mentioned below (refer Figure 1):

a User
b E-Commerce Interface
c Recommender System

This work is an attempt to closely view how to strengthen relationship between user and recommender system. In order to understand the user, Recommender System does user profiling. Categorically, we can say that most of the work is being done in Recommender System side to improve it.

In the initial days when e-commerce sites were evolving, it is being assumed that users of e-commerce site are not educated/ trained enough, so all the intelligence were being employed on the Recommender System to understand the user. Since then, the major efforts are made in the direction to understand the user better. In technical terms, it is called *user profiling*. All the data mining and statistical techniques that are employed in recommender system are primarily focused to understand the user better in order to give her appropriate recommendations.

The primary reason that researchers focused only in the direction of Recommender System to understand the user was they assumed that "User cannot express herself". So, all the prevalent techniques devised intelligence in Recommender System's side to understand the user. But today users are well versed with the internet; they are educated enough and understand well the dynamics of internet. They are empowered and they want a custom made solution for themselves, rather than a generic solution. In technical terms, it is called *personalization*. Moreover, they want their inputs being incorporated in any decision making (in this context recommendations). After going through extensive literature review, we found that following work are being done in Recommender System (refer figure 1).

The user can see the recommendation on an e-commerce website. There is almost no transparency between recommender system and the user. A user always has following questions in her mind with respect to recommender system.

i. Why she needs to create multiple profiles on different e-commerce sites? Is there any way to write preference only once and use everywhere (across multiple e-commerce websites)?
ii. How the recommendations are generated for her? Is it based on her preference or based on her neighbors?
iii. Who all/ (how many) people have rated the product before being recommended to her?
iv. In determining the overall rating of the product, how many people have actually bought the product?
v. Are the profiles who have rated are genuine profiles or attackers?

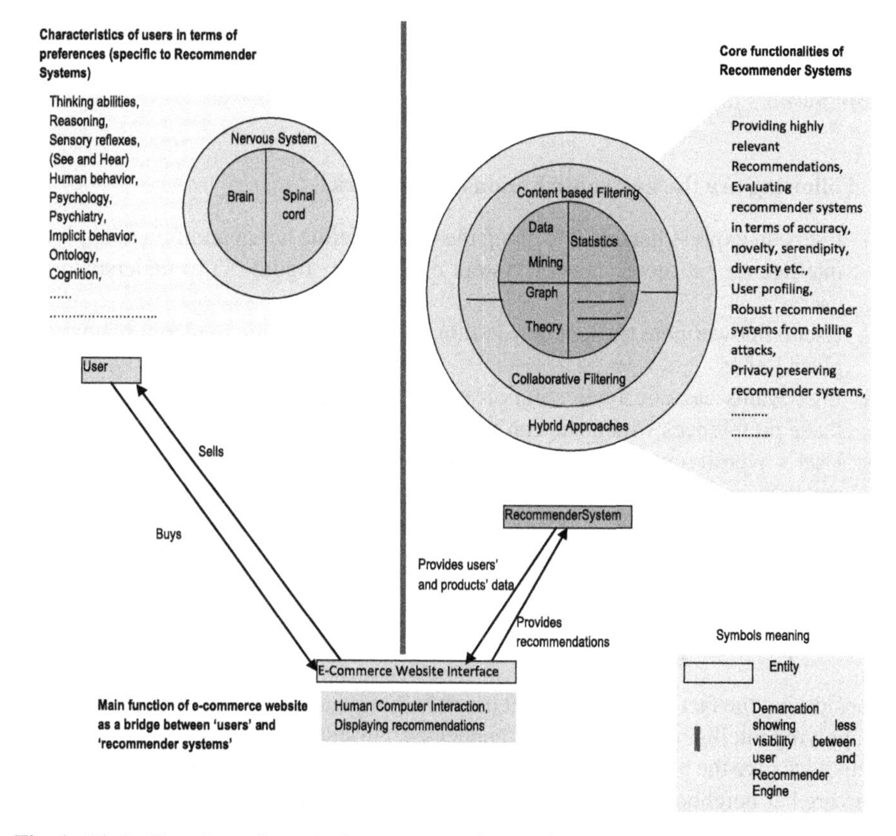

Fig. 1 Bird's Eye view of a typical e-commerce interaction

Our research questions are also pointing to the above user concerns.

Definition of the problem

Empowering a user to receive more personalized recommendations by giving her handle to create and share her profile at her own discretion using "user agent". Hence augmenting the existing intelligence mechanism by incorporating users' intelligence.

This work mainly focuses to answer the following research questions:

RQ1. *How to make users of e-commerce sites speak about her preferences?*

RQ2. *How to centralize and store users' preferences which are collected at different e-commerce websites and synchronize it with each e-commerce website? i.e how can we facilitate write once use everywhere paradigm?*

RQ3. *How to make recommendation process more user-oriented (in terms of her satisfaction on seeing the recommendations as well as her involvement in final recommendations)?*

RQ4. *Are user's neighbours found through collaborative filtering actually determines her preferences well? Or the users preferences captured explicitly contribute more towards profiling her?*

Following are the key contributions of the work:

i. User can express herself. We equip the user to define her in addition to applying intelligence (at recommender system entity, refer figure 1) to understand the user.
ii. User empowerment by her contribution in decision making in recommendations given to her.
iii. User wants custom made solution which is provided through collecting even those preferences which she can't put forward verbally (implicit preferences).
iv. User's overall transactions on all e-commerce are considered to build her profile and the same will be propagated to every e-commerce website through synchronization by user agent.
v. Offering serendipitous yet relevant recommendations.

2 Proposed Approach

Considering the fact that now internet users *"can express themselves"*, so, in addition to applying intelligence at the recommender system side to understand the users, why cannot we give the handle to the user to express herself, build her own profile, decide who are her neighbors with respect to recommender system.

The second challenge and frustration for the users of an e-commerce site is that she has to build her profile and set her preferences in every e-commerce portal *(Flipkart, Jabong, Amazon* etc.) and irrespective of overall transactions that she has done only the transactions that has been done on individual e-commerce site will be counted for building and enriching her profile.

E.g. Consider a scenario in which a user has done 100 transactions overall that spans across 5 e-commerce sites. All the transactional details will not be available with each e-commerce website and despite of 100 transactions only the local transaction with respect to the website will be considered and recommendation will be based upon those local transactions. If she has done only 2 transactions on a particular e-commerce website, despite of her 100 transaction only 2 transactions will be considered for decision making (Recommendation). This is really annoying for users as their profile preferences are not synchronized.

Our proposed approach addresses above said prevailing challenges and also empowers the users to contribute in the final decision making (Recommendation) process. In this era of internet we have to come out of the stereotypical thought "A user cannot express herself" and user is not educated enough with internet. All the existing Recommender systems seem to be made on the assumptions that the "User cannot express herself" (refer Figure 2).

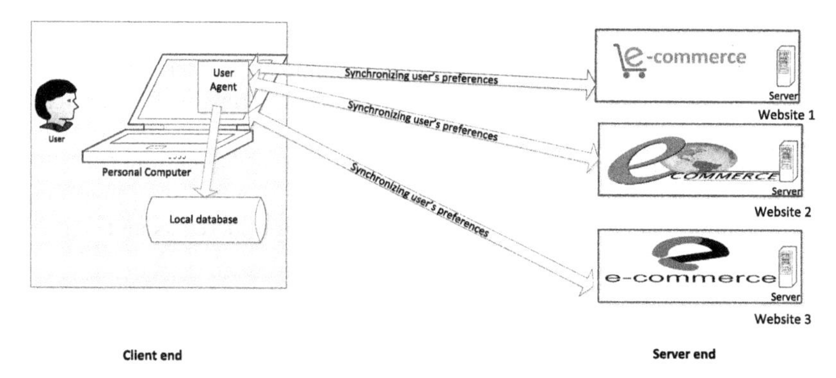

Fig. 2 User agent synchronizing user's preferences across different e-commerce websites

2.1 Typical Web Interaction of an E-commerce Site:

An e-commerce application is typically a web application and the client accessing the e-commerce application is thin client, where almost no processing is being done on the client side. Thin client is just used to render the html content which is returned from the e-commerce server.

In our proposed approach, we are advocating for a "thick client" to do better user profiling and empowering the user to enrich her profile, preferences, control the kind of recommendation she wants. In a nutshell empowering the user in recommendations and augmenting the intelligence of recommender system for appropriate recommendations to the user.

In addition to empowering the user, these thick clients will address the issue of building multiple profiles, preferences and transactional history in every e-commerce site. The user need to build her profile and preference object only once on thick client and then she can synchronize these profile and preferences object with every e-commerce website with which she wants to interact. In this way a rich transactional history will be available to each e-commerce website, hence improved decision making with respect to recommendations.

2.2 User Agent at Client's end

The user agent installed at the client end will capture the user's browsing patterns, implicit behavior based preferences and her explicit preferences. This will ensure every minute preference of a user to be captured. It is not possible for the user to express all her preferences explicitly, so some intelligent techniques are employed to capture implicit preferences as well. For example, user can explicitly say she likes a particular product, but when she sees it, she comes to know whether she wants that. Similarly, she may like something she has not explicitly expressed. User unconscious preferences are required to be captured for better user profiling. These preferences when used to recommend are serendipitous as well as useful to the user.

Sharing the preferences by this process would be at users' discretion. Users who need enhanced personalization can be benefited from this approach.

At present, the recommender system, tries to *pull* the information from the users to understand her well. This *pull* approach is depicted in figure 3, where recommender system pulls user information by anticipating, mining techniques and intelligent algorithms. The advancements done in pull approach though offers high quality recommendations but here, the user is suppressed and considered as mute entity.

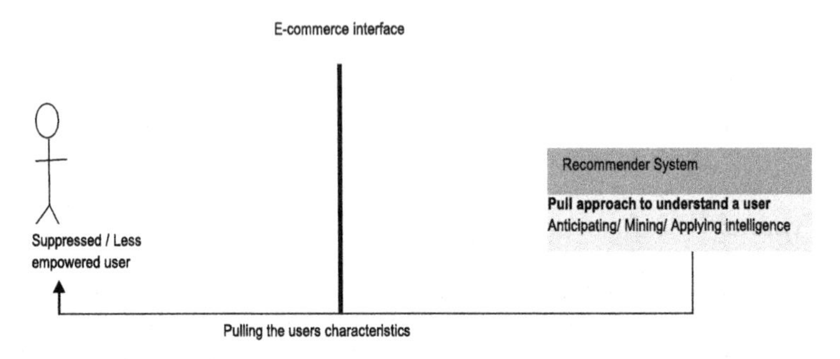

Fig. 3 Present Scenario: Pull approach by recommender systems towards user

We propose that a user is a key entity in recommendation process, which calls to make her more empowered. In other words, let her speak for herself. Let her push her preferences through user agent to the recommender system. User agent synchronizes user's characteristics with every e-commerce portal. This is a *push* approach, depicted in figure 4.

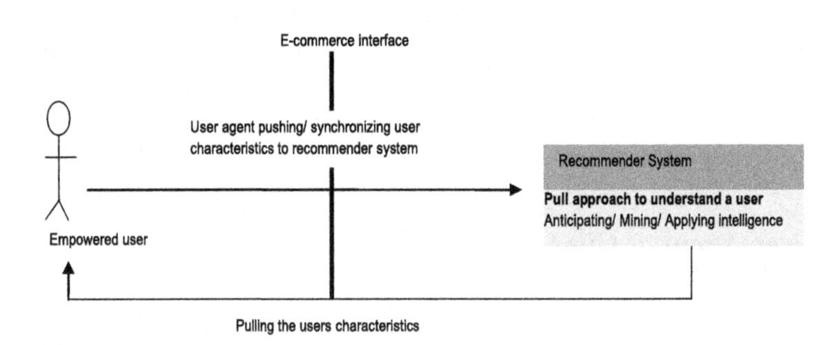

Fig. 4 Proposed solution: Push approach by user to recommender system

The existing recommender system's *pull* along with augmented user's *push* results in better recommendations.

2.3 *Proposed Solution is an Augmentation to the Existing Intelligent Approaches*

Recommender system entity (refer figure 1) is evolving and many intelligent techniques and algorithms have been developed so far. The proposed solution is an augmentation to existing intelligent mechanism in recommendation process. The proposed work highlights how a suppressed entity (a user) can be empowered to share her preferences at her discretion and the benefit of this approach.

3 Related Work

The concept of personalization plays a key role in recommender systems' designing and development. More useful recommendations must be personalized or tailor made for a user. User profiling deals with understanding users' more closely so as to provide them customized solutions. User profiling is about identifying and mining data about a user's interest domain. For accomplishing this, advanced modeling techniques are developed. User profiling models users' preferences [7].

In [2], the authors discuss that improved modeling of users and items are important extensions to current recommender systems. More closely we can capture, learn and mine users' preferences, more personalized and hence effective recommendations we can deliver. Profiling a user by her navigation patterns or browsing history is discussed in [9]. In [9], authors also discuss how recommending items based on user's implicit behavior and feedback results in better personalization.

Other user modeling techniques include modeling based on users' rating attributes of an item [4]. User profiling represented as an instance of domain ontology is discussed in [6]. Capturing users' preferences by ontology based descriptions result in better recommendations. Profiling based on concept hierarchy and item taxonomy descriptors is discussed in [8], it means users' preferences are captured on a concept level rather than an item level [8].

Adomavicius and Tuzhilin [1] propose *Recommendation Query Language (RQL)*, which can do better profiling by taking multiple dimensions into account. Value of additional information in enhancing the quality of recommendations is being acknowledged. In [5], Helocker and Konstan presented the value of additional information in recommender systems.

Other approaches to user profiling also include user-item rating matrix and keyword vectors [7]. User's tagging behavior is also used to profile a user [7, 10]. In [7], the authors have modeled users' preferences through selection agents and collection agents. Whereas collection agent shows current topic, selection agent shows single user's preferences [3].

Literature survey is an indicative of the utmost requirement to profile a user better. Since, recommender systems' inception and personalization evolution, researchers have bought up very interesting and novel ideas for user profiling to add some additional information for understanding a user. Our work is in line and towards the same goal to improve personalization and hence recommendations by profiling a user from a distinct and novel strategy.

4 Conclusions and Future Work

Collaborative filtering, probably the backbone of recommender system is based on finding preferences of highly similar users with the target user. This similarity is based on the ratings similarity of users on products bought in the past. Based on it, the target user is recommended item liked by her neighbors. This is known as user based collaborative filtering. When we employ user agent at client end and perform collaborative filtering this is called user based collaborative approach using user agent, similarly finding items which are neighbors of users' preferred items collected through user agent is called item based collaborative filtering using agent. In future, we will empirically test collaborative filtering using user agent and observe its performance improvement over traditional collaborative filtering approaches (user based collaborative filtering and item based collaborative filtering). Currently, we are in the process of building a prototype. So, in future we will be able to show the effectiveness of this approach empirically on recommendation results.

References

1. Adomavicius, G., Tuzhilin, A.: Multidimensional recommender systems: a data warehousing approach. In: Fiege, L., Mühl, G., Wilhelm, U.G. (eds.) WELCOM 2001. LNCS, vol. 2232, pp. 180–192. Springer, Heidelberg (2001)
2. Adomavicius, G., Tuzhilin, A.: Toward the next generation of recommender systems: A survey of the state-of-the-art and possible extensions. IEEE Transactions on Knowledge and Data Engineering **17**(6), 734–749 (2005)
3. Balabanović, M., Shoham, Y.: Fab: content-based, collaborative recommendation. Communications of the ACM **40**(3), 66–72 (1997)
4. Hattori, S., Mao, Z., Takama, Y.: Proposal of user modeling method and recommender system based on personal values. In: 2012 Joint 6th International Conference on Soft Computing and Intelligent Systems (SCIS) and 13th International Symposium on Advanced Intelligent Systems (ISIS), pp. 1720–1723. IEEE (2012)
5. Herlocker, J.L., Konstan, J.A.: Content-independent task-focused recommendation. IEEE Internet Computing **5**(6), 40–47 (2001)
6. Kadima, H., Malek, M.: Toward ontology-based personalization of a recommender system in social network. In: 2010 International Conference of Soft Computing and Pattern Recognition (SoCPaR), pp. 119–122. IEEE (2010)
7. Liang, H., Xu, Y., Li, Y., Nayak, R.: Collaborative filtering recommender systems using tag information. In: IEEE/WIC/ACM International Conference on Web Intelligence and Intelligent Agent Technology, WI-IAT 2008, vol. 3, pp. 59–62. IEEE (2008)
8. Nadee, W., Li, Y., Xu, Y.: Acquiring user information needs for recommender systems. In: 2013 IEEE/WIC/ACM International Joint Conferences on Web Intelligence (WI) and Intelligent Agent Technologies (IAT), vol. 3, pp. 5–8. IEEE (2013)
9. Oh, J., Lee, S., Lee, E.: A user modeling using implicit feedback for effective recommender system. In: International Conference on Convergence and Hybrid Information Technology, ICHIT 2008, pp. 155–158. IEEE (2008)
10. Tso-Sutter, K.H., Marinho, L.B., Schmidt-Thieme, L.: Tag-aware recommender systems by fusion of collaborative filtering algorithms. In: Communications of the ACM, pp. 1995–1999. ACM (1994)

Document Classification with Hierarchically Structured Dictionaries

Remya R.K. Menon and P. Aswathi

Abstract Classification, clustering of documents, detecting novel documents, detecting emerging topics etc in a fast and efficient way, is of high relevance these days with the volume of online generated documents increasing rapidly. Experiments have resulted in innovative algorithms, methods and frameworks to address these problems. One such method is Dictionary Learning. We introduce a new 2-level hierarchical dictionary structure for classification such that the dictionary at the higher level is utilized to classify the K classes of documents. The results show around an 85% recall during the classification phase. This model can be extended to distributed environment where the higher level dictionary should be maintained at the master node and the lower level ones should be kept at worker nodes.

1 Introduction

Applications like Novel document detection and clustering, Topic modeling, Spam filtering etc are still some challenging areas due to the ever increasing volume of on-line documents. The commercial value of classifying text documents have lead to many machine learning techniques in the area. The performance of such techniques can be improved by exploring domain-specific term features. Classification models have to be generated out of a set of labeled data and frequencies of class relevant terms of each document. Here, a basic necessity is to represent these extracted class specific information efficiently such that it can be used to express or approximate any input data belonging to its corresponding class at a later stage.

In areas like signal processing, image processing etc, any data can be represented or approximated in terms of a *dictionary* (a pre-determined set of data points) which is learned from the training data itself [1]. The same concept has been adopted for

R.R.K. Menon(✉) · P. Aswathi
Amrita Vishwa Vidyapeetham, Amritapuri Campus, Kollam 690525, Kerala, India
e-mail: ramya@am.amrita.edu, aswathipv45@gmail.com

S. Berretti et al. (eds.), *Intelligent Systems Technologies and Applications*,
Advances in Intelligent Systems and Computing 385,
DOI: 10.1007/978-3-319-23258-4_34

text documents representation by some works [2][3][6]. The dictionary learned out of the data is then used for classification and novelty detection [2]. Given a set of data $Y = \{y_i\}_{i=1,..m}$ in R^n where n is the number of relevant terms and m is number of documents in the dataset, the goal of dictionary learning is to find a dictionary $D \in R^{n \times k}$ such that each data point y in the set can be represented as a sparse linear combination of its columns. This can be represented as

$$Y \approx DX \quad satisfying \quad \|Y - DX\|_F \leq \varepsilon \;\&\; \|x\|_0 \leq \tau \,, \tag{1}$$

where $x \in R^k$ contains the ordered coefficient values and $X \in R^{k \times m}$ is the sparse representation matrix.

Any dictionary learning technique basically deals with two problems - Sparse coding and Dictionary composition. Sparse coding is the technique with which sparse representational values x for y with the atoms of D is obtained. The atoms to form the linear combination and its coefficient values are to be chosen well to obtain the optimal minimal representation[1][9]. There are many optimization algorithms to calculate x like Pursuit algorithms. The simplest one among them are Matching Pursuit algorithms that follows a greedy approach where the dictionary atoms are chosen sequentially and its commonly used variations are *Basis Pursuit* and *Orthogonal Matching Pursuit*(OMP). OMP solves x by using l_0 norm, as expressed below.

$$\min_x \|x\|_0 \quad subjectto \quad \|Y - DX\|_F \leq \varepsilon \tag{2}$$

Given X after the sparse coding phase, with retrieval error $\|Y - DX\|_F$, the aim of the update step is to update the dictionary D or to find a better dictionary D that can accommodate the error. There are many techniques for this like the generalized k-means, MOD (Method of Optimal Directions), Maximum A-Posteriori Probability approach, K-SVD etc. All these methods prove to learn dictionaries well but varies in their convergence, efficiency and implementation effort. In K-SVD algorithm [9, 10, 11] this is done by finding SVD of the error matrix and the matrices D and X are updated simultaneously using the factors U and V.

Once a dictionary is generated, a major question to be considered is whether a new set of data of the same class can be represented with D. The given dataset may contain documents of different classes or labels. When there are K distinct classes then K dictionaries are to be learned [2]. The dependency and relationship among the classes should be considered while learning these dictionaries such that performance is better during classification and/or clustering.

Our contribution includes hierarchically structured dictionaries and a method to classify a document into any of the K classes, based on this structure. The hierarchy consists of 2-levels. The lower level consist of K *sub-dictionaries* corresponding to K classes and the higher level has a single dictionary referred to as *parent-dictionary*(refer Fig. 1 with value of K as 3).

The structure is generated in a bottom-up fashion and classification is done with the parent dictionary. Parent dictionary is constructed out of all the sub-dictionaries,

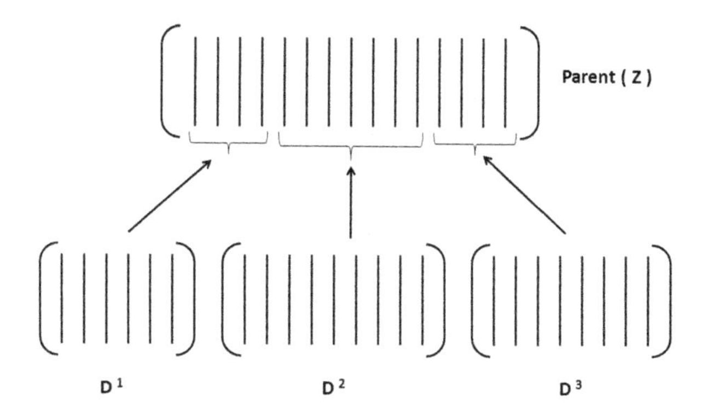

Fig. 1 2-level Hierarchical Dictionary Structure

using Principal Component Analysis(PCA), and hence contains a set of representative atoms from each of the classes. Experiments show that the parent dictionary alone is enough for classification among K classes. We perform our experiments on two different datasets those when evaluated gives around 85% recall and precision.

2 Related Work

Dictionary learning is an area which has scope for exploration in the domain of text mining. Few research has been done with regard to the usage of learning dictionaries from the vast amount of text piling up in the web/database. Sparse representation of data through sparse coding is a byproduct of this approach.

Dictionary learning has been utilized for novel document detection both for static data [2] as well as streaming data [3] using a distributed environment. It has also been used for clustering [1] but by the signal processing community. In their work, a dictionary is constructed for each cluster using which the signals are best reconstructed using the sparse code. Data points belonging to the same class can belong to different subspaces based on the generated sparse code. A dictionary incoherence term is added to the general condition of reconstructed error to make the dictionaries to represent a specific class while at the same time differentiate it from other classes. Sharing of atoms between dictionaries of two distinct classes can also occur [5]. Emerging topic detection [2] proposes an Alternative directions method (ADM). ADM is applied by representing the l_1-norm reconstruction error in the augmented Lagrangian form and then approximating it. ADM has been applied for the detection stage as well as the learning stage.

Online l_1-dictionary learning [4] uses the l_1-norm for the reconstruction error based on a variation of the ADM for novel document detection. The paper proposes to obtain bounds on the regret for updating dictionary. Experiments have been conducted on the Twitter data.Fast on-line $l1$-dictionary learning [7] compares 3 algorithms for on-line dictionary learning - dual averaging scheme (D_A), projected gradient scheme

(P_G) and online ADM. The paper claims that D_A has better predictive power were as the other two algorithms score in terms of running time and parameter stability. Kasiviswanathan et.al. [3] extends the novel document detection to a distributed environment by choosing the SPMD (Simple Program Multiple Data) model which uses the MPI concept for message passing. Here also, dictionary learning is done using the ADM approach.

Utilization of dictionary learning for classification especially in the document/text domain is yet to be well explored. Unlike signal and image domains, application of dictionary learning to the text domain will have a varied set of problems like maintainability of document structure.

3 Notations Used

Notation	Explanation
y	an input document as a vector
D	general representation for any dictionary
$D^c \in R^{n \times m}$	sub-dictionary corresponding to c^{th} class where n - number of terms m - number of examples in class c
d_i^c	i^{th} atom of cth class sub-dictionary
$Z \in R^{n \times Q}$	Parent dictionary with n rows and Q atoms together from all classes
$Z^c \in R^{n \times q_c}$	atoms of parent that represents c^{th} class q_c is number of vectors/atoms taken from D^c
z_i^c	i^{th} atom of Z^c
T	coefficient matrix obtained after sparse coding with parent Z
T^c	part of T that contains values of cth class
t_i^c	i^{th} coefficient value of T^c

4 Hierarchical Dictionary Structure

4.1 Structure of the Dictionaries

We introduce a new 2-level hierarchical structure of dictionaries which can be used later for classification purposes. The lower level consists of K *sub-dictionaries* and the higher level consist of a parent-dictionary. Given a set of labeled documents the system generates K *sub-dictionaries* for the K distinct classes in bottom up fashion. Applying PCA on each of these sub-dictionaries individually gives the most relevant vectors from each classes. These features or vectors from K classes are appended together to form a master/parent dictionary that suffices all the K classes.

The entire work is designed as 3 modules - Pre-processing data, Generating the dictionaries, Classification of documents.

4.2 Preprocessing

Given a set of documents each labeled with any of the K classes, this module processes them to generate K classes of input matrices containing numerical values. Relevant terms of each class is derived from its corresponding documents and their TF-IDF value is calculated to form the input Y^c matrix. Before the calculation, stemming is done along with removal of stop words and duplicate words to obtain more accurate values.

4.3 Procedure of Dictionary Learning

The mentioned dictionary structure is generated in 2 stages - learning K sub-dictionaries and learning the parent dictionary out of the K sub-dictionaries.

$$Y^c \approx D^c X^c \quad suchthat \ \|Y^c - D^c X^c\|_F \leq \varepsilon \tag{3}$$

As a new sub dictionary is generated, PCA is applied over it to generate the sub part Z^c of the parent dictionary Z.

$$Z = Z^1 \cup Z^2 \cup \ldots \cup Z^K \tag{4}$$

4.3.1 Learning a Sub-dictionary

An input matrix $Y \in R^{n \times m}$ consists of n terms and m number of sample labeled documents. This matrix is generated using the TF-IDF measure and contains data of a single class. As we have K classes we have K such input matrices. We have adopted the K-SVD algorithm along with considering the dependency among the sub-dictionaries. (3) is solved in this section by implementing algorithm 1.

The input matrix itself is initially considered as the dictionary D^c and is used for sparse coding in the first iteration. The K sub-dictionaries have same number of rows but varying numbers of columns. Once the sparse representation X^c is obtained, it is used to update the dictionary to accommodate the error term $\|Y^c - D^c X^c\|_F$ using SVD decomposition. Considering each atom individually, all documents that uses the atom is identified from X. Excluding the contribution of this atom in the representation, the error in retrieval is obtained on which SVD is applied. While U and V are the factors obtained during SVD, the first column of U is used to replace the selected atom of D^c and the column of V along with its corresponding singular value is used to update the corresponding row of X^c that uses the atom. This procedure is done repeatedly until convergence.

Before generating a parent dictionary from these sub-dictionaries, we have to ensure that the dictionary is as independent as possible from each other such that the classifier can distinguish the classes well. To obtain this, we add a incoherence inducing term that can be expressed as

Algorithm 1. generating a sub-dictionary

Input: Y, initial dictionary D_0, τ := number of sparse coefficients , K:= no of labels ,η := weight value
Output: Sub-dictionaries $\{D^1, D^2, .., D^K\}$ and sparse matrices $\{X\}$
for $c = 1 : K$ **do**
    $D^c := D_0$
    for $q = 1 : h$ **do**
        $[i, x_i^c] := Argmin(x)\|y_i - D^c x_i^c\|_2^2$ Subject To $\|x^c\|_0 \leq \tau$
        if $c! = 1$ **then**
            for $p = 1 : c - 1$ **do**
                $coh := coh + getIncoherence(D^c, D^p)$
            end for
            $coh := coh * \eta$
            $D^c[n, m] := D^c[n, m] + coh \quad \forall n = 1, 2, .., Nandm = 1, 2, .., M$
        end if
        for $j = 1 : k$ **do**
            $(D^c)' = (D^c)^{(q-1)}$
            $I :=$ { indices of the docs in Y uses atom $(D^c)'$}
            $E_I := Y - D_j^c{}' x_T^j$
            $[D^c, X^c] = \text{SVD}(E)$
        end for
    end for
end for

$$D_i = D_i + \eta \sum_{j=1}^{i-1} \|D_i^T D_j\|_2^2 \qquad \forall\, i = 1, 2, .., K \tag{5}$$

Each element in the dictionary is added with this value making the dictionary matrix incoherent from others[5].

4.3.2 Dictionary Learning - Learning the Parent Dictionary

Given the sub dictionaries, we generate the Parent Dictionary there by solving (4). By applying PCA on a sub-dictionary D^c we obtain a set of eigenvalues and their corresponding eigenvectors, that constitute the basis of the dictionary(refer algorithm 2). PCA projects the data to a lower dimensional representation with least error while preserving largest variances. As all the eigenvalues are not necessary for an approximation, we choose those eigenvalues greater than δ. The corresponding eigenvectors of the chosen eigenvalues forms the set $Z^c \in R^{n \times q_c}$. q_c indicates the number of eigenvectors included to the parent. The final parent dictionary $Z \in R^{n \times Q}$ is obtained by vertically concatenating $Z^1, Z^2, .., Z^K$ where Q = $q_1 + q_2 + .. + q_k$ as given in algorithm 2.

Algorithm 2. Generating Parent Dictionary

Input: D - Set of all k dictionaries, δ - threshhold for eigen value
Output: $DParent$ - Parent dictionary
for $a = 1 : k$ **do**
  $[P, Ev] := PCA(D_a)$
  for $\forall e \in Ev \geq \delta$ **do**
    $DParent_a := DParent_a \cup (P_e)$
  end for
end for

4.4 *Classification*

4.4.1 Parent-dictionary Based Classification

The parent dictionary Z contains set of vectors from all the sub-dictionaries and it represents all the K classes together. Parent based classification algorithm(refer algorithm 3) finds the sparse coding T of the new test document with parent Z. Based on the atoms being used and its coefficient values the algorithm identifies the class. Z^c indicates the part of the parent that is derived from class c^{th} dictionary. T^c indicates the coefficient values corresponding to Z^c. Thus $\sum_{i=1}^{q_c} z_i^c t_i^c$ calculates the linear combination of all atoms of cth class and then its frobenious norm is taken. The document belongs to class c for which the representation error is minimum (6).

$$\arg\min_{c} \|y - \sum_{i=1}^{q_c} z_i^c t_i^c\|_F^2 \tag{6}$$

By bounding this error with a threshold, this can be used for classification of a document into multiple labels.

5 Experimentation and Analysis

5.1 *Datasets and Preprocessing*

Experiments are conducted using two datasets dataset I and dataset II. The training samples of dataset I contains 200 documents belonging to a 'Question Bank' on the topics of computer science, collected from our academic collection. It consists of 4 labels ($K = 4$) - 'Computer Organization And Architecture', 'Computer Networks', 'Programming', and 'Database Management System'. The training samples of dataset II includes 75 documents belonging to two entirely distinct labels ($K = 2$)- 'Indian Climate' and 'Dictionary Learning'.

Algorithm 3. Classification with parent-dictionary

Input: = parent dictionary, testY, no_sparse_coef
Output: list of novel documents , labels for rest of documents
errors = error from closer dictionaries
Init: N =No.of documents in testY, $r = ()$, $doclabels = ()$, $noveldocs = ()$, $errors = ()$
for $i = 1 : N$ **do**
  $y = testY_i$
  $T := sparseCoding(Z, y, no\_sparse\_coef)$
  for $c = 1 : K$ **do**
    $res := \| \sum_{i=1}^{q_c} Z_i^c T_i^c \|_F^2$
    if $res <= \alpha$ **then**
      $r := r \cup c$
    end if
  end for
  if r is empty **then**
    alert("Document to not belong to any class !")
  end if
end for

Each document in the set of training documents taken as input belongs to any one of the classes in its domain. These labeled documents becomes the input to the preprocessing module were they are processed to generate an output file corresponding to each class/label. These files contains the TF-IDF values, to form the input Y^c matrix corresponding to each classes. Each ith column of the matrix represents i^{th} training document in the class. Removal of duplicate words and stopwords followed by stemming ensures that only relevant terms are considered in this calculation. The relevant terms identified for each class in the dataset and their corresponding TF-IDF weight is written into a file.

5.2 *Dictionary Learning*

The obtained tf-idf value-based vectors are given into the sub-dictionary learning module. Initially the input matrix itself is taken as the dictionary. Then the algorithm identifies initial coefficient values(X) by using OMP, implemented with cholesky decomposition. Once X is obtained, it is used to update the dictionary atoms along with its coefficient values using SVD factorization over error matrix. The dictionary is updated iteratively until the error criterion is met. Frobenius norm is used here to calculate the error $\|Y - DX\|_F \leq \varepsilon$. The error obtained lies between the interval 1.5 and 3.5.

With dataset II first dictionary comes with a dimension of 10×35 (10 terms and 20 documents). Similarly the dimension of second dictionary is 10×40. The parent dictionary is created from the sub dictionaries using PCA. It identifies the independent basis vectors as eigenvectors and in this experiment we consider those with their corresponding eigenvalues that are greater than δ. For dataset II, $\delta = 0.005$.

Table 1 Sample Result with Dataset I (9 out of 22 test documents)

Actual class	Documents	COA (class 1)	Networks (class 2)	Programming (class 3)	DBMS (class 4)	classified to
1	Doc 1	0.08284	0.15171	0.14925	0.20732	1
1	Doc 2	0.04193	0.07972	0.11042	0.08160	1
1	Doc 3	0.04598	0.08545	0.09380	0.14708	1
2	Doc 4	0.07379	0.04642	0.04796	0.09865	2
2	Doc 5	0.11858	0.04978	0.12538	0.06616	2
2	Doc 6	0.09460	0.04962	0.08535	0.10455	2
3	Doc 7	0.44099	0.26587	0.16008	0.32346	3
4	Doc 8	0.11906	0.09827	0.15469	0.21753	2
4	Doc 9	0.14588	0.11382	0.10693	0.08081	4

Table 2 Confusion matrix for Dataset I

Cases	Predicted Negative	Predicted Positive
Negative cases	53	3
Positive Cases	3	19

Table 3 Confusion matrix for Dataset II

Cases	Predicted Negative	Predicted Positive
Negative cases	8	1
Positive Cases	1	8

From 1st and 2nd sub-dictionaries, we get 10 and 9 eigenvectors respectively. i.e., $q_1 = 10, q_2 = 9$. On concatenating them vertically we obtain total $Q = 19$ vectors in parent. Number of rows in parent remains as n.

Similarly with dataset I, the sub-dictionary dimensions are such that $Y^1 \in R^{20\times35}$, $Y^2 \in R^{20\times23}$, $Y^3 \in R^{20\times26}$ and $Y^4 \in R^{20\times24}$. After applying the algorithm 2 over them, we obtain $Z \in R^{20\times26}$.

5.3 *Classification*

Generally during classification, each document is to be matched with all the sub-dictionaries to identify the better one. In this work, we use the parent dictionary alone for classification. To find the resultant class, equation (7) is used and results are given in Table 1 and Table 4. With dataset I, the representation error $\| \sum_{i=1}^{q_c} z_i^c t_i^c \|_F^2$ of some of the test documents, with respect to 4 classes are given in Table 1. Similarly with dataset II, the error of the test documents, with respect to 2 classes are given in Table 4.

There are 22 documents in the test set for dataset I and 19 out of 22 are correctly classified. With 9 documents in the test set for dataset II, eight documents are classified correctly. Once *confusion matrix* is calculated (as given in Table 2 and Table 3), evaluation of classification results is performed with the measures of accuracy, precision and recall. Accuracy is the correctness of results, calculated as the total number

Table 4 Result with Dataset II (9 test documents)

Actual class	Documents	Dictionary-Learning (class 1)	Climate (class 2)	classified to
2	Document 1	0.03472702	0.01276724	2
2	Document 2	0.08271272	0.08000903	2
2	Document 3	0.45815999	0.21802396	2
2	Document 4	0.02490772	0.00474065	2
1	Document 5	0.05320239	0.43174093	1
2	Document 6	0.34723903	0.30103366	2
1	Document 7	0.05320239	0.23015204	1
1	Document 8	0.35465621	0.00542857	2
1	Document 9	0.05320239	0.41254825	1

Table 5 Evaluation with Datasets

Evaluation Criteria	Dataset I	Dataset II
Accuracy - % of correct prediction	92.3 %	88.89 %
Precision - % of positive predictions that are correct	86.36 %	88.89 %
Recall - % of correct prediction of positive cases	86.36 %	88.89 %

of correct classifications divided by total number of classifications. Precision is a measure calculated as number of true positives divided by sum of true positive and false negative. Precision is a measure calculated as number of true positives divided by sum of true positive and false positive. The evaluation results of experiments with both dataset is given in Table 5. Both experiments results in an effective classification with more than 85% of accuracy, precision and recall.

6 Conclusion

Classification is a well-saught after mechanism especially in this era when data gets piled up with no bounds and hinders the way for data analysis. So our aim in this paper is to harness the potential of dictionary learning for the process of classification. We have demonstrated that our proposed method that uses a hierarchical dictionary structure will pave the way for improving the classification process. This structure can be utilized for a distributed environment where the parent dictionary can act as the master node and the sub-dictionaries can be the slave nodes. So the parent dictionary can determine the corresponding child nodes to get the complete classification done.

Acknowledgments We express our fullest gratitude to the Department Chairman, Dr. M.R. Kaimal, who supported us throughout with his valuable comments. We thank our teachers and friends for giving us valuable information and our family for their moral support.

References

1. Bruckstein, A.M., Donoho, D.L., Elad, M.: From sparse solutions of systems of equations to sparse modeling of signals and images. SIAM Rev. **51**, 34–81 (2009)
2. Kasiviswanathan, S.P., Melville, P., Banerjee, A., Sindhwani, V.: Emerging topic detection using dictionary learning. In: Proceedings of the 20th ACM International Conference on Information and Knowledge Management, CIKM 2011, pp. 745–754. ACM, New York (2011)
3. Kasiviswanathan, S.P., Cong, G., Melville, P., Lawrence, R.D.: Novel document detection for massive data streams using distributed dictionary learning. IBM Journal of Research and Development **57**(3/4), 9 (2013)
4. Kasiviswanathan, S.P., Wang, H., Banerjee, A., Melville, P.: Online l1-dictionary learning with application to novel document detection. In: Bartlett, P. L., Pereira, F.C.N., Burges, C.J.C., Bottou, L., Weinberger, K.Q. (eds.) NIPS, pp. 2267–2275 (2012)
5. Ramrez, I., Sprechmann, P., Sapiro, G.: Classification and clustering via dictionary learning with structured incoherence and shared features. In: CVPR, pp. 3501–3508, IEEE (2010)
6. Menon, S.R., Nair, S.S.: Sparsity-based representation for categorical data. In: Recent Advances in Intelligent Computational Systems (RAICS). IEEE (2013)
7. Kasiviswanathan, S.P.: Fast online l 1-dictionary learning algorithms for novel document detection. In: 2013 IEEE International Conference on Acoustics, Speech and Signal Processing (ICASSP), pp. 8585–8589. IEEE (2013)
8. Berry, M.W., Drmac, Z., Jessup, E.R.: Matrices, vector spaces, and information retrieval. Society for Industrial and Applied Mathematics **41**, 335–362 (1999)
9. Aharon, M.: Overcomplete dictionaries for sparse representation of signals. PhD thesis, Technion-Israel Institute of Technology, Faculty of Computer Science (2006)
10. Aharon, M., Elad, M., Bruckstein, A.: Svdd: An algorithm for designing overcomplete dictionaries for sparse representation. Trans. Sig. Proc. **54**, 4311–4322 (2006)
11. Rubinstein, R., Zibulevsky, M., Elad, M.: Efficient implementation of the k-svd algorithm using batch orthogonal matching pursuit. CS Technion **40**(8), 1–15 (2008)
12. Jolliffe, I.: Principal component analysis. Wiley Online Library (2002)

Efficient User Profiling in Twitter Social Network Using Traditional Classifiers

M.A. Raghuram, K. Akshay and K. Chandrasekaran

Abstract Any discussion in social media can be fruitful if the people involved in the discussion are related to a field. In a similar way to advertise an event, it is useful to find users who are interested in the content of the event. In social networks like Twitter, which contain a large number of users, the categorization of users based on their interests will help this cause. This paper presents an efficient supervised machine learning approach which categorizes Twitter users based on three important features(Tweet-based, User-based and Time-series based) into six interest categories - Politics, Entertainment, Entrepreneurship, Journalism, Science & Technology and Healthcare. We compare the proposed feature set with different traditional classifiers like Support Vector Machines, Naive-Bayes, k-Nearest Neighbours, Decision Tree and Logistic Regression, and obtain upto 89.82% accuracy in classification. We also propose a design for a real-time system for Twitter user profiling along with a prototype implementation.

1 Introduction

Twitter is a popular micro-blogging site that allows millions of users to communicate, stay in touch, establish connections and more. Users via Twitter, can post messages called as "tweets" which are limited to 140 characters containing only text or hyperlinks. The rising popularity of social networking sites like Twitter, Facebook, LinkedIn, Tumblr etc. has produced vast resources of user-generated content. As of March 2015, Twitter reports a monthly usage of 288 million active users with more than 500 million tweets exchanged per day [2].

In this context, the problem of automatically identifying user interests [15] and

M.A. Raghuram(✉) · K. Akshay · K. Chandrasekaran
Department of Computer Science and Engineering, National Institute of Technology Karnataka, Mangalore, India
e-mail: mar.11co54@nitk.edu.in, akshaysaja44@gmail.com, kchnitk@ieee.org

S. Berretti et al. (eds.), *Intelligent Systems Technologies and Applications*,
Advances in Intelligent Systems and Computing 385,
DOI: 10.1007/978-3-319-23258-4_35

user profiling [18] has gained significant attention. Twitter's Streaming API methods provide easy and programmatic access to the vast amount of data generated in the social network [3]. This has made Twitter an active hub for user personality and profile related research. Some of the studies that have been carried out include identifying user's demographic information [17], predicting brand-related events from user's tweets [14] and tweet topic identification [19]. Researches have also been carried in finding out finely-tuned features like predicting the type of Twitter account reporting an event(individual, news organization or other) [11].

Research on Indian Twitter users is rare although Indian users constitute for more than one-third of the Twitter population [5]. In our experiments, we primarily concentrated on tweets from Indian users and obtained the handles of active Twitter users using web directory listing services like Twellow [1]. Since we make use of Tweet-based features to identify the dynamic interest of the user, we assume that the user tweets about topics that he/she is interested in. We also make use of the Weka machine learning library for various feature selection & machine learning tasks and also for implementing the real-time application in Java [4].

In this paper, we predict the user's dynamic interests based on three important characteristics - the user's static profile, tweet content [12] and simple time series features of the user's tweets. We explore the supervised learning model with several classifiers - Support Vector Machines(SVM), Naive-Bayes(NB), Decision Tree(DT), k-Nearest Neighbour(kNN) and Logistic Regression(LR) with different combinations of our proposed features. We explore the impact of principal component analysis(PCA) on our proposed feature set. We also develop a prototype to evaluate our classification scheme based on the suggested features and propose a model for implementing a real-time user profiling application.

The rest of the paper is organized as follows. In the next section, we discuss some of the related works with respect to Twitter user classification and user profiling. In Section III, we present our methodology. Section IV presents the experiments, their results and analysis. In Section V, we conclude the paper along with directions for future work.

2 Related Work

In recent years, several attempts have been made on finding out the preferences of users in the Internet. The researches primarily focused on the Internet activity of the user to determine the user's interests. Also, there are various studies on sentiment analysis and tweet analysis [8, 9] in Twitter social network alone.

De Choudhury et al. [11] categorized Twitter accounts reporting worldwide events into journalists, organizations and ordinary users, based on the analysis of their Twitter time-line. The Twitter data of 1850 users was collected and the supervised machine learning model was explored. Pennacchiotti and Popescu [16] tried to build a machine learning model to determine the political affiliation of a Twitter user. The research was aimed to find whether a user is democratic or republic and also tells whether a user is attracted towards particular brand of business.

Table 1 Distribution of Twitter User Profiles collected

Interest Class	User Instances
Politics	97
Entertainment	142
Entrepreneurship	108
Journalism	72
Science & Technology	75
Healthcare	96
Total Instances	**593**

Table 2 Specific Twitter accounts collected for each class

Interest Class	Twitter account handles
Politics	@narendramodi,@NandanNilekani, @ShashiTharoor, @arunjaitley, @SushmaSwaraj
Entertainment	@iHrithik, @arrahman, @SrBachchan, @iamsrk, @AnilKapoor
Entrepreneurship	@dharmesh, @FareedZakaria, @hnshah, @kiranshaw, @Aishwarya_N
Journalism	@ndtv, @zeenews, @sardesairajdeep, @BDUTT, @timesofindia
Science & Technology	@elakdawalla, @Atul_Gawande, @sanjayguptaCNN, @IndianScience, @pallavbagla
Healthcare	@GEHealthIndia, @drsanjaygupta, @3MHealthcare_In, @fortis_hospital, @INDHEALTHCARE

Some researches focused on identifying the subjects that a user was interested in. Thongsuk et al. [19] used the topic model for identifying businesses of particular interest which match the interests of a user. The Latent Dirichlet Allocation(LDA) algorithm was used to construct the topic model. User-based features were used to identify businesses which might interest the user. Medvet and Bartoli [14] used precompiled topic description and unsupervised machine learning algorithm to detect popular themes, events and associated sentiment polarity. They focused on creating a customized user profile for Twitter users. The Time-series based features were introduced by Yang et al. [21] on a large corpus of Twitter data containing information from political and sports domains. They classify users on periodicity of activities and the patterns in which users express their opinions and also applied their proposed methods to both binary and multi-class classification of sports and political interests of Twitter users. Siswanto et al. [18] used the supervised learning model and lexical features and tried to determine users interest based on bio data and collection of users tweets. They propose two approaches for dynamic interest prediction of a Twitter user- The first method was based on classification of user's tweets using multi-label classification method and the other approach used specific accounts for classification.

A participant-based event summarization approach was proposed in Chakrabarti and Punera [10]. The proposed approach studies Twitter event streams at the participant level. Their work finds the sub-events that are associated with each participant involved in that event. They use a mixture model, combining burstiness and cohesiveness found in tweet properties related to the event which was used to generate event summaries. They collect the event related information from the tweets and re-tweets of the participants involved in the Twitter event and with time generate the

Table 3 Most commonly used words for each class of users

Interest Class	Frequent Terms(Stemmed words)
Politics	'india', 'peopl', 'pm', 'nation', 'wish', 'state', 'govern', 'parti', ,'congress', 'develop'
Entertainment	'love', 'thank', 'like', 'watch', 'great', 'happi', 'show', 'night', 'amaz'
Entrepreneurship	'busi', 'time', 'come', 'help', 'design', 'work', 'need', 'know'
Journalism	'polic', 'year', 'report', 'live', 'kill', 'world', 'video', 'week', 'right', 'court', 'presid', 'offici', 'attack', 'women', 'march'
Science & Technology	'scienc', 'think', 'research', 'data', 'find', 'interest', 'brain', 'post', 'human'
Healthcare	'drug', 'medic', 'care', 'risk', 'heart', 'learn', 'share', 'check', 'prevent', 'futur', 'food'

summaries about the event and sub-events related to it. Kouloumpis et al. [12] evaluated the usefulness of the linguistic features which capture data about the informal languages used by the users in activities like blogging. Sentiment detection based on linguistic studies of messages exchanged over Twitter can also be viewed as a supervised machine learning model.

An et al. [7] carried out a novel study on the media and information broadcasting landscape in Twitter. They use the Twitter data and the user accounts followed by people to reveal the relationship between different the classes of media and its diversity in content. Their work aimed at explaining how news related to a particular field spreads in a social network. This study helps to find the behaviour of users who read news and how publishers interact with their readers. They also explain why Twitter users follow multiple news sources.

3 Methodology

The aim of this work is to propose a generic and scalable model for automatically classifying Twitter users based on their dynamic interests given the user attributes and tweet history for a large set of Twitter users. Our approach consists of the following sequence of steps - building a large corpus of Twitter data, extracting proposed features and comparing the performance of traditional classifiers along with the effect of principal component analysis and finally develop a real-time application for Twitter user categorization.

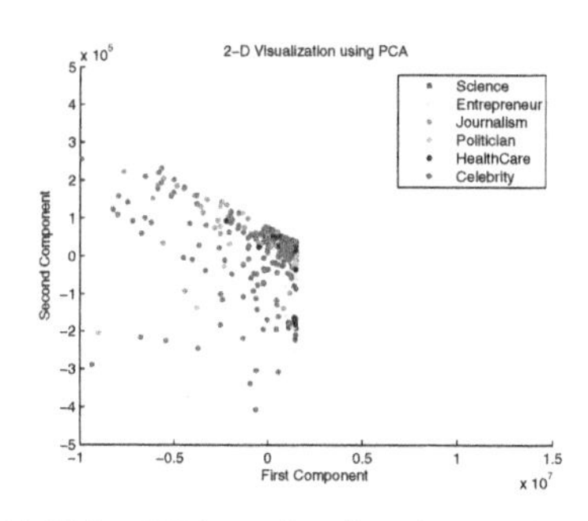

(a) Using PCA to visualize data set in 2D space

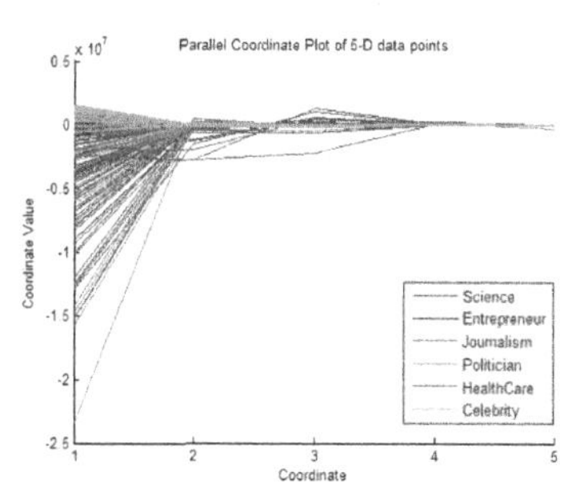

(b) Parallel Coordinates Plot of 5-dimensional feature space obtained after dimensionality reduction using PCA

3.1 Data Collection

The labelled data set of Twitter users was obtained in a semi-automatic manner wherein we first manually gather Twitter account handles for a specific class over the Internet and then use the Twitter4J Java library for extracting the Twitter feed of the user. We first focused on Indian Twitter handles that were available in the Internet(services like Twellow) and then filled the remaining instance slots with users across the globe so as to balance the data-set. For every Twitter User we collect up to last 500 tweets made by the user and pre-process the tweet data. This includes replacing URL, number with standard texts, removing special symbols/emoticons and stop words and finally performing word stemming using the Porter stemmer.

The Table 1 gives an overview of the data collected for each type of dynamic interest class. Table 2 shows some of the Twitter accounts used for collecting the data. These Twitter handles are manually labelled after inspecting the Twitter account and the tweets posted from the account. Also, only those Twitter handles that could be clearly identified with respect to a single interest class were used. Figure 1a and Figure 1b are obtained after applying principal component method for dimensionality reduction [20]. They are helpful in visualizing our data set. The first component after PCA accounts for more than 99% of the total variance which suggests that most of the data points are along a hyper-plane of higher dimensions. Figure 1a suggests a tendency of close relation among the data points although clustering of individual interest classes is not clearly visible.

3.2 Feature Selection

We propose three types of features for efficient classification of Twitter users: User-based features(Static profile), Tweet-based features and simple Time-series based features.

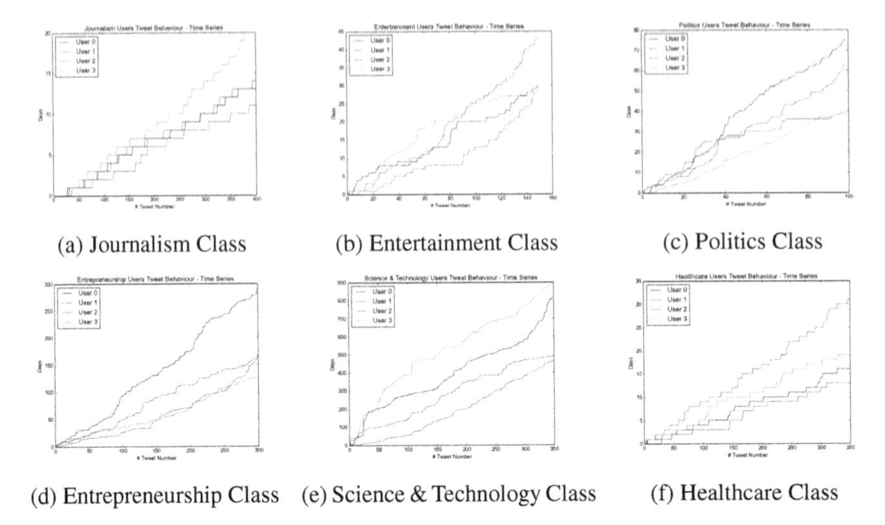

(a) Journalism Class (b) Entertainment Class (c) Politics Class

(d) Entrepreneurship Class (e) Science & Technology Class (f) Healthcare Class

Fig. 2 Tweet Behaviour Plots for each category of users

1. **User-based features**, representing the user's static profile i.e. user information that is very unlikely to undergo rapid changes. This includes user's gender, location, timezone and friends, followers and favourites count. These kind of features are also helpful in spam detection, for example Twitter's spam and abuse policy considers users having small amount of followers compared to the number of people followed by the user to be spam [13]. We consider the reputation score of the user, which is defined by Equation 1,

$$Reputation(user) = \frac{F_o(user)}{F_o(user) + F_r(user)} \tag{1}$$

where F_o and F_r represent the number of followers and friends respectively.
2. **Tweet-based features**, are extracted from the tweets posted by the user. *TF-IDF(Term Frequency - Inverse Document Frequency)* is a commonly used feature in data mining. It is used to provide a weighting method to the usual term frequency of specific dictionary words. Specifically, the weight assigned to each term is determined as shown in Equation 2,

$$Weight(t) = -\log \frac{df(t)}{U} \tag{2}$$

where t represents a term, df(t) is the document frequency - no. of users whose tweet contains the term t, U is the total number of users and weight(t) represents the corresponding weight that is multiplied to the term frequency. Other Tweet-based features include, No. of Hash-tags, No. of replies/mentions, No. of Sensitive tweets and No. of Hyperlinks/URLs etc. In Table 3, we show the most

commonly occurring words in the tweets of the users of each class. Note that these words are the stemmed words obtained using the Porter Stemmer which is used in text normalization.
3. **Time Series features** are used to represent the temporal behaviour of a user's tweets. In our experiments, we use simple statistical features of the user's time series like average, maximum/minimum, standard deviation and derivative of the time series. This is different from the features considered in [21], where class-specific keywords were used to measure the temporal behaviour of different types of users. In Figure 2, we show the tweet behaviour of different types of users. We record the time of tweeting of at-least last hundred tweets and use it to plot the graphs. We can observe that users belonging to a specific interest class tend to have almost similar tweeting patterns. Also, some class of users can be easily characterized with respect to the time series of its users. For example, the users of Journalism class have a step function pattern of tweeting which is due to the periodic news-related information posted by these type of users.

3.3 Classification Methods

We briefly summarize the classification principles of the traditional classification algorithms that were used for training/testing our data set.

1. **Support Vector Machines** is a popular classification technique which tries to determine a large-margin hyperplane that can act as decision boundary. It is therefore, a non-probabilistic linear binary classifier, which performs an implicit mapping from the input to high-dimension feature space for identifying a clear margin. The polynomial kernel(Equation 3) is commonly used for measuring the similarity between feature vectors and computing the cost function.

$$K(x, y) = (\sum_{i=1}^{N} x_i y_i + c)^d \tag{3}$$

2. **Naive Bayes** classifier is one among the many different classifiers that are based on the Bayes Theorem and is useful particularly when the input feature space is of high dimensionality. Given a set of features $X = f_1, f_2, ...f_d$ extracted from a user and a set of classification categories $c_1, c_2, ...c_k$, the Naive Bayes classifier assigns that class c_i that has the maximum posterior probability i.e. $c_i = c_j | P(C_j|X)$ is maximum
3. **Decision Tree** algorithm works on the principle of information gain i.e. at each node of the decision tree, the attribute that splits the data-set effectively is chosen. This process is then repeated on the smaller data-sets obtained by splitting the original data-set. The decision tree classifier can work efficiently with independent features and also even in the presence of outliers but does not scale suitably with large number of features compared to other classifiers like SVM or Naive Bayes.

4. **K-Nearest neighbours** is a simple classification method that does not require training / model fitting. It assumes that points that are close in the feature space are more likely to belong to the same class. Here, K represents the number of nearest neighbours to be considered for classifying the user. The most common mechanism for aggregating the k-points is to use a voting scheme where the class with highest votes is assigned as the predicted class. There are different measures that are used to determine the distance between two points, the most common being the Euclidean distance $D(x, y) = \sqrt{(x - y)^2}$
5. **Logistic regression** is a probabilistic statistical classification scheme which uses probabilistic scores to infer the relation between the classification variable(dependent) and the set of features(independent variables). The logistic function $\sigma(t)$ which is used to determine the probability score, is defined as follows

$$\sigma(t) = \frac{1}{1 + e^{-t}} \tag{4}$$

3.4 Real-Time Twitter User Classification

We propose a simple algorithm for developing a real-time user classification system that periodically updates itself. The system naturally adapts itself to changing tweet behaviour over time, of users of different categories and hence improves in its accuracy of classification. Such a system can also be easily integrated with other applications such as recommender systems to dynamically provide user interests for better performance in the integrated applications.

4 Experiments and Analysis

In all our experiments, we use the 10-fold cross validation for testing and measuring the accuracy of classification. The feature space contains the normalized term frequencies of thousand nine hundred most commonly used words in the English dictionary in their stemmed form, eight User-based features and ten Time-series based features giving a total of 1918 attributes. The Table 4 shows the result of the testing phase with the different combinations of classifier and feature set. The highest accuracy of classification is achieved when the SVM classifier and combining User, Tweet and Time-series based features. The total number of instances are 593 out of which 528 were classified correctly and 65 were classified incorrectly yielding the best accuracy of 89.04% without performing Principal Component Analysis. From Table 5 and Table 6, we can observe that the classifier is having trouble distinguishing between the fields Journalism and Entrepreneur. The True Positive rate for Journalism class is much less compared to other classes which means that actual instances of Journalism class are not being identified correctly and instead are misclassified as Entrepreneur class, thereby increasing the False Positive rate of Entrepreneur class. This type of error occurs mainly because of Twitter users who tweet about business

Algorithm 1. Real-time Twitter User Classification

```
   Data: New Twitter User v
   Result: Interest Class Label for User v
 1 Model M = Load existing classification model;
   /* Loads a set of Twitter users U where each user is
      labelled from a set of classes C and extracted set of
      features F                                           */
 2 if M is old then
 3     t = 25;
 4     NewUsers = Read t new users from Twitter Stream;
 5     for each user u in NewUsers do
 6         f = ExtractFeatures(u) ;
 7         label = M.classify(f) ;
 8         Model M = merge (u,f,label) with model M ;
 9     end
10     Store new classification model M;
11 end
12 f = ExtractFeatures(v) ;
13 label = M.classify(v) ;
14 return label
```

Table 4 Performance comparison - Testing results of different classifiers with different feature set

Classifier	**Tweet-based features**	**User,Tweet-based features**	**User,Tweet & Time-series based features**
Support Vector Machine	81.80 %	84.68 %	89.04 %
Naive Bayes	70.71 %	76.50 %	84.14 %
J48 Decision Tree	61.11 %	65.87 %	63.91 %
K-Nearest Neighbours	36.90 %	36.90 %	64.41 %
Logistic Regression	79.16 %	81.12 %	83.98 %

related news which can be similar to tweets by Entrepreneurs(For example, users emarketer, theeconomist are examples of biz news accounts).

Another important observation that can be made from Table 5 is with respect to the Entertainment class which has the perfect True Positive rate and Recall which means that any user who belongs to the Entertainment class will always be identified. The Entrepreneurship class has the lowest False Positive rate which is because it is the class that is most commonly misclassified into. This suggests that users interested in business are also likely to be interested in other topics as well.

Table 5 Detailed classification accuracy - SVM classifier + Tweet,User & Time series features

Class	TP Rate	FP Rate	Precision	Recall	F-Measure
Politics	0.969	0.002	0.989	0.969	0.979
Entertainment	1.000	0.036	0.901	1.000	0.948
Entrepreneurship	0.880	0.052	0.792	0.880	0.833
Journalism	0.681	0.013	0.875	0.681	0.766
Science & Technology	0.787	0.014	0.894	0.787	0.837
Healthcare	0.896	0.018	0.905	0.896	0.901

Table 6 Confusion Matrix for SVM classifier + Tweet,User & Time series features

Class	A	B	C	D	E	F
A - Politics	94	3	0	0	0	0
B - Entertainment	0	145	0	0	0	0
C - Entrepreneurship	0	6	95	5	1	1
D - Journalism	1	3	14	49	3	2
E - Science & Technology	0	3	6	1	59	6
F - Healthcare	0	1	5	1	3	86

Table 7 Classification accuracy for different classifiers with different feature set size obtained after using PCA

Classifier	500 features	250 features	100 features	50 features
Support Vector Machine	51.09%	78.58%	86.68%	89.71%
Naive Bayes	84.65%	84.65%	84.65%	84.82%
J48 Decision Tree	74.04%	74.87%	75.37%	76.89%
K-Nearest Neighbours	31.70%	26.81%	53.45%	77.23%
Logistic Regression	87.18%	87.68%	88.19%	88.87%

Principal Component Analysis(PCA) is a common technique in machine learning that is used to reduce the dimensionality of the feature space(can also be used to increase the dimension but is rarely done). It is based on the principle of maximum variability along each component of the new set of dimensions. The Table 7 shows the classification accuracy with different classifiers with decreasing feature space dimension obtained using the PCA tool in Weka. The results show that the highest result for the classification task is achieved by reducing the 1918 features to 50 features and using the SVM classifier. Also, general the trend across the table is increasing accuracy of classification with shrinking feature dimensions.

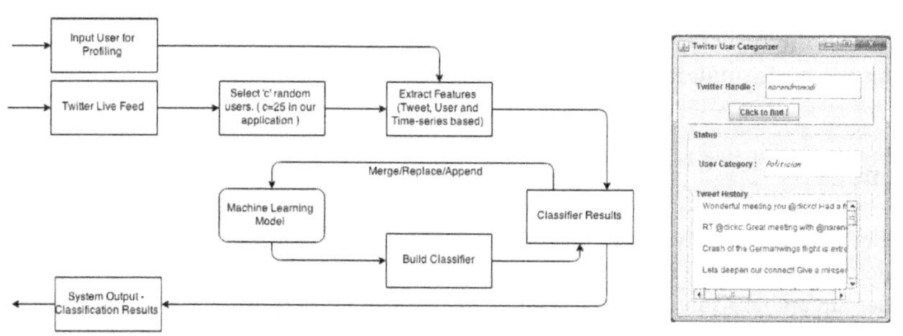

(a) Design of Real-time Twitter User Classification System

(b) Real-time Twitter User Classification Application

4.1 Real-Time Application

The Twitter user profiling application can be made more accurate and scalable by deploying it as a real-time system. We propose a simple design for a real-time Twitter user classification system along with a prototype implementation using the Weka library for machine learning in Java and python scripts for feature extraction. The application periodically collects information about a fixed number of random users from the Twitter streaming API which provides a live feed of tweets. These set of users are classified using an existing machine learning model and then these users are incorporated into the model either by replacement, addition or any other insertion scheme(We have implemented the addition scheme in our application). At any point of time, the request for user classification may be received and the machine learning model available at that time is used to classify the user. The source code and data set used for developing the application are available online [6].

5 Conclusion and Future Work

In this study we present an efficient method of categorizing Twitter users based on their interests using Tweet-based, User-based and Time-series based features. We also compared several approaches for improving the performance of the classification task along with the effect of Principal Component Analysis on our feature space. The best approach categorizes Twitter users into six interest categories namely Politics, Journalism, Entrepreneurship, Entertainment, Science & Technology and Healthcare with 89.82% accuracy. We also propose an algorithm for developing a real-time Twitter user classification system. The application we developed replaces 25 new sample users in the old model periodically and can easily scale to cater to large number of users. For future work, we plan to propose a new framework that uses category-specific keywords and also incorporates the user's social network to improve the accuracy of classification. The Multi-label approach to user interest classification is also an area that can be focused to improve performance. We also plan to increase the number of interest classes to cover a wide range of dynamic user profiles without degrading the performance of the classification task.

References

1. Twellow. https://www.twellow.com/splash/ (accessed March 10, 2015)
2. Twitter. https://about.twitter.com/company (accessed March 10, 2015)
3. Twitter Streaming APIs. https://dev.twitter.com/streaming/overview (accessed March 10, 2015)
4. Weka 3: Data Mining Software in Java. http://www.cs.waikato.ac.nz/ml/weka/ (accessed March 16, 2015)
5. India to have third-largest Twitter population by 2014: eMarketer (2014). http://indianexpress.com/article/india/politics/india-to-have-third-largest-twitter-population-by-2014-emarketer (accessed March 10, 2015)
6. Github - Twitter User Categorization (2015). https://github.com/AKSHAYH/twitterusercategorization (accessed March 19, 2015)
7. An, J., Cha, M., Gummadi, P.K., Crowcroft, J.: Media landscape in twitter: a world of new conventions and political diversity. In: ICWSM (2011)
8. Bifet, A., Frank, E.: Sentiment knowledge discovery in twitter streaming data. In: Pfahringer, B., Holmes, G., Hoffmann, A. (eds.) DS 2010. LNCS, vol. 6332, pp. 1–15. Springer, Heidelberg (2010)
9. Bollen, J., Mao, H., Zeng, X.: Twitter mood predicts the stock market. Journal of Computational Science **2**(1), 1–8 (2011)
10. Chakrabarti, D., Punera, K.: Event summarization using tweets. In: ICWSM 2011, pp. 66–73 (2011)
11. De Choudhury, M., Diakopoulos, N., Naaman, M.: Unfolding the event landscape on twitter: classification and exploration of user categories. In: Proceedings of the ACM 2012 Conference on Computer Supported Cooperative Work, pp. 241–244. ACM (2012)
12. Kouloumpis, E., Wilson, T., Moore, J.: Twitter sentiment analysis: the good the bad and the omg! In: ICWSM 2011, pp. 538–541 (2011)
13. McCord, M., Chuah, M.: Spam detection on twitter using traditional classifiers. In: Calero, J.M.A., Yang, L.T., Mármol, F.G., García Villalba, L.J., Li, A.X., Wang, Y. (eds.) ATC 2011. LNCS, vol. 6906, pp. 175–186. Springer, Heidelberg (2011)
14. Medvet, E., Bartoli, A.: Brand-related events detection, classification and summarization on twitter. In: 2012 IEEE/WIC/ACM International Conferences on Web Intelligence and Intelligent Agent Technology (WI-IAT), vol. 1, pp. 297–302. IEEE (2012)
15. Michelson, M., Macskassy, S.A.: Discovering users' topics of interest on twitter: a first look. In: Proceedings of the Fourth Workshop on Analytics for Noisy Unstructured Text Data, pp. 73–80. ACM (2010)
16. Pennacchiotti, M., Popescu, A.-M.: Democrats, republicans and starbucks afficionados: user classification in twitter. In: Proceedings of the 17th ACM SIGKDD International Conference on Knowledge Discovery and Data Mining, pp. 430–438. ACM (2011)
17. Siswanto, E., Khodra, M.L.: Predicting latent attributes of twitter user by employing lexical features. In: 2013 International Conference on Information Technology and Electrical Engineering (ICITEE), pp. 176–180. IEEE (2013)

18. Siswanto, E., Khodra, M.L., Dewi, E., Joni, L.: Prediction of interest for dynamic profile of twitter user. In: 2014 International Conference of Advanced Informatics: Concept, Theory and Application (ICAICTA), pp. 266–271. IEEE (2014)
19. Thongsuk, C., Haruechaiyasak, C., Saelee, S.: Multi-classification of business types on twitter based on topic model. In: 2011 8th International Conference on Electrical Engineering/Electronics, Computer, Telecommunications and Information Technology (ECTI-CON), pp. 508–511. IEEE (2011)
20. Van der Maaten, L.J.P., Postma, E.O., van den Herik, H.J.: Matlab toolbox for dimensionality reduction. MICC, Maastricht University (2007)
21. Yang, T., Lee, D., Yan, S.: Steeler nation, 12th man, and boo birds: classifying twitter user interests using time series. In: 2013 IEEE/ACM International Conference on Advances in Social Networks Analysis and Mining (ASONAM), pp. 684–691. IEEE (2013)

Towards Development of National Health Data Warehouse for Knowledge Discovery

Shahidul Islam Khan and Abu Sayed Md. Latiful Hoque

Abstract Availability of accurate data on time is essential for medical decision making. Healthcare organizations own a large amount of data in various systems. Researchers, health care providers and patients will not be able to utilize the knowledge in different stores unless integration of the information from disparate sources is completed. Developing health data warehouse is a complex process and also consumes a significant amount of time but it is essential to deliver quality health services. In this paper the architecture of a data warehouse model and the development process suitable for integrating data from different healthcare sources have been presented. We have developed a Star schema suitable for large data warehouse. Integrating health data requires a rigorous preprocessing and we have completed the preprocessing of national health data by applying efficient transformation techniques. Finally the knowledge discovery potentials from the data warehouse are also presented with relevant examples.

1 Introduction

Health informatics is an intersection of computer science with health domain. It focuses on resources and techniques required to optimize the collection, storage, retrieval and exploitation of information in medical research. Its application areas are health care management, diagnosis, clinical care, pharmacy, nursing and public health [1, 2]. Knowledge discovery from data (KDD) is a major process to identify valid, new, potentially useful and eventually comprehensible patterns in data. Data mining, a foremost part in KDD, consists of applying data analysis and learning algorithms to produce potential interesting patterns over the data [3, 4, 5].

S.I. Khan(✉) · A.S. Md. Latiful Hoque
Department of CSE, Bangladesh University of Engineering and Technology (BUET), Dhaka, Bangladesh
e-mail: nayeemkh@gmail.com, asmlatifulhoque@cse.buet.ac.bd

S. Berretti et al. (eds.), *Intelligent Systems Technologies and Applications*,
Advances in Intelligent Systems and Computing 385,
DOI: 10.1007/978-3-319-23258-4_36

Health data means any data that is contained in patients' health record. There are several sources to collect these data such as doctor's prescription, notes from hospital admission or diagnostic test reports. Health data comes in diverse varieties such as text or numbers (e.g., Patient ID, demography, diagnostics data, etc.), analog or digital signals (ECG, ECHO, EMG, etc.), images (radiological, ultrasound, etc.), and videotapes. A sophisticated factor of the data is that patient identification information has to be removed from other parameters [2, 6, 7, 8]. Electronic Health Record (EHR) describes the diseases, diagnosis and treatments of patients. EHRs are usually stored in hospitals or clinics where those are produced. Patients are usually treated in several hospitals and clinics. [9].

One of the main Information Technology challenges in medical practice is how to integrate several distinct and isolated information stores into a single logical repository to create consistent information for all users. A huge amount of health records and related documents created by clinical diagnostic equipments are generated daily. These valuable data are stored in various medical information systems such as HIS (Hospital Information System), PACS (Picture Archiving and Communications System), RIS (Radiology Information System) in various hospitals, departments and diagnostic laboratories. Data required to make proper medical decisions are trapped within fragmented and heterogeneous health systems that are not properly integrated. As a result healthcare suffers because medical practitioners and health care providers are unable to access this information to perform activities such as diagnostics, and treatment optimization to improve patient care [1, 6, 7].

Successful healthcare data management is a vital issue to develop a support system for clinical decision-making process. Traditional operational database does not support critical data analysis tasks of the health care providers. It contains detailed transactional data but do not incorporate important historical data, and since it is highly normalized, it performs poorly for complex queries where joining of numerous tables is required to generate various clinical reports [7, 8].

A data warehouse (DW) is a subject-oriented, integrated, non-volatile, and time-variant collection of data to support management decisions [10]. It unites the data spread throughout an organization into a single central structure. DW may be considered as a *proactive* approach for information integration, as compared to the traditional *query driven* approaches where data processing starts when a query arrives. Its main advantages are standardizing data across organizations and improved turnaround time for analysis and reporting. Long initial development time and associated high cost are treated as the major drawbacks of DW [7, 11, 12].

A health data warehouse is a data store, different from hospitals' operational databases, used to analyze the consolidated historical health data [7, 8]. Development of a Health DW contains two key phases. Firstly, a conceptual view of the DW is specified according to the user requirements in the configuration phase. Secondly, the allied data sources and the Extraction-Transform-Load (ETL) process are determined. After the initial load, during DW operation, data must be refreshed on a regular basis such that data stored in the DW reflect the current state of the operational systems [5, 8].

This paper is intended to present the current scenario of healthcare data integration. Besides a large scale general platform data warehousing model suitable for integrating fragmented data is also described. Health data of various hospitals and clinics of Bangladesh are used in this research. The terms *Health*, *Pathological* and *Medical* are used for similar meaning.

The rest of the paper is organized as follows. In Section 2 we have presented selected literature reviews on DW, Health DW and KDD techniques. Section 3 describes briefly some design issues of National Health DW (NHDW). In Section 4 we have shown the calculation of approximate size of our DW. Some preprocessing techniques that we have used are illustrated in Section 5. Section 6 gives readers ideas about how our DW will be used for knowledge discovery and mining. Finally Section 7 concludes the paper.

2 Related Works

Mullins et. al. adopt data warehouses technology for health system development in [13], where the design experience in the University of Virginia Health System is reported. In their paper DW has been used to provide clinicians and researchers with direct, rapid access to desired patients' data.

So many constraints and issues complicated data mining performed for medical datasets i.e., the way the data is collected, accuracy and ethical, privacy and social issues [2]. Several researches were carryout to observe impact of missing values and noises and how this can influence the output. Zhu et al. classified noises into class noise and attributes noise. Attribute noise include incorrect attribute values, incomplete attributes, missing attribute values and don't care values [14].

Some researchers have focused on the techniques that are applicable for medical applications and have built in mechanism to handle noise and missing values. For example decision tree, logic programs, K-nearest neighbor, and Bayesian classifiers [15, 16]. Lee et al recommended that Bayesian networks and decision trees are the primary techniques applied in medical information systems [17]. It is stated that neural networks performed better than logistic regression [18]. Medical data mining applications should follow a five stage data mining development cycle: planning tasks, developing data mining hypotheses, preparing data, selecting data mining tools, and evaluating data mining results [19].

Faisal S. used multiple imputation technique to grip missing data in Pathology Databases in [20]. Optimizing public health data gathered for KDD using feature selection is studied in [21]. Cubillas et. al. proposed a model for improvement in appointment scheduling in health care centers [22].

Hoque et. al. conferred present structure of pathological data, requirements to formulate efficient models and the necessity to reform the present structure for predicative data mining in [23].

Neural Network is used for the diagnosis of diabetes by Kumari and Singh [24]. Yilmaz et. al. presented a modified K-means algorithm based data preparation method for diagnosis of heart and diabetes diseases in [25]. Latest research using Big Data tools to analyze Health Informatics is presented in [26].

3 Architecture of National Health DW

The architecture of NHDW is shown in Fig. 1. At the initial step health data from different governmental and private sources will be collected (such as hospitals, clinics, diagnostic centers, research centers etc.). These data will be integrated into a temporary repository using Extraction, Transform and Load process. Cleaning, noise reduction, normalization techniques are applied next. Then data are loaded into the DW. Knowledge discovery be performed afterward over NHDW [27].

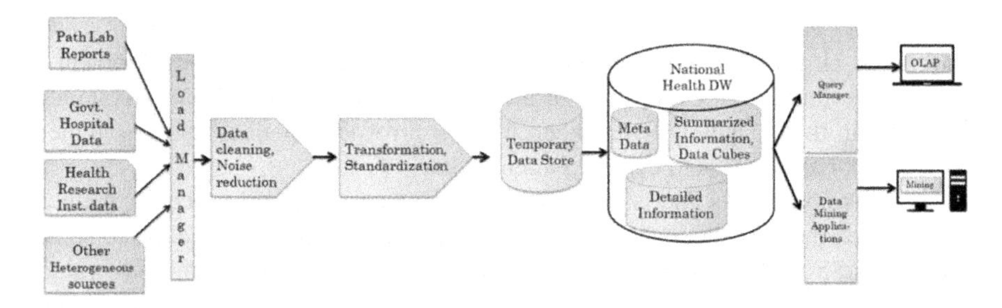

Fig. 1 Architecture of Health Data Warehouse

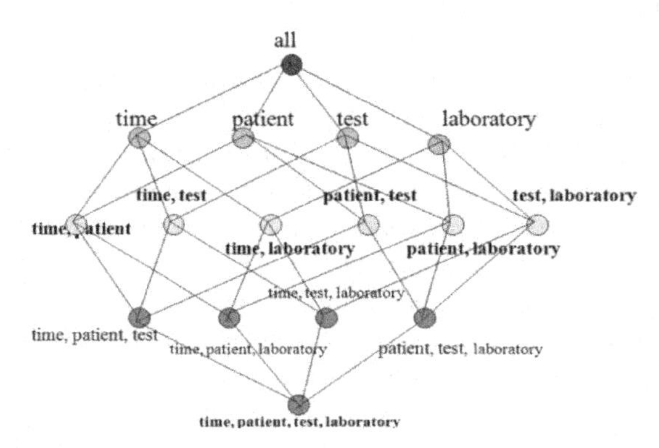

Fig. 2 Health Data Cube

Data cube used to design NHDW is shown in Fig. 2. 0-D apex cube (top red circle) provides highest level of summarization and 4-D base cube (bottom green circle) provides lowest level of summarization respectively. To reduce huge space requirement of full materialization, partial materialization is performed [9, 10].

There are many logical models like Star schema, Star Cluster schema, Snowflake schema, Fact Constellation schema etc. Star schema, Snowflake schema and Fact Constellation schema are widely used. Most suitable Logical Data Warehousing Model is the Star Schema for better Efficiency [9, 12, 13]. We have used Star Schema in our design shown in Fig. 3. The crow's feet at the end of the links indicate many-to-one relationship between the fact table and dimension tables.

By using the fact table and the available dimension tables, there are numerous ways of data aggregation. For analytical purpose of health data, frequently looked-for aggregated datasets have to be created in advance for the users to achieve better system performance. A major precept of data warehousing is to have data readily and easily available. Followings are some aggregated datasets for our DW:

- Patient count by Diagnosis, Gender, Age, Date
- Count of Procedures by Provider and Date
- Billing and discount information

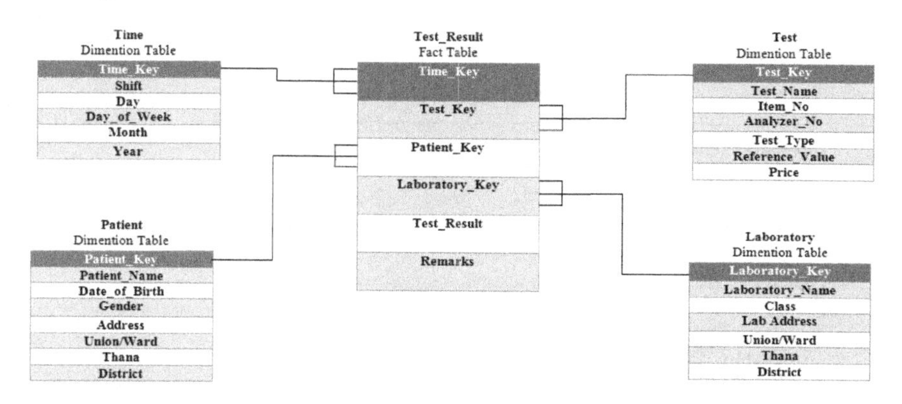

Fig. 3 Fact Table and Dimension Tables of NHDW

4 Evaluation of NHDW

4.1 Physical Size Estimation of NHDW

Total number of government hospitals under Directorate General of Health Services (DGHS) in Bangladesh is 717 [28, 29]. There are also 8,000 registered diagnostic centers under DGHS According to the Bangladesh Private Clinic and Diagnostic Owners Association (BPCDOA) data till December 2014 [30, 31]. Therefore currently there are more than 8717 authorized places where pathological tests are performed. For simplicity of calculation, if we consider average 500 patients' reports are produced every day, each report consists of 15 test attributes then putting the values in above equation we get:

T= 65377500 records (tuples /test attributes) per day.

Everyday 65 million records will likely be added in the fact tables of NHDW. If 0.25KB memory space is required to store a single record, then the DW will consume 16.25 GB memory/day. To manage this big data the Government should go for Cloud based storage and services [26, 32]. Cloud based large databases can be partitioned by available techniques such as [33, 34, 35].

4.2 Data Preprocessing

Preprocessing of data includes data cleaning, missing values imputation, normalization, transformation etc. As for NHDW, data comes from different public and private hospitals, diagnostic centers and other sources, different preprocessing steps have been performed on data. Table 1 presents a partial Random Plasma Glucose data using Min-Max Normalization technique and Fig. 4 illustrate the ratio of Glucose level among test samples.

Table 1 Metadata for Type_ID generation from Urine color

TESTRESULT_ID	Random Plasma Glucose (mmol/L)	Min-Max Normalized Value
114080000001077	8	0.28
114080000001362	5.5	0.19
114080000001955	9.3	0.33
114080000002314	17.2	0.61

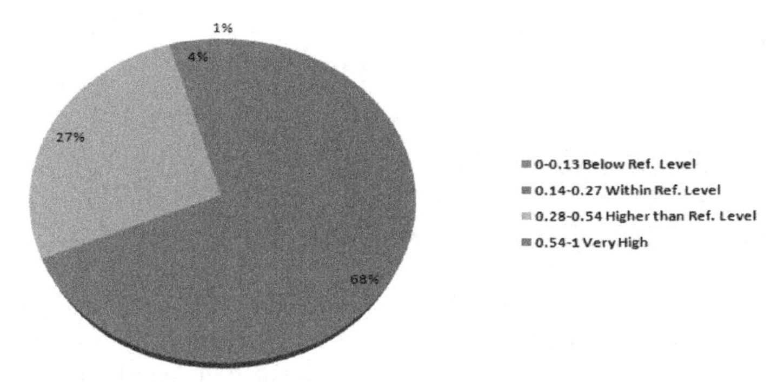

Fig. 4 Ratio of Random Plasma Glucose where 0.14-0.27 is the normalized reference range

4.3 Knowledge Discovery from NHDW Platform

Developing a common repository of nationwide health data has many ways to contribute the improvement of national health standard as well as knowledge discovery process. How healthcare providers and patients can be benefited from NHDW, are depicted in the following instance 1 and 2.

Instance 1: Temporal impact analysis

Using NHDW temporal pathological results of each citizen of the country can be tracked and recommendations can be provided. Table 2 presents the current guidelines by American Heart Association regarding Triglyceride levels [36].

Table 2 Triglyceride guidelines by American Heart Association

Triglyceride Level (mg/dl)	Interpretation
<151	Normal range
151-199	Slightly above normal
200-499	High
>499	Very high- High risk

For example, if Mr. X with date of birth 1st May 1985 has Triglyceride levels 160(mg/dl) on 1st January 2012, 185(mg/dl) on 1st January 2014 and 210(mg/dl) on 1st April 2015 respectively, the system can recommend whether this growth in his Triglyceride level is normal, risky or highly risky considering other 100 million available citizens data of the country.

Instance 2: Impact analysis of diagnostic tests

Using NHDW knowledge about impact of diagnostic tests on diseases detection can be discovered. So trend of performing unnecessary tests by some immoral doctors and diagnostic centers can be reduced. For example if the following association rule is derived from NHDW Support =70%, Confidence=97% for a Costly pathological test P_3:

$$\text{Patient\_Age}(X<18) \wedge (\text{Sex}=\text{'F'}) \Rightarrow \text{Negative}(X, P_3)$$

It can be said confidently that P_3 test has almost no impact of disease diagnosis for adolescent girls. So patients can be aware of these types of superfluous tests. In this way many other interesting patterns can be derived from NHDW by using various data mining algorithms like association or clustering.

5 Conclusions

Developing a DW by gathering huge data from available health databases ought to give easier and better access to data for health service providers, researchers and governmental authorities. In this paper we have presented the architecture of National Health Data Warehouse model and its developmental stages to process and analyze large-scale health data. We have developed a Star schema suitable for large data warehouse. We have shown some preprocessing of healthcare data by applying suitable transformation techniques. Finally the knowledge discoveries prospective from NHDW are also presented with instances. Considering quality factors Star schema is chosen to design DW. In order to get maximum benefit from the model presented in this research, there should be a cooperative mind among different health service providers to help the governmental bodies for the implementation of the DW platform by providing de-identified health data.

Acknowledgments This research is performed in Dept. of CSE, BUET and funded by the ICT Division, Ministry of Posts, Telecommunication and Information Technology, Government of the People's Republic of Bangladesh.

References

1. Roddick, J.F., Fule, P., Graco, W.J.: Exploratory medical knowledge discovery: experiences and issues. SIGKDD Explor. Newsletter **5**(1), 94–99 (2003)
2. Cios, K.: Uniqueness of medical data mining. Artificial intelligence in medicine. **26**, 1–24 (2002)
3. Fayyad, U.M., Shapiro, G.P., Smyth, P.: From Data Mining to Knowledge Discovery: An Overview. Advances in Knowledge Discovery and Data Mining, 1–36 (1996)
4. Khosla, R., Dillon, T.: Knowledge discovery, data mining and hybrid systems. In: Engineering Intelligent Hybrid Multi-Agent Systems, pp. 143–177. Kluwer Academic Publishers (1997)
5. Inmon, W.H.: EIS and the data warehouse: a simple approach to building an effective foundation for EIS. Database Programming and Design **5**(11), 70–73 (1992)
6. Stolba, N., Banek, M., Tjoa, A.M.: The security issue of federated data warehouses in the area of evidence-based medicine. In: First International Conference on Availability, Reliability and Security, ARES 2006. IEEE (2006)
7. Sahama, T.R., Croll, P.R.: A data warehouse architecture for clinical data warehousing. In: Australasian Workshop on Health Knowledge Management and Discovery, HKMD 2007 (2007)
8. Lyman, J.A., Scully, K., Harrison, J.H.: The development of health care data warehouses to support data mining. Clin. Lab. Med. **28**(1), 55–71 (2008)
9. Nugawela, S.: Data Warehousing Model For Integrating Fragmented Electronic Health Records From Disparate And Heterogeneous Clinical Data Stores, M.Sc. Thesis, Queensland University of Technology (2013)
10. Inmon, W.: Building the Data Warehouse, 4th edn., Wiley, New York (2005)
11. Jiawei, H., Micheline, K., Jian, P.: Data Mining Concepts and Techniques, 3rd edn., Elsevier (2012)
12. Kimball, R., Ross, M.: The Data Warehouse Toolkit: The Definitive Guide to Dimensional Modeling, 3rd edn., Wiley (2013)
13. Mullins, M., Siadaty, M.S., Lyman, J., et al.: Data mining and clinical data repositories: Insights from a 667,000 patient data set. Comput. Biol. Med. **36**, 1351–1377 (2006)
14. Zhu, X., Khoshgoftaar, T., Davidson, I., Zhang, S.: Special issue on mining low-quality data. Knowledge and Information Systems **11**, 131–136 (2007)

15. Brown, M.L., Kros, J.F.: Data mining and the impact of missing data. Industrial Management & Data Systems **103**, 611–621 (2003)
16. Lavrač, N.: Selected techniques for data mining in medicine. Artificial intelligence in medicine **16**(1), 3–23 (1999)
17. Lee, I.N., Liao, S.C., Embrechts, M.: Data mining techniques applied to medical information. Medical Informatics & the Internet in Medicine **25**(2), 81–102 (2000)
18. Obenshain, M.K.: Application of Data Mining Techniques to Healthcare Data. Infection Control and Hospital Epidemiology **25**(8), 690–695 (2004)
19. Wang, H., Wang, S.: Medical knowledge acquisition through data mining. In: IEEE International Symposium ITME (2008)
20. Faisal, S.: Missing Data in Pathology Databases. MSc Thesis, Australian National University (2011)
21. Partington, S.N., Papakroni, V., Menzies, T.: Optimizing data collection for public health decisions: a data mining approach. BMC Public Health **14**, 593–598 (2014)
22. Cubillas, J.J., Ramos, M.I., Feito, F.R., Ureña, T.: An improvement in the appointment scheduling in primary health care centers using data mining. J. Med. Syst., **38,** 89 (2014)
23. Hoque, A.S.M.L., Galib, S., Tasnim, M.: Mining pathological data to support medical diagnostics. In: Workshop on Advances on Data Management: Applications and Algorithms. Department of Computer Science and Engineering, BUET, Dhaka, pp. 71–74 (2013)
24. Kumari, S., Singh, A.: A data mining approach for the diagnosis of diabetes mellitus. In: IEEE 7th International Conference on Intelligent Systems and Control (2013)
25. Yilmaz, N., Inan, O., Uzer, M.S.: A New Data Preparation Method Based on Clustering Algorithms for Diagnosis Systems of Heart and Diabetes Diseases. J. Med. Syst. **38 (2013)**
26. Herland, M., Khoshgoftaar, T.M., Wald, R.: A review of data mining using big data in health informatics. J. Big Data **1**, 2 (2014)
27. Khan, S.I., Hoque, A.S.M.L.: Towards development of health data warehouse: bangladesh perspective. In: Proc. 2nd International Conference on Electrical Engineering and Information & Communication Technology (2015)
28. HEALTH BULLETIN, 2nd edn., DGHS, Ministry of Health and Family Welfare, Government of the People's Republic of Bangladesh (2014)
29. http://www.dghs.gov.bd/index.php/en/health-program-progress/hpnsdp-2011-16/84-english-root/ehealth-eservice/497-hpnsdp-2011-16-brief (Accessed February 20, 2015)
30. http://www.bpcdoa.com/clinics_and_diagnostics.html (Accessed February 22, 2015)
31. http://www.thefinancialexpress-bd.com/2014/12/15/71077/print (Accessed February 22, 2015)
32. Liang, Z., Sherif, S., Anna, L., Athman, B.: Cloud Data Management. Springer, Switzerland (2014)
33. Khan, S.I., Hoque, A.S.M.L.: A New Technique for Database Fragmentation in Distributed Systems. International Journal of Computer Applications **5**(9), 20–24 (2010)
34. Raouf, A.E., Badr, N.L., Tolba, M.F.: Dynamic distributed database over cloud environment. In: Hassanien, A.E., Tolba, M.F., Taher Azar, A. (eds.) AMLTA 2014. CCIS, vol. 488, pp. 67–76. Springer, Heidelberg (2014)
35. Harikumar, S., Ramachandran, R.: Hybridized fragmentation of very large databases using clustering. In: IEEE International Conference on Signal Processing, Informatics, Communication and Energy Systems (SPICES) (2015)
36. Triglycerides: Why do they matter? http://www.mayoclinic.org/diseases-conditions/high-blood-cholesterol/in-depth/triglycerides/art-20048186 (Accessed June 07, 2015)

Requirement of New Media Features for Enhancing Online Shopping Experience of Smartphone Users

Anuja Koli, Anirban Chowdhury and Debayan Dhar

Abstract Now-a-days consumers are using different social media (e.g. Facebook, WhatsApp etc.) to share the product quality information to take feedback about the product from their friends, family members, colleagues etc. The aim of this study was to found out the feasibility of implementation of different media features in e-retailing platform for taking feedback about the product to enhance the consumers' experience of online purchase. With this intention, user survey was conducted using a standardized questionnaire which includes items about the users' demographic information, likeliness to share product information for getting online feedback, probable acceptance of future online purchase system having new media features, willingness to use proposed online purchase system with new media features, priority to use specific media features for sharing product choice information through online. Results of the present study suggest that users would like to share product quality information with others through the proposed system which has product comparison screen share, personalized reviews and share and voice chat options in e-commerce app/ website. Therefore, these preferred media features may be integrated to create a better user experience in online purchase platform which in turn helps in quick product decisions during online purchase.

Keywords E-commerce · E-retailing · Decision making · Multi-media · Product choice

1 Introduction

Online shopping enable consumers to buy a range of products and services through the internet. At present online shopping has become hassle free since consumers, especially the working segment of the society, need not invest time and money in

A. Koli · A. Chowdhury(✉) · D. Dhar
Department of User Experience Design, MIT Institute of Design, Rajbaug,
Pune 412201, Maharastra, India
e-mail: chowdhuryanirban14@gmail.com

S. Berretti et al. (eds.), *Intelligent Systems Technologies and Applications*,
Advances in Intelligent Systems and Computing 385,
DOI: 10.1007/978-3-319-23258-4_37

going to the stores physically. Even it enables a lot of options to purchase a product. Online shopping is quick, easy and fastest growing segment of Indian economy that offers a wide variety of goods for purchase. It also engages consumers with a lot of exciting deals. Even, many e-retailers websites are very trustworthy as the money transaction is customer friendly. Even, in case of any defected item, it can be replaced and consumers duly return the defected product from their door step. Hence, educated people are observed to show preference to e-stores since a number of items are available within considerable price range.

Now, it is considered that use of smart phones are growing by both, the working and non-working population of India. Consumers can now shop from any place at any time of the day. Mobile commerce (or m-commerce) describes purchasing from an online retailer's mobile optimized online site or application. Yi-Fen and Yu-Chen (2014) said that online shopping through mobiles will become one of the services with a vast development potential and explored the influence of various factors using a technology acceptance model [4].

A study was done by Diaz (2014) to state that there are different influential factors for social networking platforms which can be used to find the right stimuli to induce customers into purchase [7]. With the development of myriad of shopping portals (such as Amazon, Ebay, Flipkart, Jabong, Yepme, Myntra and many more), the competition has also increased manifold among e-retailers. E-retailers are now employing social media such as Facebook, Twitter and even instant messengers to publicize and attract consumers towards their shopping portals so that they can retain their market share. Though these features are effective to some extent, currently, there is no online shopping platform which has social networking sites like features such as sharing product quality information with others (friends, family members etc.). It is questionable that chance of improvement of user experience of online shopping through implementation of media features for sharing product quality information. Users may or may not accept this kind of new media features for sharing the product quality information. Therefore, the present paper aims to evaluate the user experience aspects of such different social media features which can be integrated into online shopping platform to share product quality information. In this paper, the user study was conducted with an intention to improve the user acceptance of online shopping through new media-features. Before conducting the actual study, it is better to look into some previous literature regarding the users' experiences, in the context of online shopping.

2 Literature Review

2.1 Factors Influencing Online Shopping

There have been numerous studies on shopping online and the factors influencing it. Van der Heijden et al. (2003) showed that perceived risk and perceived ease-of-use of technology have a direct effect on the attitude towards online shopping. Stages like need recognition, pre-purchase search, the evaluation of alternatives, actual purchase and post purchase evaluation are differentiated in their frameworks [23]. Hong et al. (2014) reported that emotional trust plays a greater role than cognitive trust in influencing consumers' intention to purchase online [25]. It can be considered while designing ecommerce portals in future. Cynthia et al. (2003) presented a model that identified three perceptual factors that impact on-line trust: perception of credibility, ease of use and risk [6]. Mauricio and Paul (2003) found that performance based risk perceptions and perceived ease of use of the e-service reduces risk concerns affect e-services adoption adversely [9].In results of a study by Escobar and Carvajal (2014) on Spanish consumers of LCC flights, it is derived that key determinants of purchasing are trust, habit, cost saving, ease of use, performance and expended effort, hedonic motivation and social factors [8]. However, the review conducted by Ton et al. (2004) showed that attitudes toward online shopping and intention to shop online are not only affected by ease of use, usefulness, and enjoyment, but also by exogenous factors like consumer traits, situational factors, product characteristics, previous online shopping experiences, and trust in online shopping [20]. Ashraf et al. (2014) developed an extended technology acceptance model that incorporates importance of trust and perceived behavioral control, the perceived ease of use and perceived usefulness on consumers' intentions to shop online which was validated across various cultures [3]. Each and every above mentioned factor may have some effect on online purchase decision; but, Chu et al. (2014) in their study concluded that combining multiple persuasive means lead to greater persuasiveness than using any of them individually. On the whole, the persuasive power derived from the persuasive means that appeal to users' emotions and the credibility of the product pages is greater than that derived from the persuasive mean that appeals to logic [5]. From the above discussion it is reflected that most of the studies on online shopping experience considered either usability related factors or trust and other psychosocial factors to explain online shopping experience. However, studies are less considering the social networking related media features in the context of online purchase experience.

2.2 User Experience in Online Shopping: Role of Social Media Assistance

According to Almeida et al. (2014), Web 3.0 enables people and machines to connect, evolve, share and use knowledge on a greater scale and make Internet experience better in new ways [1]. These researchers predicted that stores will also change over the next twelve years more than the people have over the past five decades. Reynolds et al. (2014) argued that digital technologies could be tools that affect attitudes towards a shopping centre in a positive way, and the use of digital devices could make the overall experience better [22]. Hajli (2014) found that the users' trust is affected positively by forums and communities leading to a more definite intention of consumers to buy online [11]. Similarly, Hassanein and Head (2007) stated that higher levels of perceived social presence have positive impact on the perceived usefulness, trust and enjoyment of shopping websites, leading to positive consumer attitudes [13].

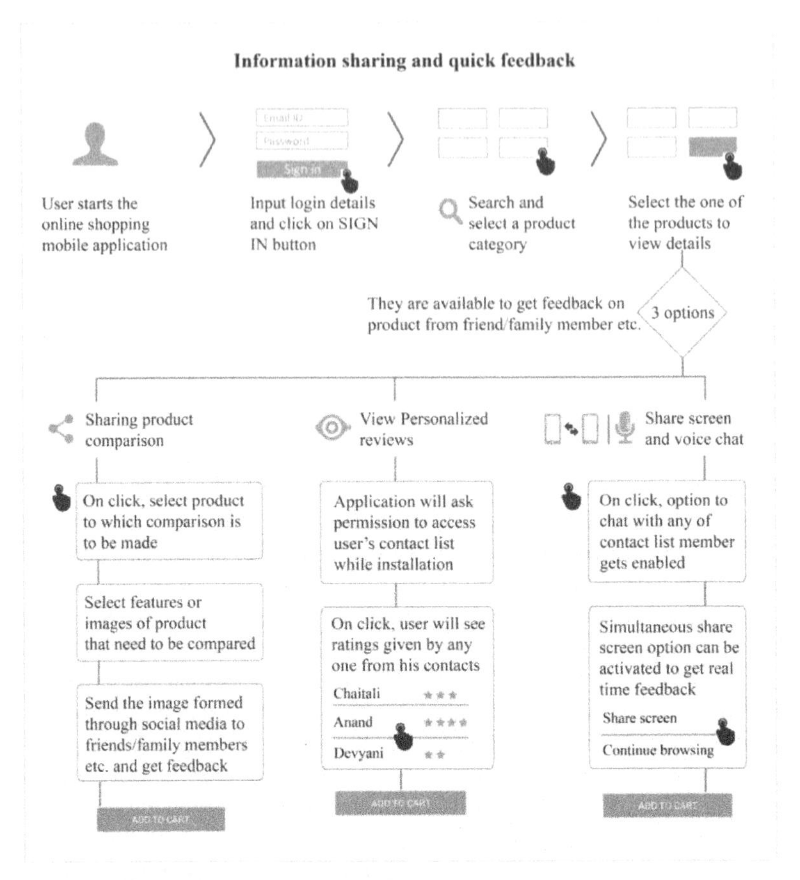

Fig. 1 Information flow of the proposed new media features

Now-a-days it is observed that people are interested in collaborative online shopping with their friends, family members etc. to have a better experience. Hassanein and Head (2005) showed in their study that in an online apparel shopping domain, increased levels of social presence through socially-rich descriptions and pictures positively impact consumer attitude [12]. Ho and Vogel (2014) confirmed that trust can be obtained through social networking sites in e-commerce through an interactive and user-centered environment [14]. Huang and Benyoucef (2014) highlighted the need of investigation of user preferences of social features implemented in e-commerce websites currently [15]. Antunes et al. (2014) have described an application framework consisting of extensive usage of visual elements and gestures and independence from specific decision-making methods, processes and tasks which is supporting collaborative handheld decision-making [2]. Further, different types of information commonly used by group-buying sites induce product purchase. It was established by Kuan et al. (2014) that the "like" information on e-retailers' website has a positive influence on consumer intention [17].

The challenge that e-retailers faced was to provide a better shopping experience than the traditional way of shopping. Features of traditional way of shopping like touch and feel factor, trust factor and decision making in groups which are not matching with the online shopping experience. Over the years with the increase in number of consumers there has been significant rise in such demands as well. So, to bridge the gap between traditional stores and online stores, features like 'product reviews' were made available to help consumers choose the desired product though the authenticity of these reviews is a question yet to be answered. The user experience of the retailer's website will thus help engage more consumers and enhance relationship between consumers and e-commerce websites. It was highlighted by Lim (2014) that one of the new and under-researched areas of electronic commerce is online group buying phenomenon [18]. Thus, there is a scope of the study of enhancement of online shopping experience of users through various social networking related media features. Inaddition, Ganesh and Agarwal (2014) reported that friend referral is also considered by online shoppers while choosing a particular brand [10]. Thus, another feature added recently is 'share' wherein consumer can share the product quality information through various popular media like Facebook, Whatsapp, email, Wechat, bluetooth etc. This feature plays a major role in getting feedback from friends, family members, relatives, colleagues and anyone who the consumer trusts when it comes to shopping goods/expensive goods in particular. In the current scenario only the link of that particular product page gets shared due to which the time spent between the consumer viewing the product and then actually buying it is much more than expected. Thus, the present study was focused on providing a viable option to enable consumers to take quick decisions thus enhancing online shopping experience from user experience perspective. Some new features like personalized reviews, sharing product comparison and screen share and voice chat can be tested on the application to make it even more personalized, engaging and real time. If those features are good enough then e-retailers can incorporate there features in their websites. Therefore, in the present study an attempt has been made to evaluate the feasibility of incorporation of different media features for taking feedback about the product, to enhance the consumers' experience of online purchase.

3 Methods

3.1 Participants

This study recruited 50 participants in which 64% users were below 25 years of age, 26% users were in 25 to 40 years age group and the remaining 10% were from 40 to 55 years age group. Around 88% participants said that they are playful with smartphones, 6% were not sure and the remaining 6% said they are not playful with the device. Maximum participants were observed to be playful, flexible and creative with their smartphones but there is no significant difference [Table 1]. Around 64.0% participants said they are likely to buy online, 10.0%

were not sure while 26.0% said they are not likely to buy online. Total 40.0% participants said they need help from friends while shopping online, 24.0% need help from family members, 2.0% from relatives and colleagues each while 30.0% said they don't need any help while shopping online. Therefore a majority of participants were found to need help from their near and dear ones while shopping online. Majority of people among participants (78.0%) use WhatsApp followed by Facebook, We-chat and others.

3.2 Description about Proposed System with New Media Features

Voice chat and screen sharing, sharing product comparison and personalized reviews are the three features which are proposed in this study to observe the feasibility to implement these features in the current scenario of online shopping experience on smartphones in India.

Voice chat and screen sharing feature was proposed for the online product purchase website/ application so that users can share their mobile screen with friends, family members or relatives keeping the control with themselves showing them the product page which they are viewing and simultaneously chat with them to get feedback on the same. It would enable the users to get real time feedback and thus take quick decisions while shopping online. Sharing a snapshot of the selected product comparison features will help consumers to take decisions quickly when it comes to expensive products. Personalized reviews are reviews given by people from the contact list of user's mobile phone that can be tracked by. So the user will know whom to contact so as to get the best possible feedback. They will give appropriate and trustworthy feedback to help consumers take quick and efficient decisions. Fig. 1 is presenting the information flow of the proposed media features for online shopping.

3.3 Measures for User Survey

First consumer behavior was studied through contextual enquiry while shopping in traditional stores as well as online stores was observed. Based on the information retrieved, a questionnaire was designed and reframed so as to collect quantitative data. With the intention to know the feasibility of incorporation of different media features for taking feedbacks about the product; likeliness to share product information for getting online feedback, probable acceptance of future online purchase system having new media features, willingness to use proposed online purchase system with new media features, priority to use new media for sharing product choice information through online, priority ratings of each media features were studied along with demographic information of participants. Items of the questionnaire are presented in Table 1.

3.4 Procedure

Participants were recruited for taking responses about their perception on variables (smartphone playfulness, likeliness to purchase online, likeliness to share product information for online feedback, priority to use new media features for sharing product choice information through online medium, willingness to use proposed online purchase system with new media features and Probable acceptance of future online purchase system). The questionnaire was presented on computer screen (Dell Inspiron 15 system; OS: Windows 7^{TM}, resolution: 1366 x 768, 60p Hz) and data were collected through Google Docs. The collected data were analyzed using SPSS20.0. The frequency distribution of scores of variables showed a mix of normal and non-normal distribution on histograms. Hence non-parametric data analysis is followed. Only one group of people is involved due to which CHI square test – Goodness of fit test is performed on the six variables.

Table 1 Items used in the questionnaire

Variables and scale items	
Mobile playfulness (7 point Likert scale used) [24]	
MP1	I am "Playful" when using smartphone
MP2	I am "Flexible" when using smartphone
MP3	I am "Creative" when using smartphone
Likeliness for online purchase (7 point Likert scale used)	
LOP1	I would like to experience online product purchase
LOP2	I would like to purchase different products from online stores
Likeliness to share product information for online feedback (7 point Likert scale used)	
LSPI1	I would like to discuss about features of product before shopping online
LSPI2	I would like to discuss about looks of product before shopping online
Priority to use new media for sharing product choice information through online (7 point Likert scale used)	
PNM1	Written chat
PNM2	Voice chat
PNM3	Photo share
PNM4	Screenshot
PNM5	Phone call
Willingness to use online purchase system with new media assistance (7 point Likert scale used)	
WOPS1	It would be helpful if I could discuss about product through sharing product comparison information before shopping online
WOPS2	I would like to discuss about product before shopping online through voice chat and screen sharing
WOPS3	I would be pleased to discuss about product before shopping online through sharing product photo or screenshots

Table 1 (*Continued*)

Variables and scale items	
Probable acceptance of future online purchase system (7 point Likert scale used)	
PAPS1	It would be more usable if I could get feedback on product quickly and easily (in real time)
PAPS2	I am willing to use the system which has share screen and voice chat, 2 and personalized reviews
Product quality evaluative assistance (Multiple choice questions)	
PQA1 Do you need any help from the following people for decision making while shopping online?	
Options: 1. Friends; 2. Family members; 3. Relatives; 4. Colleagues; 5. No one	
PQA2 How do you like to shop?	
Options: 1. Always go with friend; 2. Always go with family; 3. Always go alone; 4. Sometimes with friend; 5. Sometimes with family	
Type of online platform for sharing product quality information (Multiple choice questions)	
TOP1 Do you use any kind of platform to get friend's opinion or product review while shopping online?	
Options: 1. Facebook; 2. Whatsapp; 3. Wechat; 4. Text message; 5. E-mail; 6. None	
TOP2 If none, did you know such options are available to get feedback on product from friends/family?	
Options: 1. Yes; 2. No	
TOP3 Do you want to have a feature to get product review or feedback from friends/ family members/ colleagues?	
Options: 1. Yes; 2. No	
Priority ratings of each media features for sharing product choice information through online (7 point Likert scale used)	
PNM1	Written chat
PNM2	Voice chat
PNM3	Photo share
PNM4	Screenshot
PNM5	Phone call

4 Results

Likeliness to share product information or online feedback, probable acceptance of future online purchase system, willingness to use online purchase system with new media assistance and priority to use new media for sharing product choice information through online medium were the key factors studied through the questionnaire. Scale data was collected with regards to all the four factors.

4.1 User's Acceptance of Futuristic Online Purchase System

4.1.1 Likeliness to Share Product Information for Online Feedback

Significantly more people (86.0%) said that they will like to share product information for online feedback as compared to people who are not sure about their likeliness (2.0%) and the people who don't like to do the same (12.0%) [Please see Table 2]. Therefore, more number of people wanted to get feedback on looks and features of the product while shopping online. Total 78.0% of the participants prefer to get feedback from friends, family members, colleagues etc. using Whatsapp followed by Facebook, messaging, mail, Wechat and others.

3.1.2 Priority to Use New Media Features for Sharing Product Information

All participants of present study wanted to use new media for sharing product choice information through online medium [please see Table 2]. It was also found in this study that sharing of product comparison was the most preferred feature of the participants (56.0%) followed by getting personalized reviews (52.0%). It is followed by voice chat and screen sharing (40.0%) and others (4.0%).

3.1.3 Willingness to Use the Proposed Online System

About 94.0% people said that they are willing to use online purchase system with new media assistance, 2.0% are not sure while 4.0% are not willing to use it. However, significantly more participants are willing to use the proposed system with new media features (share product comparisons, personalized reviews, share screen and voice chat) [Please see Table 2]. Around 90.0% people said that they want more features to get that feedback. Most of the participants (84.0%) said that it would be more usable if they could get feedback on product quickly and easily i.e. in real time, 4.0% are not sure while 12.0% don't want the feedback.

3.1.4 Probable Acceptance of Future Online Purchase System

A total of 84.0% participants said that they will accept proposed online purchase system including sharing product comparison, personalized reviews and screen share and voice chat features. About 4.0% participants are not sure about it while 12.0% will not like to use it. Therefore, significantly more number of people is willing to accept new media in the current scenario [Please see Table 2].

Table 2 Important factors for enhancing online purchase experiences.

Variable name	Value < 4	Value = 4	Value > 4	χ^2 (df)	Sig. level	Effect size
Likeliness to share product information for online feedback	6(12%)	1(2%)	43(86%)	47.68 (10)	<0.001	0.954
Priority to use new media features for sharing product choice information through online medium	0	0	50(100%)	33.6 (18)	0.014	0.672
Willingness to use proposed online purchase system with new media features	2(4%)	1(2%)	47(94%)	26.2 (14)	0.024	0.524
Probable acceptance of future online purchase system	6(12%)	2(4%)	42((84%)	42.4 (10)	<0.001	0.848

5 Discussion

Now-a-days people have developed a strong liking and attraction towards social media. In the present study it was found that about most of these participants need help from friends, family members, relatives and colleagues through those media for taking a decision while shopping online. It was also observed that maximum participants were likely to discuss about features of product specially the appearance of the product before shopping online. Later Ho and Vogel (2014) confirmed that trust can be obtained through social networking sites in e-commerce through an interactive and user-centered environment [14]. Huang and Benyoucef (2014) highlighted that the need of investigation of user preferences of social features implemented in e-commerce websites currently [15]. Therefore, the India consumers would like to share product information with their friends, family members, relatives and colleagues through those media before they buy products online because trust may still be an issue in this context. Another reason for likeliness to use the proposed system as taking one of the proposed opinions can assist them to come to a decision quickly and easily if it is an expensive product. It was also observed that maximum number of participants was aware of platforms which are available to get friend's opinion or product review while shopping online. Whatsapp being the most popular messaging application these days was found to be used by 78.0% of the participants to get feedback from friends, family members, colleagues etc. followed by Facebook, messaging, mail, Wechat and others. These reasons actually explains the findings of the present study that states significantly most of the online buyers (86.0%) would like to share product information for online feedback before taking product decision during online purchase.

All participants of present study wanted to use new media for sharing product choice information through online medium because it was observed that maximum participants (92.0%) desired to get new features for getting product review or feedback from friends/family members/colleagues. It was also found in this study that sharing of product comparison was the most preferred feature of the participants (56.0%) followed by getting personalized reviews (52.0%) since personalization was found to be valuable today [16]. Voice chat and screen sharing (40.0%) was third preferred option for sharing product choice information.

In this study, most of the participants (90.0%) want more features to get that feedback. Maximum participants of the present study want easy and quick product feedback because the time gap between the consumer viewing a product and then actually buying it after getting feedback is considerably high at present. The consumer would be enabled to take quick decisions using the proposed system. These may be the reasons for willingness to use the proposed online system. Literature also supports this fact as it was stated that more personalized features enhanced the shopping experience for the consumers shopping online [19].

In the present study, share screen and voice chat, share product comparison and personalized reviews were the proposed features. Significantly more participants (84.0%) said that they will accept proposed online purchase system having these aforesaid features as it will enhance users/ consumers experience for easy product review in collaborative and reliable online shopping environment.

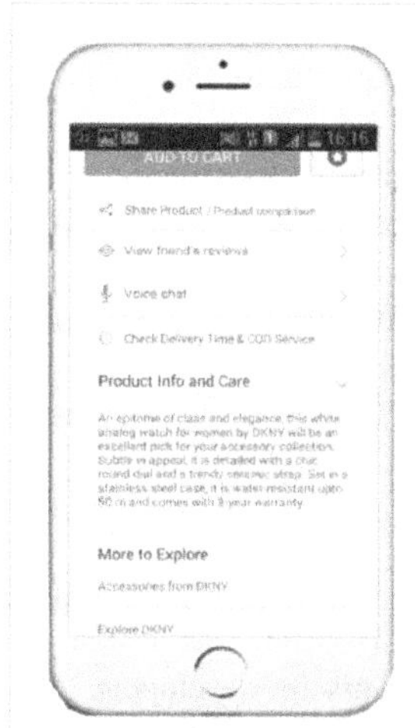

Share product comparison
This feature will enable the users to share selected options, images or information of two or more products with friend, family member etc. through social media applications available currently. It gives the user the liberty to share only that part of information on which he/she wants feedback.

Personalized reviews
This is one of the feature embedded in the shopping application that is considered in this study. These are reviews given by people from user's smartphone contact list. Thus reviews can be viewed increasing personalization. These are basically friend's /colleague's views on particular product.

Share screen and voice chat
Voice chat is another embedded feature which will enable the users to talk with any of their smartphone contacts while browsing through the shopping application. It will also provide an option to share screen with that contact so that a friend /family member can view what the user is viewing and get real time feedback.

Fig. 2 Screenshots of proposed media features

6 Conclusion

Currently, online consumers are sharing product information for getting feedback about the product separately through using social networking sites. No e-retailers' website/ application have their own product information sharing media which is providing the aforesaid / proposed facilities together though there are needs of this kind of facilities by users. Therefore, this kind of system features may be integrated in e-retailer's websites/ applications for providing better user experience. It was stated by Hassanein and Head (2005) that the increased levels of social presence through socially-rich descriptions and pictures positively impact consumer attitude [12]. In 2007, further they reported that higher levels of perceived social presence have positive impact on the perceived usefulness, trust and enjoyment of shopping websites, leading to positive consumer attitudes [13]. Thus, the proposed kind of system features will be helpful for e-retailers to attract their consumers and to build trust among users as they are getting feedbacks about products from their known persons (family members, colleagues, friends etc.) who already purchased/ used the product. Fig. 2 highlighted about the new media features which are preferred by consumers for group purchase decision while shopping online. Proposed new media features can also be used in similar fashion in other context when information sharing and online real time feedbacks are required to take a group decision.

This study was mainly conducted to check the feasibility of new media in online shopping environment. Though, inclusions of new media features are feasible according to the current results, still the effect of other social factors (such as age gender, educational qualification, culture etc.) on adaptation of the proposed media features need to be studied in the context of online shopping.

References

1. Almeida, F., Santos, J.D., Monteiro, J.A.: E-commerce business models in the context of web3.0 paradigm. Int. J. Adv. Info. Tech. **3**(6), 6102 (2014)
2. Antunes, P., Zurita, G., Baloian, N.: An application framework for developing collaborative handheld decision-making tools. Behaviour & Information Technology **33**(5), 470–485 (2014)
3. Ashraf, A.R., Thongpapanl, N., Auh, S.: The Application of the Technology Acceptance Model Under Different Cultural Contexts: The Case of Online Shopping Adoption. Journal of International Marketing **22**(3), 68–93 (2014)
4. Chen, Y.F., Lan, Y.C.: An empirical study of the factors affecting mobile shopping in Taiwan. Int. J. Tech. Hum. Interaction **10**(1), 19–30 (2014)
5. Chu, H.L., Deng, Y.S., Chuang, M.C.: Investigating the Persuasiveness of E-Commerce Product Pages within a Rhetorical Perspective. Int. J. Business and Manage. **9**(4), 31 (2014)
6. Corritore, C.L., Kracher, B., Wiedenbeck, S.: On-line trust: concepts, evolving themes, a model. Int. J. Hum.-Comput. St. **58**(6), 737–758 (2003)

7. Diaz, E.: Factors that influence Impulsive Buying on Social Networking Sites Platforms: The Case of Facebook for Social E-commerce. Dissertation, International Management Institute, Success University (2014)
8. Escobar-Rodríguez, T., Carvajal-Trujillo, E.: Online purchasing tickets for low cost carriers: An application of the unified theory of acceptance and use of technology (UTAUT) model. Tourism Management **43**, 70–88 (2014)
9. Featherman, M.S., Pavlou, P.A.: Predicting e-services adoption: a perceived risk facets perspective. Int. J. Hum.-Comput. St. **59**(4), 451–474 (2003)
10. Ganesh, L., Agarwal, V.: E-shopping: An Extended Technology Innovation. J. Res. Mark. **2**(1), 119–126 (2014)
11. Hajli, M.N.: Social Commerce for Innovation. Int. J. Innovation Mgt. (2014). doi:10.1142/S1363919614500248
12. Hassanein, K., Head, M.: The impact of infusing social presence in the web interface: An investigation across product types. Int. J. Elect. Com. **10**(2), 31–55 (2005)
13. Hassanein, K., Head, M.: Manipulating perceived social presence through the web interface and its impact on attitude towards online shopping. Int. J. Hum.-Comput. **65**(8), 689–708 (2007)
14. Ho, R., Vogel, D.: The impact of social networking functionalities on online shopping: an examination of the web's relative advantage. Int. J. Business Information Syst. **16**(1), 25–41 (2014)
15. Huang, Z., Benyoucef, M.: User preferences of social features on social commerce websites: An empirical study. Technol. Forecast. & Soc. Change (2014). doi:10.1016/j.techfore.2014.03.005
16. Kobsa, A.: Privacy-enhanced web personalization. In: Brusilovsky, P., Kobsa, A., Nejdl, W. (eds.) The adaptive web. Springer Verlag, Heidelberg (2007)
17. Kuan, K.K., Zhong, Y., Chau, P.Y.: Informational and normative social influence in group-buying: Evidence from self-reported and EEG data. J. Manage. Information Syst. **30**(4), 151–178 (2014)
18. Lim, W.M.: Sense of virtual community and perceived critical mass in online group buying. J. Strategic Mark. **22**(3), 268–283 (2014)
19. Nah, F.F.H., Davis, S.: HCI Research issues in Electronic Commerce. J. Electronic Commerce Research **3**(3), 98 (2002)
20. Pereay Monsuwé, T., Dellaert, B.G., De Ruyter, K.: What drives consumers to shop online? A literature review. Int. J. Serv. Ind. Manag. **15**(1), 102–121 (2004)
21. Ravasan, A.Z., Rouhani, S., Asgary, S.: A Review for the Online Social Networks Literature (2005-2011). European J. Business and Management **6**(4), 22–37 (2014)
22. Reynolds, J., Sundström, M.: Digitalisation, retail transformation and change: what will European consumers want from their future shopping centre experience? In: Proc: The 4th Nordic Retail and Wholesale Conference Hosted by Center for Retailing, Stockholm School of Economics, Sweden, 5–6, 2014. http://hdl.handle.net/2320/14372
23. Van der Heijden, H., Tibert, V., Marcel, C.: Understanding online purchase intentions: contributions from technology and trust perspectives. Eur. J. Inform. Syst. **12**, 41–48 (2003)
24. Webster, J., Martocchio, J.J.: Microcomputer Playfulness: Development of a Measure with Workplace Implications. MIS Quart **16**, 201–226 (1992)
25. Zhang, H., Lu, Y., Gao, P., Chen, Z.: Social shopping communities as an emerging business model of youth entrepreneurship: exploring the effects of website characteristics. Int. J. Technol. Manage. **66**(4), 319–345 (2014)

Formal Architecture Based Design Analysis for Certifying SWS RTOS

Yalamati Ramoji Rao, Manju Nanda and J. Jayanthi

Abstract In recent times Formal Techniques have been strongly recommended in the engineering life-cycle of safety -critical systems. With this, Architecture Analysis & Design Language (AADL) is a widely spectrum accepted architecture modeling language that can be wrap with Formal Modeling techniques, that proficiently helps in the design of a safety-critical system and circumscribes various analytical features for modeling the hardware and software architecture/s, against the required as per the guidelines set aside in RTCA DO-178C (333- Formal Based Modeling). This paper discusses the use of architecture modeling language along with formal based techniques for the analysis of RTOS architecture which is important in the correct implement of the given requirements. The architecture of the RTOS is expressed and analyzed using AADL. A suitable case study such as Stall Warning System/Aircraft Interface Computer (SWS/AIC), RTOS scheduler is modeled and analyzed. The analysis of results are mapped to the workflow prescribed in RTCA DO-178C for generating the certificate artifact and establishing the effectiveness of architecture based design analysis in the software engineering process.

Keywords Safety-Critical system · Multi-function · RTOS · Formal method · Architecture analysis & design language (AADL) · SWS/AIC · Certification artifacts

1 Introduction

Avionics architecture means set of various electronics operated in aircrafts. Since 70's avionics architecture is composed of several digital and communication

Y.R. Rao
Institute of Science and Technology, JNTUK-Kakinda University,
Kakinada, Andhrapradesh, India
e-mail: ramojirao999@gmail.com

M. Nanda(✉) · J. Jayanthi
CSIR- National Aerospace Laboratories, Bangalore, India
e-mail: {manjun,jayanthi}@nal.res.in

S. Berretti et al. (eds.), *Intelligent Systems Technologies and Applications*,
Advances in Intelligent Systems and Computing 385,
DOI: 10.1007/978-3-319-23258-4_38

systems which support more and more avionics applications such as flight controls, and flight management system's to name a few. Over the years as the demand is increasing, the avionics system architecture is becoming complex and central component of an aircraft. In case of civil aerospace 30 to 40 % of development cost is due to avionics system whereas in case of military it goes to 40 to 50% [1]. The higher cost of avionics system is because it involves more R&D work for each and every system that has been installed in aircraft. This cost can be reduced if we analyze the system earlier in the phase i.e. before the design level which is the Architecture level. Architectural analysis not only helps in detecting the design flaws earlier but it also helps in understanding the system requirements.

RTOS plays a critical role in realizing the real time functionalities of the Safety-Critical Systems. The scheduling feature of the RTOS provides the timeliness of the embedded systems. In this paper, we propose an approach to analyze the architecture of the RTOS. Analysis of RTOS helps in identifying potential risks of the scheduler against the requirements [2]. Developing a reliable RTOS architecture for a complex system is a critically important step for ensuring that the system satisfies its principal objectives. Examples of safety critical RTOS includes QNX, VxWorks, DEOS, and LynxOS [3].

The paper provides an approach to analyze the RTOS scheduler. This is done by means of a case study of the SWS/AIC RTOS scheduler. The RTOS scheduler is modeled using the AADL using the OSATE plug-in in Eclipse IDE. The scheduler timeliness, processor utilization and response time features are analyzed as per the project requirements. The analysis results are mapped to the RTCADO-178C/ DO-333 workflow to establish certification of RTOS using formal methods.

2 Literature Survey

2.1 Complex Safety-Critical Systems

Advanced avionics systems are computer centric systems which are used to achieve the desired functionality. Failure in these systems endangers the human life and large scale of economics [5][6]. Functionality within the avionics software continues to expand. Additional software capabilities bring many more lines of code, and generate opportunity for error. As the complexity requirements and criticality of avionics software grow, innovative tools are used to test, verify and secure the architecture of such systems." Safety-Critical Systems has to be bug free, and it has to work according to specifications", that's where the FAA's RTCA DO-178B(C) standards comes in. Static analysis software tool analyze source code to derive properties that can helps in detecting the errors that might not be apparent to the programmer, while dynamic analysis tool helps in showcasing which code is executed by the test suite. "All have their place in safety critical development", for a complex safety critical avionics software, the requirements for testing and validation are higher than most of other software [7][8][9][10][11].

2.2 RTOS

Real-Time Operating Systems (RTOS) are required to provide a predictable platform for the execution of multiple software tasks on single microprocessors.Every RTOS has a kernel, which serves as its core. The core is surrounded by shell layers, which provide protection and access authorizations. The RTOS manufacturer provides a set of APIs, which are used to access this kernel to perform tasks. The kernel contains a scheduler to execute these threads in a sequential manner. RTOSs are of three types; Hard real-time, Firm real-time and Soft real-time. In a hard real-time system the tasks and interrupt requests must be completed within their required deadlines. Failure to meet a single deadline may lead to a critical catastrophic system failure. Firm-real time system tolerates the missing of a few deadlines; however, missing more than a few may lead to complete disaster. A soft real-time system is one in which deadlines are mostly met, but this constraint is not very stringent. RTOS must have sufficient number of priority levels and must avoid priority inversion. RTOS Scheduler is the most critical component as it ensures the timeliness of the tasks as per the requirements [3].

2.3 Formal Methods

Nowadays the advantages of formal methods for the development and certification of safety-critical software are widely recognised by both Software Engineering community and standard organisations [18]. Some safety critical domains specific standards explicitly either highly recommend or mandatory require the use of formal methods. The specific benefit recognised to formal methods is that they allow "complex behaviour" to be analysed (by means of proofs or state exploration), reviewed, and analysed in their totality, rather than merely sampled as by testing or simulation. Thus, the major benefit derives from a double application of formal methods that is by formal requirements specification coupled with formal verification. From the certification point of view, formal methods are recognised to increase the degree of confidence in achieving software of high integrity levels [23]. Formal methods that are able to address real-time issues offer a unique opportunity for the specification of such time critical operating systems[20].The advantage of using formal methods in avionics software is because of

- *Reduced Cost* : Early detection/Elimination of defects
- *Increase Confidence*: Complete examination of models and requirements
- *Satisfy certification Objectives*: RTCA DO-178C [19]

3 Proposed Methodology

3.1 RTOS Kernel Overview

In this paper we propose a methodology to validate the RTOS architecture of a proven safety-critical system, Stall Warning System / Aircraft Interface Computer

(SWS/ AIC)kernel architecture. The proven system is selected in order to prove the effectiveness of architecture analysis and its certification impact with respect to the civil aerospace software standard RTCA DO-178C [16].

SWS/AIC system kernel is known asAPM2000 kernel. The scheduler schedules and dispatches the application tasks as per their grouping into fast, slow, and background tasks[21]. The complete processing of application is carried out through foreground and background tasks. The foreground task consists of fast, and slow tasks, fast task is executed every 25 msec. The slow task is executed every 100 msec. The fast task has higher priority than the slow task and background task. Background task has the least priority. At the startup sequence, the software manages all the interrupts to schedule and execute the application. The scheduling of SWS/AIC kernel is shown in Figure 1.

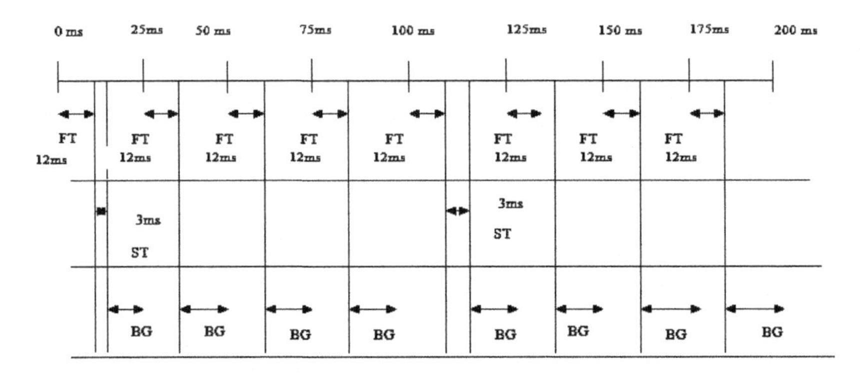

Fig. 1 Scheduling in the Stall Warning System

Note - *FT: Fast Task; ST: Slow Task; BG: Background Task*

In order to model the SWS/AIC kernel architecture as per the requirements we use AADL based architecture analysis approach. To design the kernel architecture detailed understanding of the FT, ST& BG was performed.

3.2 Fast Task Requirements and AADL Implementation

Fast task is given control by the kernel when the 25ms RTC interrupt occurs. This task has a higher priority than the slow and background task. The maximum task duration for the foreground task is 25msec. Only 60% of the foreground task will be utilized which is 60% of 25msec = 15msec. The design for the fast task duration is 80% of the 15msec, which is 12msec.In the fast task, excessive context switch and associated over-head is reduced by limiting the nesting of calls. The fast task utilization is such that the processor throughput is not affected. The activities in the Fast task include: Acquire inputs, CCDL Tx/Rx, ARINC output processing, Compute FT, Average Time, Discrete data validation, PTT on ground/air, WDT Kernel call, CBIT and Shaker stall Validation. These tasks are shown in Figure 2. The AADL model of the Fast Task is shown in Figure 3.

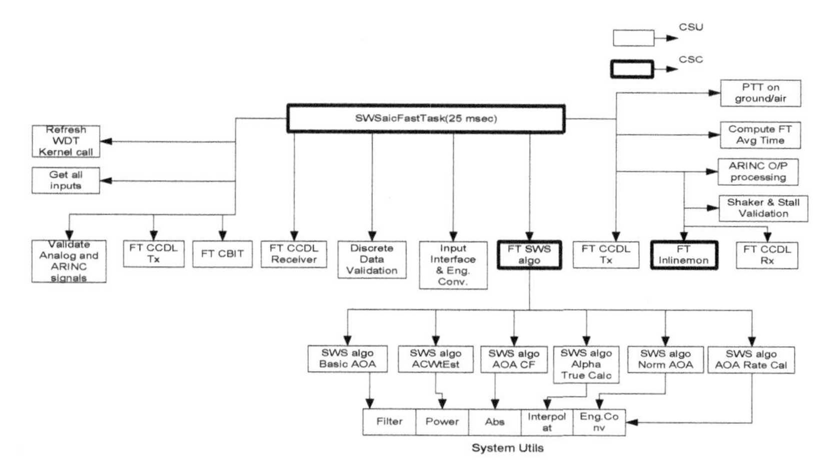

Fig. 2 Fast task Process Flow

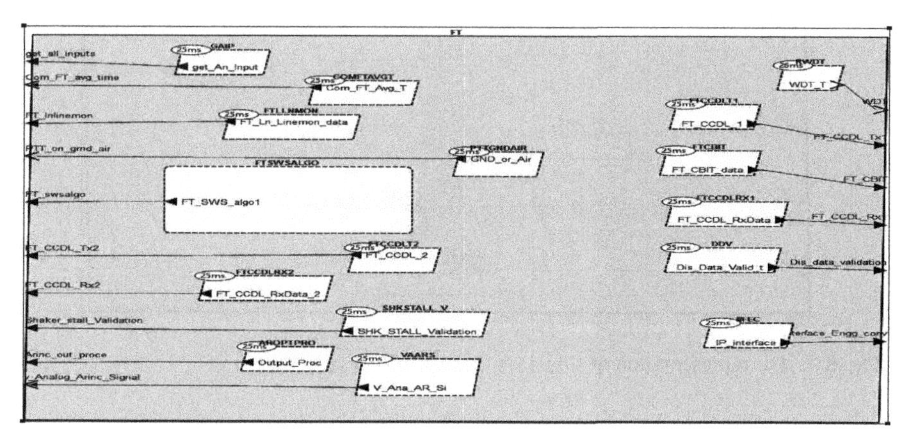

Fig. 3 AADL representation of fast task

3.3 *Slow Task Requirements and AADL Implementation*

Slow task in the foreground is given control by the kernel when the 100ms RTC interrupt occurs. This task has a higher priority than background task but lower than the fast task. The design for slow task duration is about 20% of 15msec = 3msec. The Fast Task interrupt can preempt this 100ms slow task. The slow task utilization is such that the process or throughput is not affected. The tasks in the slow task are: Output validation, output processing, Compute average time, & aic algo. These tasks are shown in Figure 4 and the AADL representation of slow task is shown in Figure 5.

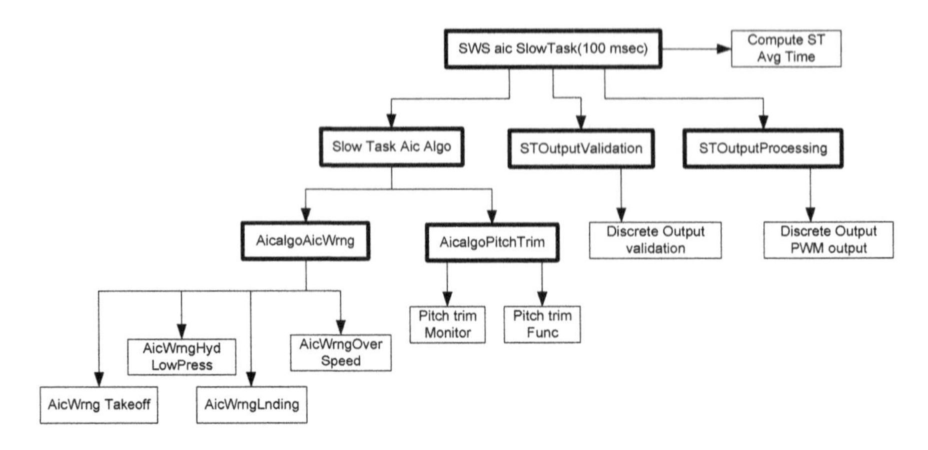

Fig. 4 Slow task Details

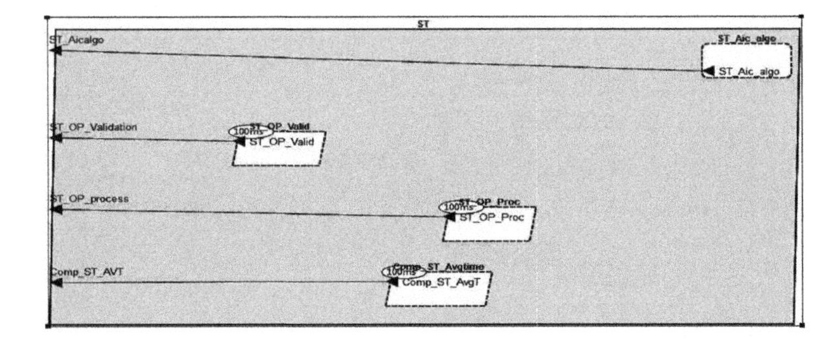

Fig. 5 AADL representation of slow task

3.4 Background Task Requirements and AADL Implementation

Background task is given control by the kernel after Power On and during the initialization of the system. It remains in this task until the 25ms fast task or 100ms slow task interrupt occurs. The fast task has the higher priority than the slow task. After the execution of fast task the kernel passes the control to slow task. After the completion of slow task, if no fast task/ slow task interrupt occurs only then the kernel passes the control to the background task. The tasks in the background task are: PBIT startup, CBIT, and Failed mode. The tasks are shown in Figure 6 and the AADL representation of background task is shown in Figure 7.

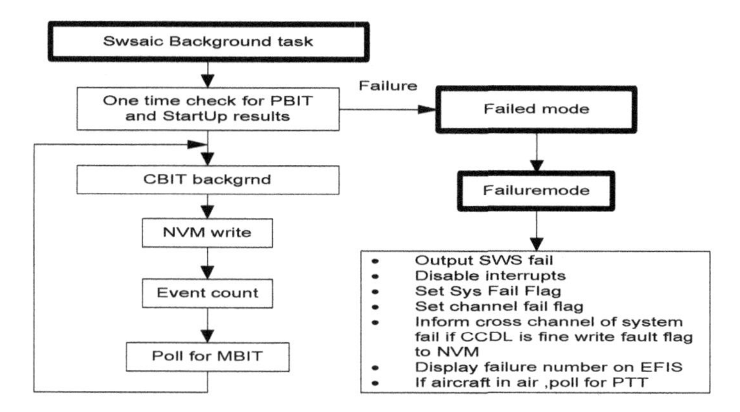

Fig. 6 Background task details

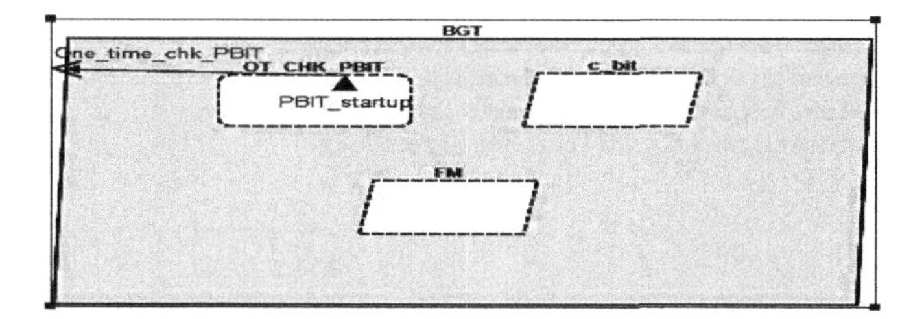

Fig. 7 AADL representation of Background task

4 Kernel Scheduler Simulation and Analysis

The representation of the architecture using AADL becomes easier to understand and implement as compared to the manual approach which gets finalized after multiple reviews.

The analysis of the AADL architecture provides model, application and execution statistics. The modeling and analysis using AADL is carried out using the OSATE Tool, Version 2.2[4].

The entire SWS/AIC kernel architecture is justified with evaluation of metrics such as: The *model statistics, application statistics, and execution platform. Model statistics involves* various aspects of model such as the component types, their declaration instances, and data flow types. The *application statistics* provides brief description about the software architecture involving thread instances, process instances, semantic connections between them, and end-to-end flow instances. The *execution platform* statistics provides clarification about the hardware instances involved in supporting the execution of the above justified instances such as processors and memory units that are bound during the execution of the software instances. The metrics generated by the tool is shown in Figure 8.

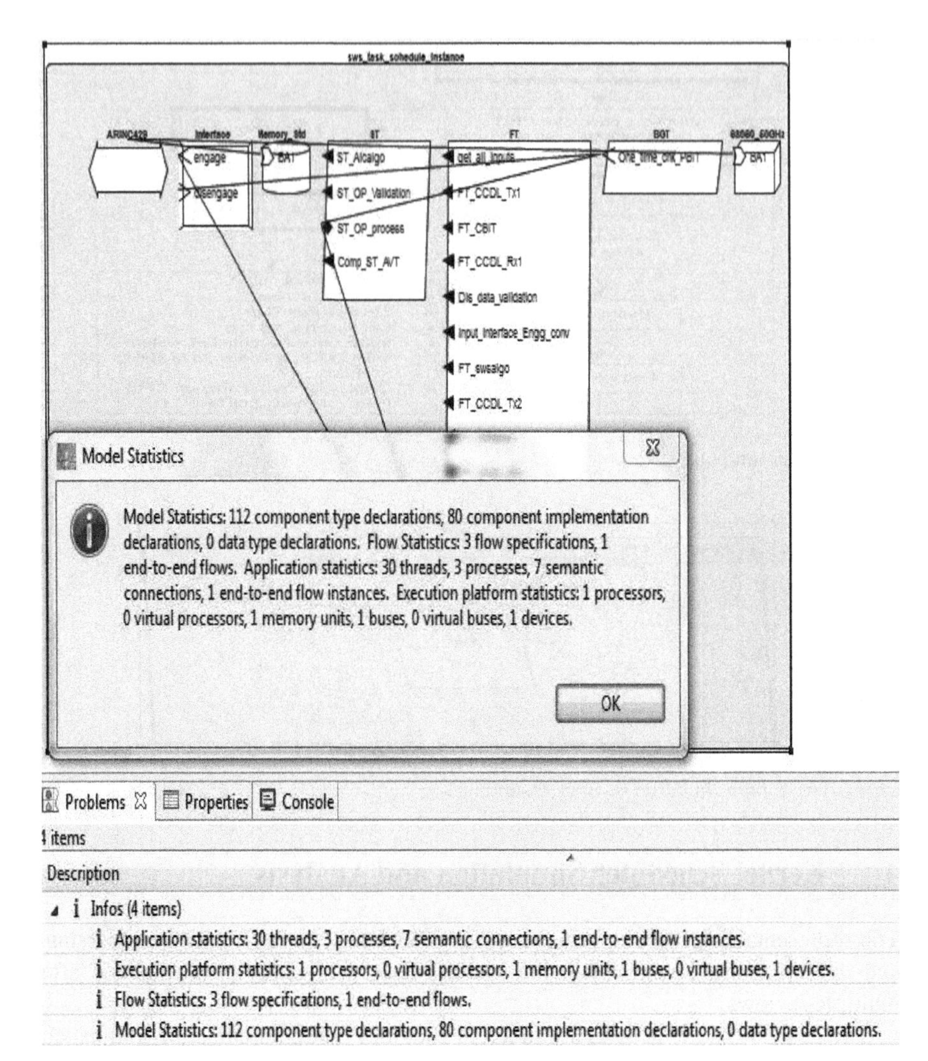

Fig. 8 Model Statistics of SWS/AIC Kernel Architecture

4.1 Kernel Scheduling Analysis

Analysis is the art of planning your activities so that you can achieve your goals and priorities as per the project requirements. Analysis helps in:

- Understand what you can realistically achieve with your time.
- Make sure you have enough time for essential tasks.

The kernel scheduling analysis using AADL Inspector provides cheddar schedulable table, Theoretical schedulable test and simulation schedulable test. Cheddar schedulable table provides the scheduling analysis of each and every task. The cheddar schedulable table for the SWS/AIC scheduler is in Figure 9.

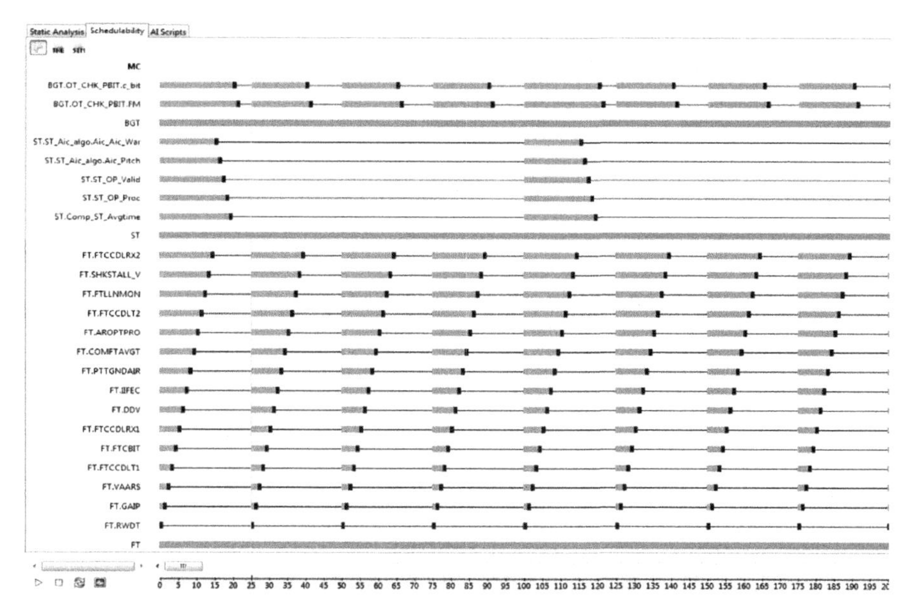

Fig. 9 Scheduling analysis of SWS/AIC Kernel

In Figure 9: Orange color represents thread is in ready state, Black color represents thread is in running state, Solid black line represents thread is in suspended state, and Green color represent process is in running.

Theoretical scheduling provides report on processor utilization factor with period & worst case task response time. The theoretical scheduler for the SWS/AIC is shown in Figure 10

test	entity	result
processor utilization factor	root.mc	Can not apply the feasibility test on processor utilization factor on this scheduler.
base period	all	100.00000
processor utilization factor with deadline	all	0.73000
processor utilization factor with period	all	0.73000
worst case task response time	root.mc	All task deadlines will be met : the task set is schedulable.
response time	root.mc.bgt.ot_chk_pbit.fm	22.00000
response time	root.mc.bgt.ot_chk_pbit.c_bit	21.00000
response time	root.mc.st.comp_st_avgtime	20.00000
response time	root.mc.st.st_op_proc	19.00000
response time	root.mc.st.st_op_valid	18.00000

Fig. 10 Theoretical Scheduling of SWS/AIC Kernel

$$\textit{Processor Utilization}: \mathrm{U} = \sum_{i=1}^{n} \frac{\mathrm{c_i}}{\mathrm{p_i}} \leq \mathrm{n}.(2^{\frac{1}{\mathrm{n}}} - 1)$$

n = No. of Tasks; C_i= Execution Time i^{th} task; P_i = Period of ith task.

Fast task has C=15 msec, P = 25 msec; Slow task has C=5 msec, P = 100 msec; Background task C = 2msec, P = 25 msec; So U= 0.73 &B = n.$(2^{1/n}$-1) = 0.77 since U $\leq$B deadline will be met[24].

Simulation scheduling provides task response time computed from simulation, number of preemption and number of context switches. The simulation scheduler for the SWS/AIC is shown in Figure 11.

test	entity	result
Task response time computed from simulatio	root.mc	No deadline missed in the computed scheduling : the task set is schedulable if you co
Number of preemptions	root.mc	0
Number of context switches	root.mc	72
Task response time computed from simulatio	root.mc.bgt.ot_chk_pbit.c_bit	worst = 21, best = 0 and average = 13.80000
Task response time computed from simulatio	root.mc.bgt.ot_chk_pbit.fm	worst = 22, best = 0 and average = 14.60000
Task response time computed from simulatio	root.mc.ft.aroptpro	worst = 11, best = 0 and average = 8.80000
Task response time computed from simulatio	root.mc.ft.comftavgt	worst = 10, best = 0 and average = 8.00000
Task response time computed from simulatio	root.mc.ft.ddv	worst = 7, best = 0 and average = 5.60000
Task response time computed from simulatio	root.mc.ft.ftcbit	worst = 5, best = 0 and average = 4.00000
Task response time computed from simulatio	root.mc.ft.ftccdlrx1	worst = 6, best = 0 and average = 4.80000

Fig. 11 Simulation Scheduling of SWS/AIC Kernel

Worst Case Response Time: $W_i^n = C_i + \sum_{j \in hp(i)} \left(\frac{C_i}{P_j}\right) . C_j$

C_i= Execution Time i^{th} task, P_i = Period of i^{th}task, P_j = period of high priority task, C_j= Execution time of priority task, hp(i)=the set of tasks which have a higher priority than task i.
W_1^0=15, W_2^0=5,W_2^1=8,W_2^2=9.8,W_3^0=2,W_3^1=4,W_3^2=5.....etc[24].

The cheddar schedulable table, theoretical scheduling and the simulation scheduling for the SWS/AIC system satisfies the requirements and can be used as a proof for certification.

5 Mapping of Kernel Architecture Analysis to Certification Artifacts as per RTCA DO-178C Workflow

Architecture analysis not only helps in understanding the requirements better and uncovering the design flaws earlier in the process but it also helps in the addressing the certification issues. RTCA DO-178C[16][17] is the civil aerospace software engineering guideline. The ADDL analysis report can be used to address the objectives provided in the 178C guidelines. The objectives addressed by this analysis are:

- *Software architecture is compatible with high level requirements.*
- *Software architecture is consistent.*
- *Software architecture is compatible with target computer.*
- *Software architecture is verifiable.*
- *Software architecture is conforms to standards.*
- *Software partitioning integrity is confirmed.*

In our case study, the software for SWS/AIC system is designed, developed and qualified as per the highest criticality level of the 178C guidelines. With the AADL approach for the kernel architecture, 6 objectives of the 71 objectives are addressed. Out of 71objectives, 6 are related to software architecture, in which 3 are independent. Table 1 shows the software criticality and the architecture level objectives.

Table 1 The objectives related to software architecture based on criticality level.

Criticality Level	Failure condition	Objectives	With independence
A	Catastrophic	6/71	3/33
B	Hazardous	6/69	0/21
C	Major	4/62	0/8
D	Minor	0/26	0/5
E	No Safety Effect	0/0	0/0

The analysis results by using AADL Inspector can be mapped to satisfy the objectives as shown in Table 1. Analysis report generated by the tool can be used as an artifact to verify the effectiveness of the scheduler as per the functional, operational and safety requirements.

6 Conclusion and Future Scope

The work focuses on the importance of RTOS architecture analysis in case of safety critical system and mapping the analysis to RTCA DO-178C compliance. AADL based architecture analysis helps in analyzing the RTOS architecture in terms of model statistics, resource allocation with respect to processor/s and threads, latency analysis & scheduling analysis of various processes and threads. This is demonstrated using the case study of the proven SWS/AIC kernel architecture. AADL-based approach proposed helps in developing the confidence of the kernel architecture for its functionality, operation and safety. The analysis outcome provides evidence which can be mapped to the certification objectives of RTCA DO-178C. Hence incorporating the AADL based approach in the engineering process make the process effective as it will uncover the requirements and design flaws earlier in the life cycle and increase the safety feature of the software.

Future work involves implementation of this approach for certification of highly complex RTOS used in the advanced avionics systems, such as Integrated Modular Avionics which uses the hard-real time space and times partition operating system.

Acknowledgments Authors would like to thank the Director, CSIR-NAL for providing them the opportunity to work in this area.

References

1. Bieber, P., Boniol, F., Boyer, M., Noulard, E., Pagetti, C.: New Challenges for Future Avionic Architectures (4), May 2012. pp10
2. Dobrica, L., Niemelä, E.: A Survey on Software Architecture Analysis Method. Ieee Transactions on Software Engineering **28**(7), July 2002. pp10
3. Dhage, S.: Qualification of RTOS for safety critical systems using formal methods. INDIAcom 2015. pp12
4. Singhoff, F., Legrand, J., Nana, L.: Scheduling and memory requirements analysis with AADL. In: International Conference Proceedings, November 13–17, 2005. pp11
5. Designing Safety-Critical Avionics Software Using Open Standards. http://www.google.com.unpublished
6. Howard, C.E.: Safety- and security-critical avionics software, February 1, 2011. http://www.militaryaerospace.com/articles/print/volume-22/issue-2/technology-focus/safety-and-security-critical-avionics-software.html
7. Donini, R., Marrone, S., Mazzocca, N., Orazzo, A., Papa, D., Venticinque, S.: Testing complex safety- critical systems in SOA Context, November 12, 2007. pp8
8. Alexander, R., Alexander-Bown, R., Kelly, T.: Engineering Safety-Critical Complex Systems. http://www.cs.york.ac.uk/nature/tuna/outputs/finalreport.pdf, pp27
9. Correa, T., Becker, L.B., Farines, J.-M.: Supporting the design of safety critical systems using AADL. In: 2010 15th IEEE International Conference on Engineering of Complex Computer Systems, pp. 331–336. pp6
10. Knight, J.C.: Safety critical systems: challenges and directions. In: Proceedings of the 24rd International Conference on ICSE 2002 (2002). ieeexplore.ieee.org, pp4
11. Nordhoff, S.: DO-178C/ED-12C the new software standards for avionics industry: goal, changes and challenges. http://www.sqs.com, pp26
12. Feiler, P.H., Gluch, D.P., Hudak, J.J., Lewis, B.A.: Embedded System Architecture Analysis Using SAE AADL, June 2004. Technical Note, CMU/SEI-2004-TN-005, pp45
13. Adalog, J.-P.R., Axlog, J.-F.T.: AADL Workshop, October 17–18, 2005. 2005 Overview of AADL Syntax
14. Casteres, J., Ramaherirariny, T.: Aircraft integration real-time simulator modeling with AADL for architecture tradeoffs, pp. 346–351 (2009). pp6
15. Rammig, F., Ditze, M., Janacik, P., Heimfarth, T., Kerstan, T., Oberthuer, S., Stahl, K.: Basic Concepts of Real Time Operating systems. Hardware-Dependent Software, Springer Science + Business Media B.V., 16–44 (2009). pp28
16. RTCA DO-178C Software Consideration in Airborne Systems and Equipment Certification
17. RTCA DO-333 Formal Method Supplement to DO-178C and DO-278A
18. Formal Methods for Software Architectures: Software Architectures, SFM 2003, Bertinoro, Italy, September 22–27, 2003. http://www.springer.com
19. Cofer, D.: "DO-178C", High Confidence Software & Systems Conference, May 8, 2012. pp33
20. Wang, Y., Ngolah, C.F.: Formal Description of a Real-Time Operating System using RTPA, vol. 2, pp. 1247–1250, May 4–7, 2003. pp3
21. CSIR-NAL: Software Design Description (SWDD) of SARAS aircraft
22. Noll, T.: Safety, dependability and performance analysis of aerospace systems. In: Third International Workshop on Formal Techniques for Safety-Critical Systems (FTSCS 2014) November 1–5, 2014
23. Fisher, K.: Using Formal Methods to Enable More Secure Vehicles: DARPA's HACMS Program, September 16, 2014
24. Hugues, J., Singhoff, F.: AADLv2: an Architecture Description Language for the Analysis and Generation of Embedded Systems. ISAE, France

A Metaheuristic Approach for Simultaneous Gene Selection and Clustering of Microarray Data

P.S. Deepthi and Sabu M. Thampi

Abstract Cancer subtype discovery from gene expression data is a daunting task due to the relatively large number of genes associated with the samples. Selecting important genes to identify the underlying phenotype structure has been addressed in this work. The proposed work integrates gene selection using Firefly algorithm into K-means clustering. The algorithm guarantees a suboptimal subset of genes necessary for disease subtype clustering. Experiments were conducted on publicly available cancer gene expression datasets and an accuracy of 70-80% was obtained. Comparison with the previous work on Particle Swarm Optimization based feature selection shows that the proposed work achieved better convergence and accuracy.

1 Introduction

Genomic research has reached new frontiers with the advent of microarray technologies like oligonucleotide arrays and cDNA microarray. Gene expression data arising from these experiments measure thousands of genes in parallel and data mining tasks have proven to be useful in gaining insights into various disease patterns and patient survival rate. In particular, clustering and classification have been employed to detect natural groupings of patient samples and genes[7].

Gene expression data clustering falls into three categories - gene clustering, sample clustering and subspace clustering. In gene based clustering, genes having similar functions are assigned to same group. In sample based clustering, tissue samples are grouped to identify the phenotypes of diseases. Subspace clustering treats both

P.S. Deepthi(✉) · S.M. Thampi
Indian Institute of Information Technology and Management - Kerala(IIITM-K), Trivandrum, India
e-mail: deepthisath@gmail.com, smthampi@ieee.org

P.S. Deepthi
LBS Centre for Science and Technology, Trivandrum, India

S. Berretti et al. (eds.), *Intelligent Systems Technologies and Applications*,
Advances in Intelligent Systems and Computing 385,
DOI: 10.1007/978-3-319-23258-4_39

genes and samples symmetrically and hence clusters in both dimensions[7]. Sample based clustering has been found efficient in discovering new subtypes of diseases like cancer which is essential for better prognosis and treatment.

Traditional clustering techniques like k-means, self organizing maps and hierarchical clustering have been employed for gene clustering. But applying these methods for sample clustering, taking into account the complete set of genes as features will degrade the clustering reliability. Hence there is a need for selecting the important genes. Dimensionality reduction techniques like Principal Component Analysis[15] and many feature selection techniques have been found in literature to deal with the "curse of dimensionality"[9]. Most of the feature selection techniques employ sequential search or random search which is inefficient as they generate local optimal solutions. Several gene selection approaches have also been suggested for microarray sample classification. However, unsupervised gene selection is challenging due to the lack of class information.

In this scenario, a metaheuristic approach for simultaneous gene selection and clustering of disease subtypes is presented. The proposed work integrates informative gene selection into clustering for categorising cancer subtypes. Firefly algorithm is used to generate near optimal subset of genes by choosing the objective function as squared error criteria of the K-means algorithm. The proposed method can produce compact clusters by escaping the local optimal solutions.

The rest of the paper is organized as follows. Section 2 discusses the related works. Section 3 describes the firefly algorithm. Section 4 presents the proposed work and algorithm. Section 5 discusses experimental results and section 6 concludes the paper.

2 Related Works

The problems related to high dimensional data spaces are referred to as the "curse of dimensionality" - a term coined by Richard Bellman. The clustering models help to identify the functional dependencies in the dataset and to find interesting insights into the domain that the dataset represents. As the number of attributes increases these functions become more complex. It is difficult to define a measure of proximity, distance or neighbourhood as dimensions increase. Dimensionality reduction techniques like Principal Component Analysis[15] has been employed for projecting data into lower dimensional space. These methods perform a feature transformation which may degrade the clustering result. Hence research began to shift towards developing feature selection algorithms.

A survey of feature selection techniques for clustering and classification is presented in [9]. The categorising framework defined in this paper shows that most of the techniques adopt complete, sequential or random search for exploring the feature space to select relevant attributes. Two schemes - a filter and a hybrid filter-wrapper approach are proposed in [11] for model based learning. It uses Sequential Forward Selection and Sequential Backward Elimination strategy for unsupervised learning of Naive Bayes. Ranking by SVD entropy suggested in [13] also follows sequential search. These strategies are inefficient for gene expression data as search space is large.

Feature selection by gene shaving method based on principal components is presented in [6]. Informative gene selection using cosine silmilarity to improve clustering is suggested in [12]. However, these schemes are based on variance of data and hence are suited only for those data that have large variances. A multiobjective approach for gene selection and clustering using genetic algorithm is proposed in [10]. But it involves complex mutation operations. Gene selection using Independent Component Analysis and clustering using extensions of Non negative Matrix Factorization is putforth in [16]. Difficulty to determine the number of independent components is a problem of this method.

Meta-heuristic approaches have been employed for gene selection for sample classification. Gene selection by combining feature relevance with genetic algorithm and Particle Swarm Optimization is suggested in [1]. The framework for feature selection in [3] employs feature ranking followed by a genetic algorithm based wrapper approach for supervised feature selection. Although there is a multitude of algorithms for supervised gene selection, few works focus on gene selection for sample clustering.

In [4] authors developed a Particle Swarm Optimization based algorithm for informative gene selection of microarray data. The proposed work adopts firefly algorithm to generate near optimal subset of genes required for sample clustering. The squared error criteria of k-means algorithm is the objective function fed to the firefly for optimization. The proposed work achieved better accuracy and faster convergence than the previous work.

3 Firefly Algorithm

Firefly algorithm [14] is a nature inspired technique proposed by Xin-She Yang which models the flashing behaviour exhibited by fireflies. It is built upon the following three assumptions-

i) A firefly can be attracted to any other firefly as fireflies are unisex.

ii)A firefly's attractiveness is proportional to its brightness seen by other fireflies, hence for any two fireflies, the dimmer firefly moves towards brighter ones, but if there are no brighter fireflies nearby, a firefly moves randomly.

iii) The brightness of a firefly is proportional to the value of its objective function.

Variation of attractiveness is defined by

$$\beta = \beta_0 e^{-\gamma r^2} \tag{1}$$

where β is the attractiveness parameter and γ is the light absorption coefficient.

Update formula for any two fireflies i and j is

$$x_i(t+1) = x_i(t) + \frac{\beta}{1+\gamma r_{ij}^2}(x_j(t) - x_i(t)) + \alpha(r - 0.5) \tag{2}$$

where second term is due to attraction and third term is for randomization, α is the randomization parameter and r is a random number in the range [0,1].

4 Proposed Work

4.1 Proposed Model

Sample based clustering of gene expression data requires informative gene selection. An exhaustive search of the complete gene dimension is infeasible. Hence a hybrid Firefly-K-means algorithm has been proposed for generating near optimal subset of genes and subsequent clustering. Randomization of firefly algorithm is exploited to reduce the search space. The proposed model is given in Figure 1.

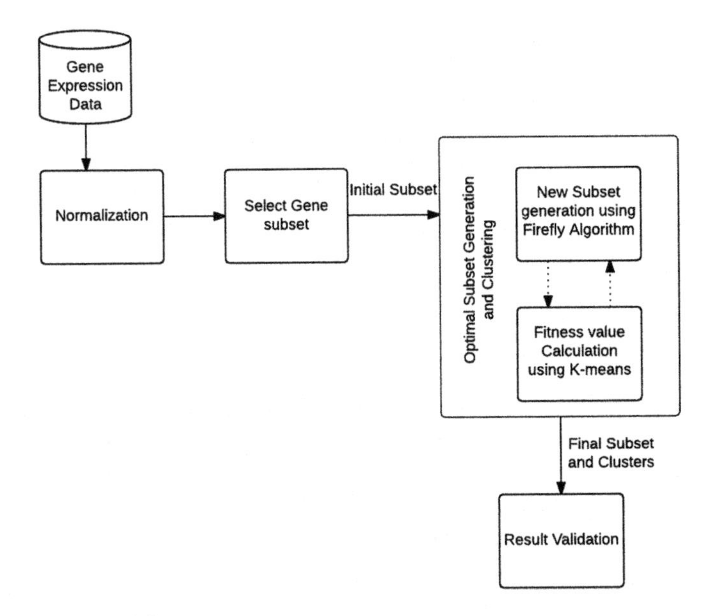

Fig. 1 Proposed Model

The mapping of firefly and objective function in terms of microarray data is given below.

Firefly : Each firefly represents a candidate solution. Hence a gene subset is modelled as a firefly.

Objective function : Objective function defines the value that the firefly algorithm tries to optimize. Since goodness of clustering is judged by the compactness of clusters, squared error in k-means algorithm is chosen as the objective function.

In each iteration, the new subsets of genes are generated based on equation(2). Each subset is evaluated for objective function using within cluster sum of squared error given in equation(3)

$$Error = \sum_{i=1}^{k} \sum_{o \epsilon C_i} (o - m_i)^2 \tag{3}$$

where o=(o1, o2, ...on) are the objects, C=(C1, C2...Ck) are the k clusters and mi is the mean of ith cluster

The subset that minimizes the criteria defined in equation(3) describes the best subset of genes. In this context, a subset of genes is represented as a firefly and the squared error measure in k-means algorithm is adopted as the objective function that the firefly tries to optimize.

4.2 Algorithm

The proposed Firefly-Kmeans hybrid algorithm is given in Algorithm 1.

Algorithm 1. Firefly-Kmeans Algorithm

Input : Dataset D with N instances and M features, Number of clusters k
Output: Feature subset m, k clusters

1: Initialise *POPULATIONSIZE*, the number of subset of genes(fireflies)
2: **for** i=0 **to** *POPULATIONSIZE*-1 **do**
3: Initialise position of gene subset
4: Perform Kmeans clustering with each gene subset as features
5: Set fitness value I_i of each firefly as defined in equation(3)
6: **end for**
7: **repeat**
8: **for** i=0 to *POPULATIONSIZE*-1 **do**
9: **for** j=0 to *POPULATIONSIZE*-1 **do**
10: **if** $I_j > I_i$ **then**
11: Change position of subset i towards j using equation(2)
12: **end if**
13: Perform Kmeans clustering with updated position
14: Update light intensity of the subset using equation(3)
15: **end for**
16: **end for**
17: Sort the intensity values and find current best
18: **until** Maximum number of iterations
19: Return current best as best subset of genes
20: Return the cluster results corresponding to current best

Initialization

For initializing the population of fireflies, subsets of genes are randomly chosen from uniform distribution. Attractiveness corresponds to squared error criteria which is calculated by performing k-means algorithm on the selected subsets of genes.

Centroid Initialization
The initial centroids of the k-means are found using k-means++ algorithm[2].

Update Process
The intensity values corresponding to each gene subset is found by performing k-means algorithm with the centroids initialised above. Let I_i and I_j be the squared error obtained on clustering using two subsets i and j. If I_j is greater than I_i then the subset j is updated to move towards subset i using equation(1). Once the position of subset j is updated, k-means is executed to find the updated squared error. This process is repeated till maximum number of iterations is reached. At this stage there are n fireflies with different objective function values which are sorted to get the best value. The firefly with the least sum of squared error value represents the best solution, ie the best subset of genes and the cluster results are retained.

5 Experimental Results

The proposed algorithm was implemented in Java and experiments were conducted on a PC with Intel core i5-3210M Processor, 2GB RAM and 500GB hard disk.

5.1 Datasets

Five publicly available gene expression datasets, namely Leukemia, MLL(Mixed-Lineage Leukemia), DLBCL(Diffuse Large B-cell Lymphoma), SRBCT (Small Round Blue Cell Tumors) and Lymphoma are chosen for the study. These microarray datasets are accessed from http://orange.biolab.si/datasets.psp (available on June 6, 2014) and details are given in Table 1.

Table 1 Datasets

Dataset	#Samples	#Genes	Class
Leukemia	72	5148	2
MLL	72	12534	3
DLBCL	77	7071	2
SRBCT	83	2309	4
Lymphoma	62	4026	3

5.2 Preprocessing

Normalization is performed to pre-process the data. Scale factor and translation factor are set as -2 and 1 and all entries were made in the range (-1,1). This process was conducted for all experiments performed.

5.3 Parameter Tuning

Performance and convergence of the algorithm depends on the parameters of the firefly algorithm. Table 2 gives the parameters chosen for conducting the experiments.

Table 2 Parameter Settings

Parameter	Value
POPULATIONSIZE	20
α	0.20
$\beta 0$	0.25
γ	0.1

5.4 Results and Discussion

The number of features were kept varying and experiments were repeated. Comparison with other feature selection methods like *Random Projection*, *Ranker* followed by *PCA* and *Correlation Feature Selection*(CFS) using *Sequential Forward Search*(SFS) were also conducted. These methods were implemented using weka [5]. Random Projection reduces the feature space by projecting it into a lower dimensional space. Variation in data is preserved. Principal Component Analysis followed by Information gain based ranking was also performed. Result of k-means algorithm on features selected by these methods is analysed.

The performance of all the above methods on the cancer gene expression datasets were evaluated using two criteria -*within cluster sum of squared error* and *cluster purity*. Within cluster sum of squared error measures the cohesion of objects in a cluster. It is the sum of squared distance of the objects from the cluster centroid. The lesser the value of this value, the more close are the objects to the centroid.

Purity is another metric used to evaluate the clustering results. It measures the accuracy of clustering and ranges from 0 to 1. Accurate clustering results in purity close to 1.

Performance analysis on leukemia dataset for 6, 8 and 11 attributes are given in Table 3. A maximum accuracy of 80% was obtained for 11 attributes using the proposed approach. Among all other methods, CFSSubset gave an accuracy of 76%.

Table 3 Result on Leukemia dataset

Method	Features	WC Error	Purity
Random Projection	6	14.00	0.45
	8	19.34	0.63
	11	26.08	0.52
Ranker+PCA	6	12.91	0.65
	8	14.61	0.67
	11	24.53	0.68
CFSSubset+SFS	6	13.20	0.68
	8	17.01	0.72
	11	23.73	0.76
Proposed Method	6	7.70	0.71
	8	11.44	0.74
	11	19.26	0.80

Results on MLL dataset with 15, 19 and 22 attributes are tabulated in Table 4. An accuracy of 77% was attained for 22 attributes. CFSSubset gave an accuracy of 71%. For all other methods accuracy was below 70%.

Table 4 Result on MLL dataset

Method	Features	WC Error	Purity
Random Projection	15	33.72	0.27
	19	43.68	0.60
	22	51.48	0.69
Ranker+PCA	15	32.82	0.40
	19	38.85	0.65
	22	45.60	0.67
CFSSubset+SFS	15	34.64	0.67
	19	41.28	0.69
	22	45.40	0.71
Proposed Method	15	24.45	0.72
	19	33.05	0.73
	22	38.85	0.77

For DLBCL, results on 17, 20 and 23 genes are given in Table 5. 78% accuracy was achieved for 23 genes whereas CFSSubset gave 71% accuracy. Error was also improved.

For SRBCT, 17, 20 and 23 genes were chosen. The purity and within cluster sum of error are tabulated in table 6. The proposed method obtained an accuracy of 75% for 12 attributes.

The results for Lymphoma with genes are summarized in table 7. The proposed method obtained an accuracy of 75% for 12 attributes.

Table 5 Result on DLBCL dataset

Method	Features	WC Error	Purity
Random Projection	17	39.86	0.62
	20	45.45	0.64
	23	51.69	0.66
Ranker+PCA	17	43.22	0.68
	20	47.92	0.63
	23	52.23	0.68
CFSSubset+SFS	17	36.21	0.69
	20	42.02	0.49
	23	51.62	0.70
Proposed Method	17	21.78	0.72
	20	24.38	0.76
	23	32.73	0.78

Table 6 Result on SRBCT dataset

Method	Features	WC Error	Purity
Random Projection	8	20.42	0.45
	10	23.44	0.62
	12	31.25	0.67
Ranker+PCA	8	20.42	0.45
	10	23.44	0.62
	12	31.25	0.67
CFSSubset+SFS	8	15.26	0.63
	10	21.40	0.66
	12	24.99	0.68
Proposed Method	8	14.38	0.72
	10	18.14	0.73
	12	21.52	0.75

From the experimental results, it is clear that the proposed method achieves better results than other methods in terms of within cluster sum of squared error and cluster purity. The improvement in accuracy is attributed to the fitness function chosen. As each gene subset is evaluated using the squared error criteria of k-means algorithm, it is guaranteed that a near optimal solution is achieved. The randomized behavior of firefly algorithm not only reduces the search space but also bypasses the local optimal solutions. An accuracy of 70-80% has been achieved for all the datasets. Among all the methods compared, CFSSubset gave considerably good results. Maximum purity obtained for leukemia with number of genes selected m=11, MLL with m=22 and DLBCL dataset with m=23, SRBCT with m=12 and Lymphoma with m=13 is plotted in Figure 2 and Figure 3.

Table 7 Result on Lymphoma dataset

Method	Features	WC Error	Purity
Random Projection	10	18.37	0.46
	11	20.65	0.52
	13	24.41	0.66
Ranker+PCA	10	19.62	0.54
	11	22.32	0.60
	13	29.94	0.68
CFSSubset+SFS	10	19.06	0.64
	11	21.14	0.66
	13	29.73	0.70
Proposed Method	10	15.66	0.70
	11	17.14	0.72
	13	19.22	0.73

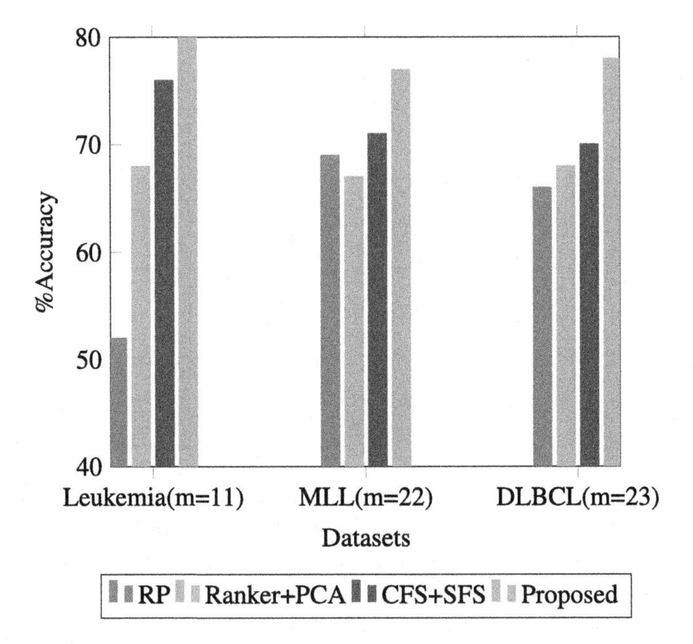

Fig. 2 Comparison of accuracy

5.5 *Comparison with Previous Work*

In the previous work [4], authors have presented a Particle Swarm Optimization (PSO) based feature selection method for optimal subset generation. Performance analysis and comparison of PSO based method and firefly algorithm is summarised in Table 8. The best number of genes found for leukemia is 11, MLL is 23, DLBL is 22, SRBCT is 12 and Lymphoma is 13. Error, accuracy and CPU time in seconds for

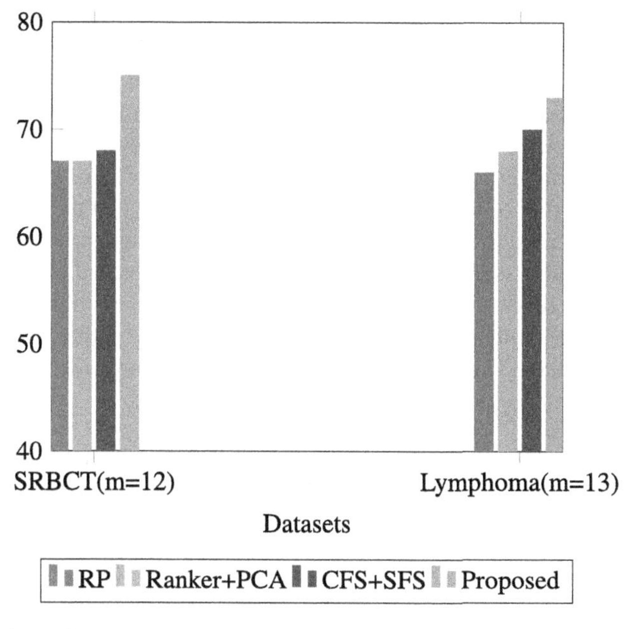

Fig. 3 Comparison of accuracy

these subset of genes are tabulated. It is observed that the firefly based method attained better accuracy than PSO based method. The proposed method also converged faster than the previous algorithm.

Table 8 Results on PSO and Firefly

Method	Dataset	Error	Purity	CPU time
PSO based	Leukemia	19.87	80	52.714
	MLL	41.48	75	172.716
	DLBCL	35.16	76	109.798
	SRBCT	23.22	73	31.047
	Lymphoma	19.86	72	40.326
Firefly Based	Leukemia	19.26	80	49.204
	MLL	38.85	77	170.043
	DLBCL	32.73	78	101.342
	SRBCT	21.52	75	29.506
	Lymphoma	19.22	73	36.203

6 Conclusions

A metaheuristic algorithm that employs firefly algorithm and k means for simultaneous gene selection and clustering of microarray data is proposed. Informative gene selection is achieved through the randomized behavior of firefly algorithm and quality of clustering is assured by incorporating error criteria of k-means algorithm as objective function of firefly. Experimental results show that the proposed method outperforms other feature selection techniques. Comparison with the previous work on PSO based feature selection shows that the method achieved better accuracy and improved convergence. A clustering accuracy of 70-80% was obtained for the chosen datasets. The number of optimal genes was set by carrying out the experiments with different values. The algorithm can be modified to automatically determine this value.

Acknowledgments This work is supported by the SPEED-IT research fellowship by Kerala State IT Mission, Department of Information Technology, Govt of Kerala and Indian Institute of Information Technology and Management - Kerala.

References

1. Alba, E., García-Nieto, J., Jourdan, L., Talbi, E.G.: Gene selection in cancer classification using PSO/SVM and GA/SVM hybrid algorithms. In: IEEE Congress on Evolutionary Computation, pp. 284–290 (2007)
2. Arthur, D., Sergei, V.: k-means++: The advantages of careful seeding. In: ACM-SIAM Symposium on Discrete Algorithms, pp. 1027–1035 (2007')
3. Cannas, L.M., Dessí, N., Pes, B.: A hybrid model to favor the selection of high quality features in high dimensional domains. In: Intelligent Data Engineering and Automated Learning, pp. 228–235 (2011)
4. Deepthi, P.S., Thampi, S.M.: Unsupervised gene selection using particle swarm optimization and k-means. In: Proceedings of the Second ACM IKDD Conference on Data Sciences, pp. 134–135 (2015)
5. Hall, M., Frank, E., Holmes, G., Pfahringer, B., Reutemann, P., Witten, I.H.: The WEKA data mining software: an update. ACM SIGKDD Explorations Newsletter, 10–18 (2009)
6. Hastie, T., Tibshirani, R., Eisen, M.B., Alizadeh, A., Levy, R., Staudt, L., Brown, P.: Gene shaving as a method for identifying distinct sets of genes with similar expression patterns. Genome Biology, 1–21 (2000)
7. Jiang, D., Tang, C., Zhang, A.: Cluster analysis for gene expression data: A survey. IEEE Transactions on Knowledge and Data Engineering, 1370–1386 (2004)
8. Kim, Y., Street, W.N., Menczer, F.: Feature selection in unsupervised learning via evolutionary search. In: Proceedings of the Sixth ACM SIGKDD International Conference on Knowledge Discovery and Data Mining, pp. 365–369 (2000)
9. Liu, H., Yu, L.: Toward integrating feature selection algorithms for classification and clustering. IEEE Transactions on Knowledge and Data Engineering, 491–502 (2005)
10. Mukhopadhyay, A., Maulik, U., Bandyopadhyay, S.: Simultaneous informative gene selection and clustering through multiobjective optimization. In: 2010 IEEE Congress on Evolutionary Computation, pp. 1–8 (2010)

11. Sondberg-Madsen, N., Thomsen, C., Pena, J.M.: Unsupervised feature subset selection. In: Proceedings of the Workshop on Probabilistic Graphical Models for Classification, pp. 71–82 (2003)
12. Tajunisha, N., Saravanan, V.: A new approach to improve the clustering accuracy using informative genes for unsupervised microarray data sets. International Journal of Advanced Science and Technology, 85–94 (2011)
13. Varshavsky, R., Gottlieb, A., Linial, M., Horn, D.: Novel unsupervised feature filtering of biological data. IEEE Transactions on Knowledge and Data Engineering, e507–e513 (2006)
14. Yang, X.S.: Nature-Inspired Metaheuristic Algorithms. Luniver Press, Frome (2008)
15. Yeung, K.Y., Ruzzo, W.L.: Principal component analysis for clustering gene expression data. Bioinformatics, 763–774 (2001)
16. Zheng, C.H., Huang, D.S., Zhang, L., Kong, X.Z.: Tumor clustering using nonnegative matrix factorization with gene selection. IEEE Transactions on Information Technology in Biomedicine, 599–607 (2009)

Author Index

Printed in the United States
By Bookmasters